Symposium:
Seed Proteins

other AVI books

Food Chemistry and Microbiology

PHOSPHATES IN FOOD PROCESSING   *deMan*
MICROBIOLOGY OF FOOD FERMENTATIONS   *Pederson*
FOOD ANALYSIS: THEORY AND PRACTICE   *Pomeranz and Meloan*
FOOD ENZYMES   *Schultz*
PROTEINS AND THEIR REACTIONS   *Schultz and Anglemier*
CARBOHYDRATES AND THEIR ROLES   *Schultz, Cain and Wrolstad*
LIPIDS AND THEIR OXIDATION, SECOND PRINTING   *Schultz, Day and Sinnhuber*
SOYBEANS: CHEMISTRY AND TECHNOLOGY, VOL. 1, PROTEINS   *Smith and Circle*
PRACTICAL FOOD MICROBIOLOGY AND TECHNOLOGY, SECOND EDITION   *Weiser, Mountney and Gould*

Food Engineering

FUNDAMENTALS OF FOOD ENGINEERING, SECOND EDITION   *Charm*
ENCYCLOPEDIA OF FOOD ENGINEERING   *Hall, Farrall and Rippen*
FOOD PROCESSING OPERATIONS, VOLS. 1, 2, AND 3   *Joslyn and Heid*
BAKERY TECHNOLOGY AND ENGINEERING, SECOND EDITION   *Matz*
FOOD DEHYDRATION, VOLS. 1 AND 2   *Van Arsdel and Copley*
HANDBOOK OF REFRIGERATING ENGINEERING, FOURTH EDITION, VOLS. 1 AND 2   *Woolrich*

Food Technology

THE TECHNOLOGY OF FOOD PRESERVATION, THIRD EDITION   *Desrosier*
ECONOMICS OF NEW FOOD PRODUCT DEVELOPMENT   *Desrosier and Desrosier*
ECONOMICS OF FOOD PROCESSING   *Greig*
HANDBOOK OF COMMERCIAL SUGARS   *Junk and Pancoast*
QUALITY CONTROL FOR THE FOOD INDUSTRY, THIRD EDITION, VOL. 1   *Kramer and Twigg*
CEREAL TECHNOLOGY   *Matz*
BREAD SCIENCE AND TECHNOLOGY   *Pomeranz and Shellenberger*
FRUIT AND VEGETABLE JUICE PROCESSING, SECOND EDITION   *Tressler and Joslyn*
FREEZING PRESERVATION OF FOODS, FOURTH EDITION, VOLS. 1, 2, 3, AND 4   *Tressler, Van Arsdel and Copley*
FOOD OILS AND THEIR USES   *Weiss*

Nutrition

MILESTONES IN NUTRITION   *Goldblith and Joslyn*
NUTRITIONAL EVALUATION OF FOOD PROCESSING, SECOND PRINTING   *Harris and Von Loesecke*
PROTEINS AS HUMAN FOOD   *Lawrie*
PROGRESS IN HUMAN NUTRITION   *Margen*

Food and Field Crops Technology

CEREAL SCIENCE   *Matz*
POULTRY: FOODS AND NUTRITION   *Schaible*
POTATOES: PRODUCTION, STORING, PROCESSING   *Smith*
PEANUTS: PRODUCTION, PROCESSING, PRODUCTS   *Woodroof*

Symposium:
Seed Proteins

Editor   **G. E. Inglett, Ph.D.**

Chief, Cereal Properties Laboratory,
Northern Marketing and Nutrition
Research Division, Agricultural
Research Service, U.S. Department
of Agriculture, Peoria, Illinois

WESTPORT, CONNECTICUT

THE AVI PUBLISHING COMPANY, INC.

1972

Contributors

DR. AARON M. ALTSCHUL, Professor, Georgetown University Medical School, Washington, D.C.

DR. MARIANNA B. ANDERSON, Graduate Assistant, Horticulture Department, Purdue University, Lafayette, Indiana.

MR. BYRON L. BERNTSON, Program Specialist, Nutrition and Agribusiness Group, Foreign Economic Development Service, USDA, Washington, D.C.

DR. W. BUSHUK, Director of Technology, Canadian International Grains Institute, Winnipeg, Manitoba, Canada.

MRS. GLORIA B. CAGAMPANG, Assistant Chemist, International Rice Research Institute, Los Baños, Laguna, The Philippines.

DR. JOE H. CHERRY, Professor, Horticulture Department, Purdue University, Lafayette, Indiana.

DR. SIDNEY J. CIRCLE, Director, Protein Research, Anderson Clayton Foods, Clayton Research Center, Richardson, Texas.

DR. WALTER L. CLARK, Associate Director of Research-Exploratory, John Stuart Research Laboratories, The Quaker Oats Company, Barrington, Illinois.

MR. W. EMILE COLEMAN, Research Chemist, U.S. Environmental Protection Agency, Solid Wastes Office, Cincinnati, Ohio.

DR. JOSE MADRID CONCON, Assistant Professor, Nutrition and Food Science, College of Home Economics, University of Kentucky, Lexington, Kentucky.

DR. ARTHUR DALBY, Associate Professor, Department of Botany and Plant Pathology, Purdue University, Lafayette, Indiana.

DR. IOLO AB I. DAVIES, Lecturer in Biochemistry, School of Pharmacy, City of Leicester Polytechnic, Leicester, England.

DR. JULIUS W. DIECKERT, Professor, Department of Biochemistry and Biophysics, Texas A and M University, College Station, Texas.

MRS. MARILYNE C. DIECKERT, Technician, Department of Biochemistry and Biophysics, Texas A and M University, College Station, Texas.

MR. H. K. GARDNER, JR., Acting Head, Oilseed Products Investigations, Engineering and Development Laboratory, Southern Marketing and Nutrition Research Division, USDA, New Orleans, Louisiana.

DR. GEORGE E. INGLETT, Chief, Cereal Properties Laboratory, Northern Marketing and Nutrition Research Division, USDA, Peoria, Illinois.

DR. WITOLD JACHYMCZYK, Assistant Professor, Institute of Biochemistry, Warsaw University, Warsaw, Poland.

DR. G. RICHARD JANSEN, Head of Department, Food Science and Nutrition Department, Colorado State University, Fort Collins, Colorado.

DR. VIRGIL A. JOHNSON, Research Agronomist, Agronomy Department, Agricultural Research Service, USDA, University of Nebraska, Lincoln, Nebraska.

DR. BIENVENIDO O. JULIANO, Head, Chemistry Department, International Rice Research Institute, Los Baños, Laguna, Philippines.

DR. JAN KANABUS, Graduate Assistant, Horticulture Department, Purdue University, Lafayette, Indiana.

DR. CONSTANCE KIES, Professor, Department of Food and Nutrition, University of Nebraska, Lincoln, Nebraska.

DR. MELVIN P. KLEIN, Associate Director, Laboratory of Chemical Biodynamics, Lawrence Radiation Laboratory, University of California, Berkeley, California.

MR. LEO N. KRAMER, Physicist, Laboratory of Chemical Biodynamics, Lawrence Radiation Laboratory, University of California, Berkeley, California.

MRS. M. G. LAMBOU, Research Chemist, Southern Marketing and Nutrition Research Division, USDA, New Orleans, Louisiana.

MR. PAUL J. MATTERN, Associate Professor, Agronomy Department, University of Nebraska, Lincoln, Nebraska.

DR. EDWIN T. MERTZ, Professor, Department of Biochemistry, Purdue University, Lafayette, Indiana.

DR. LARS MUNCK, Chemist in charge, Nutrition Laboratory, Swedish Seed Association, S-268 00 Svalof, Sweden.

DR. JOHN J. MURPHY, Research Biochemist, Stauffer Chemical Company, Mountain View, California.

DR. ROBERT L. ORY, Head, Protein Properties Investigations, Oilseed Crops Laboratory, Southern Marketing and Nutrition Research Division, USDA, New Orleans, Louisiana.

Contributors

DR. GEORGE C. POTTER, Manager, Biochemistry, John Stuart Research Laboratories, The Quaker Oats Company, Barrington, Illinois.

MISS M. L. ROLLINS, Head, Microscopy Investigations, Cotton Properties Laboratory, Southern Marketing and Nutrition Research Division, USDA, New Orleans, Louisiana.

DR. DANIEL ROSENFIELD, Director, Nutrition and Technical Services, Food and Nutrition Service, USDA, Washington, D.C.

DR. ALLEN J. ST. ANGELO, Research Chemist, Southern Marketing and Nutrition Research Division, USDA, New Orleans, Louisiana.

DR. JOHN W. SCHMIDT, Professor, Agronomy Department, University of Nebraska, Lincoln, Nebraska.

DR. ALLAN K. SMITH, Oilseeds Protein Consultant, New Orleans, Louisiana.

MR. H. L. E. VIX, Chief, Engineering and Development Laboratory, Southern Marketing and Nutrition Research Division, USDA, New Orleans, Louisiana.

DR. WALTER J. WOLF, Head, Meal Products Investigations, Oilseed Crops Laboratory, Northern Marketing and Nutrition Research Division, USDA, Peoria, Illinois.

Preface

This book contains the proceedings of a Symposium on Seed Proteins held at the American Chemical Society meeting in Los Angeles, California, on March 28-April 2, 1971. Reviewed at this Symposium, sponsored by the Protein Subdivision of the ACS Agricultural and Food Chemistry Division, were the important aspects of seed proteins—their synthesis, properties, and products after processing. Areas of current interest and the latest technological progress received particular emphasis.

This book represents a multidisciplinary study of seed proteins, including the disciplines of agronomy, biochemistry, chemistry, economics, engineering, food science, genetics, industrial technology, molecular biology, and nutrition. Contributors to this volume are authorities having a close and continuing acquaintance with plant science. Not all authors participated in the symposium, however, but were invited by the editor to contribute chapters to enrich the organization of this book. The subjects selected give a broad coverage of the areas and disciplines concerned with the science and technology of seed proteins. Another editor and symposium organizer could have easily selected differently. My intention was to provide information on seed proteins needed by industrialists, scientists, students, and technologists.

Contributors from various Federal Agencies and Departments are expressing their own opinions in their articles and not those of the U.S. Government.

G. E. Inglett

July 1971

Acknowledgments

Sincere appreciation goes to those symposium participants who were willing to meet the challenge and to their companies, government agencies, and universities who allowed their participation.

I am indebted to the officers of the Agricultural and Food Chemistry Division of the American Chemical Society; namely, Drs. K. Morgareidge, I. Hornstein, E. L. Wick, and R. J. Magee, who allowed us the forum for this symposium. Dr. A. M. Altschul not only served as chairman of a half-day session, but was also active during early organizational stages of the symposium. Other chairmen of half-day sessions were Drs. B. S. Schweigert, W. J. Hoover, and H. L. Wilcke.

Gratitude is also expressed to Dr. J. Baudet, who read three short comments from the French Ministry of Agriculture, INRA organizations. Similar appreciation is acknowledged for comments given by Drs. W. H. Martinez, R. C. Pickett, J. W. Finley, and G. D. Kapsiotis.

Finally, I should like to thank the Director of the Northern Regional Research Laboratory, Dr. R. J. Dimler, for his encouragement; as well as support from the following staff members: Messrs. C. W. Blessin, G. N. Bookwalter, J. F. Cavins, and D. D. Christianson; Drs. H. W. Gardner, U. Khoo, J. A. Rothfus, J. H. Sloneker, J. S. Wall, M. J. Wolf, and W. J. Wolf.

Contents

Section V. Methodology

George E. Inglett | **Seed Proteins in Perspective**

A seed is considered the propagative portion of a plant. It contains a fertilized and ripened ovule comprising a miniature plant usually accompanied by a supply of food as endosperm or perisperm. A protective coat, often accompanied by auxiliary structures (as an aril or caruncle), is another feature of a seed. Under suitable conditions, it will develop into a plant similar to the one that produced it. In this symposium we will consider only some of the major seed crops that are used for food, feed, and processing. These are primarily wheat, rice, barley, corn, oats, soybean, cottonseed, and peanut. Other seeds will be mentioned as they relate generally to the cereal or oilseed presentations or as invited short comments.

As the population of our civilization continues to explode, increased world-wide seed production with improved nutritive value is a mandatory objective. The seed crops represent food—sometimes the seeds are used in their near native form for the less sophisticated diet. Foods prepared from processed seeds in more highly cultured societies usually give little resemblance to the original grains.

Cereals and oilseeds are the mainstay of the world's food supply. Considerable amounts of these grains, however, are used only indirectly through the feeding of animals which, in turn, supply meat products. As the population of the world continues to increase, there will be a greater demand to increase not only production but also the direct food utilization of cereals and oilseeds. Some 70% of the world's protein supply comes from vegetable sources and 30% from animal sources. Grains directly provide nearly 1/2 of man's total supply of protein; and more protein is provided from livestock products which derive much of their nutritional value from cereals. Cereals will continue to be the main source of protein in the foreseeable future. They will also provide the bulk of the calorie supply. Oilseeds, including pulses and nuts, are the main protein-rich vegetable crops. They play a vital role in protein nutrition particularly in areas of the world where animal products are too expensive or not readily accepted. Oilseeds and their products will be even more important sources of protein for direct consumption by humans in the future. Presently in the developed nations the oilseed meals are used mainly for animal feeds.

World production of the most important seed crops are shown in Table 1.1. All cereals produced in 1968 amounted to 1.18 billion metric tons. Wheat, the major seed crop, contributed 28% of this amount followed by rice (24%), corn (21%), barley (11%), oats (5%), sorghum (4%), rye (3%), and others (4%). Of the oilseed crops, soybeans produced in 1968 amounted to 44 million metric tons, which was more than twice the amount of cottonseed (21 million metric tons).

TABLE 1.1

WORLD PRODUCTION OF THE MOST IMPORTANT SEED CROPS

Cereals	Production, 1968 (Million Metric Tons)	Oilseeds	Production, 1968 (Million Metric Tons)
Wheat	333	Soybeans	44
Rice, paddy	284	Cottonseed	21
Corn	251	Peanuts	15
Barley	131	Sunflower	10
Oats	54	Rapeseed	6
Rye	33	Linseed	3
Sorghum and millet		Sesame	2
Sorghum	42		
Millet	16		
Unspecified	27		

Source: Production Yearbook: 1969, Vol. 23, FAO, Rome, Italy.

Peanuts (15 million metric tons) and sunflower (10 million metric tons) were the next largest oilseed crops.

Maximization of our seed protein resources either for human food or animal feed requires more sophisticated basic and applied research efforts of many scientific disciplines. In this symposium we will cover both basic and applied research projects. Some of the more basic research is concerned with the genetic control of protein synthesis in the seed crops.

SYNTHESIS AND DEPOSITION OF SEED PROTEINS

The synthesis of proteins by seeds—cereals and oilseeds—is important because the utilization of seeds can depend largely upon the quantity and kinds of proteins produced. This will be illustrated in this symposium by the discussions on high-lysine corn proteins in contrast to the proteins of ordinary corn. The nutritive value of *opaque-2* corn proteins is virtually equivalent to milk protein whereas ordinary corn proteins are deficient in the essential amino acids, lysine and tryptophan. The *opaque-2* gene is responsible for the smaller quantity of zein in the endosperm and the larger quantity of more nutritious proteins, glutelins. Microscopy was important in showing the structural difference between the various genotype endosperms. The biochemical separations of the endosperm proteins have also revealed the differences in the relative proportions of zein and glutelins that account for the remarkable nutritional values of high-lysine corn in contrast to ordinary corn.

A gene for altering the amino acid composition in barley seeds has also been found. The Hily (high lysine) gene of barley gives seeds with increased lysine values. Genetic studies on wheat genotypes have not yet revealed a gene that gives a large improvement in high-lysine protein content. However, there are good prospects for wheat with higher protein content. There is a concerted effort by the International Rice Research Institute to improve the protein content of rice by at least 2% by crossing IR8 with high-protein varieties. In recent years an oat variety with a protein content of nearly 25% has been discovered.

The synthesis of larger quantities of proteins with improved nutritive value of cereals offers great promise for alleviating hunger problems in many segments of our civilization. Rapid methods of determining seed protein, quality, and quantity will expedite the successful breeding and commercialization of these crops. Present-day methods of amino acid analyses are too slow to meet the vast number of samples to be screened for genetic improvement. A potential method for determining protein value *in situ* is by the use of X-ray photoelectron spectroscopy. A report is included in the symposium. Also, more research concerned with the methods of determining proteins synthesized and stored in seeds is needed.

PLANT PROTEIN PROPERTIES

The kinds of proteins that are in seeds determine the chemical and physical properties of the seed products. Wheat proteins of flour have suitable elasticity for forming dough, an essential step in breadmaking. Soybean proteins of soy meal are very versatile; they can be used as emulsifying agents for water and fat in meat products or spun into food fibers used to construct new food items. It becomes more important as time moves along to determine the types of proteins in seeds and relate them to properties. Only when the kinds of proteins and methods of their detection are known can adequate progress be made for new and improved seed sources.

One kind of storage protein in seeds is known as protein bodies. Protein bodies, their synthesis and deposition in cereals and oilseeds, have been studied in the peanut, cottonseed, soybean, shepherd's purse, hempseed, barley, and corn. Protein bodies are not found in mature wheat kernels, so perhaps the protein bodies in the seeds are not as important to the functional properties as the interstitial protein components. However, much more information is needed, not only on existing protein composition of seeds but also on genetic studies to improve seed proteins for functional properties and nutritive value. Since genetic improvement of our seed crops is a long-range project, some of the more immediate problems of improving nutritive value of seed-derived foods may be met by fortification.

Treatments that plant proteins receive during processing, of course, greatly influence properties and nutritional value of the seed products. Microscopy of the grain components has opened up new avenues of processing. Cottonseed processed by a liquid cyclone procedure developed at the USDA Southern Regional Research Laboratory will be used to illustrate the use of microscopy on a new high-protein product. The glands containing the gossypol of ordinary cottonseed were removed by the cyclone process.

There are many facets of protein synthesis, properties, and processes associated with our large worldwide seed crops. It was our intention in this symposium to discuss some of the major seeds and to illustrate where basic and applied research could be used for the betterment of mankind.

CHAPTER 2

Daniel Rosenfield
and
Byron L. Berntson

Economics and Technology
of Cereal Fortification

An approach to nutrition improvement not primarily dependent upon socio-economic development is needed. Food fortification in general and cereal fortification in particular constitute such an approach. The thrust of our paper will be directed towards the fortification of wheat, rice, and corn with amino acids to increase their biologically utilizable protein. Such fortification involves the addition of other nutrients such as vitamins and minerals. After reviewing the economics, physiological impact, and technology of fortifying cereals with amino acids, we will conclude with a discussion of protein versus calorie shortages and the food versus diet approach to combatting malnutrition (Rosenfield 1970, 1971).

POVERTY-CAUSED MALNUTRITION

In discussing poverty-caused malnutrition, it is first necessary to define the term malnutrition. It is generally defined as some measurable degree of ill health due to inadequate nutrition that can be prevented or cured by improved nutrition. With such a broad definition, malnutrition can include obesity from an intake of too many calories in relation to those expended in physical activity, coronary and cerebral atherosclerosis when related in part to excessive quantities of saturated fats and cholesterol in the diet, hypertension when due in part to excessive intakes of sodium, anemias due to lack of iron, folic acid and vitamin B_{12}, and severe lack of calories. Since there are some 50 different nutrients in foods and water, there can be just as many different types of malnutrition.

The type with which our work is most concerned is underconsumption of specific nutrients such as vitamin A, iron, or protein due to food consumption patterns of people in general and poor people in particular. For poor people, inadequate nutrient intake occurs as they have little choice but to consume only a few types of food day after day with monotonous regularity. The less the variety of foods in the diet, the greater the risk of not consuming sufficient amounts of all the needed nutrients. As family income increases, variety of foods consumed increases and malnutrition generally decreases. Because of this direct relationship between income and variety, it cannot be expected that malnutrition can be significantly overcome in poor people by teaching them to eat more varied diets.

We believe that protein malnutrition is the most prevalent and critical form of poverty-caused malnutrition. There is not unanimous agreement on this since protein malnutrition can manifest itself in a variety of ways which are not readily measured or visually detected, i.e., clinically observed. Our understanding today of protein malnutrition appears to be somewhat analogous to the general feeling about atherosclerosis 20–30 yr ago. Previously, one usually had to have a heart attack to know one's heart was undergoing deterioration. Now, blood analysis for cholesterol serves as a most useful early indicator of potential heart trouble. What is needed in the protein area is a biochemical test for the early detection of protein malnutrition, particularly for humans other than infants. This needed indicator would give warning that one is not in a healthy protein nutritional state and is therefore susceptible to other illnesses. Biochemical indices such as serum albumin levels which are in use today generally require drastic changes in protein nutrition to be of value. By the time protein malnutrition is clinically evident other diseases may occur which obscure the fact that protein deprivation was the primary cause of sickness. Were more sensitive biochemical indicators available today, the endemic character of subclinical protein malnutrition for people other than special groups, such as infants and pregnant and lactating women, would be more widely recognized.

In contrast to the status of biochemical markers for protein nutrition, techniques for the measurement of vitamin nutrition are relatively hypersensitive. Consequently, the appraisals of nutritional status which are currently being done in this and other countries reflect the state of technology. Vitamin deficiency is overdiagnosed in nutrition surveys on the basis of saturation tests and protein deficiency is underdiagnosed on the basis of inadequate tests. Development of techniques for the evaluation of protein malnutrition would redress this imbalance and give the physician, public health worker, and government policy planner a better view of the health status of populations.

Parenthetically, it should be noted that protein intake is the major food variable influenced by income. As animal protein consumption is the first to rise with income and go down with loss of income, it is a good guess that poor people suffer from protein shortage with subsequent malnutrition. This does not mean that animal protein is necessary for good protein nutrition; plant mixtures properly selected and fortified can serve equally well. It is just that it is simpler from a nutrition standpoint to meet the physiological needs for protein of an entire community with animal protein. With plant protein, the risk of inadequate nutrition is greater since the selection of plant protein foods generally takes place without knowledge of their complementary effects.

PROTEIN NEEDS AND RESOURCES

Having reviewed briefly the problem of poverty-caused malnutrition, let us now look at some of the facts about protein uses and supplies. Table 2.1 presents protein consumption data from 1968 (these should be taken as estimates, not precise values). It is seen that cereals provide over 50% of the protein in the human diet; fish, 3%; and animals, 23%. The term "other" refers to sources such as legumes (beans, peas, lentils), tubers, and nuts.

TABLE 2.1

1968 PROTEIN CONSUMPTION

Source	Consumption (Million Metric Tons)	
Cereals	60.5	
Fish	3.4	
Animals	26.4	
Milk		12.5
Meat		12.7
Eggs		1.8
Other	23.2	
Total	113.5	

The surprising fact to many is that this value of over 100 million metric tons is sufficient to feed today's world population estimated to be 3.5 billion if there is no concern for diet aesthetics and variety. For example, if it is assumed that the crude protein requirement per person is 65 gm per day,[1] a population of 3.5 billion needs about 83 million metric tons per year. Even if this value underestimates the need by 50%, there would still be, on a theoretical basis, adequate supplies of protein from a nutritional point of view to meet the world's requirements. Thus, in spite of all the publicity, the protein problem is not one of overall shortage; it is rather one of inequitable distribution of total supplies and of shortage of particular sources such as meat.

In our judgment, it is not realistic to expect significant long-term solutions to come from efforts to redistribute protein supplies or to increase animal production. If the protein requirements of animals are included and the affluent model of protein consumption is followed, protein supplies would have to go up by at least a factor of five. Another way to look at this is to compare the plant calorie equivalents per capita per day currently required to maintain various food consumption patterns and levels. Altschul (1971) has made such calculations for selected countries as shown in Fig. 2.1. Note that a country such as India would have to increase its available plant calories from 3100 to 11,000 if it were to follow the U.S. model. Tunisia would have to more than double its plant supplies if it wishes to emulate its Western European neighbor, Italy. The less developed world as a whole would have to increase plant calorie supplies by 2½ times if it wishes to follow the developed world.

---

[1] This number is used for purposes of calculation only; protein requirements are subject to many factors such as age, sex, and stress.

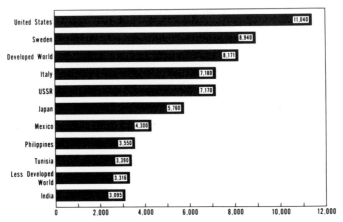

FIG. 2.1. PLANT CALORIE EQUIVALENTS PER CAPITA PER DAY

This is not to say that redistribution and animal husbandry efforts should be stopped. The U.S. Food for Peace Program, which is essentially a program of redistribution, is going to be needed for a long time to come. And there are some situations where it makes much sense to undertake animal husbandry ventures. But poor countries cannot afford to remain permanently in debt to rich countries via continuing importation of substantial amounts of foods; and the economics of livestock production negate the growing and purchase of significant amounts of animal protein by poor people.

TABLE 2.2

EDIBLE PROTEIN PRODUCTION BY
ANIMALS AND PLANTS

Item	Edible Protein Extracted/Acre of Land/Year (Lb)
Alfalfa, U.S. avg	600
Soybean culture	452
Legume culture	372
Wheat culture	272
Milk-forage, silage fed to cows	82
Beef-forage, silage fed to steers	38

Tables 2.2 and 2.3 present some information on protein production by animals and plants. While the numbers may differ slightly the point is clear that much more plant protein compared to animal protein can be obtained per unit of land. This fact of life shows up in the following data for protein costs per

TABLE 2.3

RELATIVE PROTEIN PRODUCTION

Commodity	Avg Yield per Acre (Lb)	Protein per Acre (Lb)
Soybeans	1440 (24 bu)	508
Shelled corn	3584 (64 bu)	323
Wheat	1500 (25 bu)	180
Milk	2800	97
Beef	342	58

food source as calculated by Kermit Bird (1971) in Table 2.4. Note that his costs are per pound of utilizable protein, not total protein. From a nutritional standpoint, cost of biologically utilizable protein is a more meaningful concept than cost of total protein since some of the total protein content of a food has no biological value. Protein costs, in turn, manifest themselves in the quantities of grain which are consumed directly and that which is consumed in the form of meat, milk, and eggs. Figure 2.2, showing per capita grain consumption versus per capita income in developed and low income countries, illustrates and reinforces the conclusion that increased animal supplies just cannot be the answer for countries with realistic aspirations for making adequate protein available for the desires and requirements of their populations.

TABLE 2.4

RELATIVE COSTS OF PROTEIN

Food Source	Cost (¢/Lb)	Total Protein (%)	Avg NPU[1]	Utilizable Protein ($/Lb)
Beef	48.7	19.5	76.7	3.26
Milk, whole, fluid	6.7	3.5	81.6	2.34
Milk, skim, powder	22.4	35.6	79.6	0.79
Casein	24.0	99.0	72.1	0.34
Eggs, medium	25.0	12.8	93.5	2.09
Soybean flour, low fat	8.5	44.7	61.4	0.31
Soybeans, extruded	28.0	52.5	58.0	0.92
Corn meal, whole	8.5	9.2	51.1	1.81
Wheat, whole grain	3.3	12.2	65.2	0.41
Cottonseed meal	35.0	42.3	57.7	1.57
Rice, whole	9.0	7.5	70.2	1.71

Source: Bird (1971).
[1] Net protein utilization.

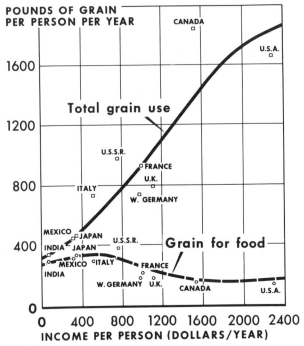

FIG. 2.2. INCOME AND PER CAPITA GRAIN CONSUMPTION,
TOTAL AND FOR FOODS

GREEN REVOLUTION

No discussion of food proteins would be complete without reference to the so-called Green Revolution. It is reasonable to ask: can the Green Revolution alleviate the problem of poverty-caused malnutrition? Certainly, the specter of famine has been postponed as shown by increases in agricultural production wherever adequate supplies of water, fertilizer, pesticides, and modern equipment are coupled with the new high yielding varieties (Brown 1969). But it costs money to apply science and technology to the centuries-old agricultural systems of low income countries (Wharton 1969; Ladejinsky 1970). For example, the lack of money has restricted the construction of irrigation facilities which, in turn, has been a constraint on the spread of the high yielding varieties. The question of where the poor countries are going to get the necessary capital to advance the Green Revolution may be rhetorically asked. The general conclusion can be made that in the long-run the Green Revolution can spread only at the speed at which the whole process of economic development takes place.

As the Green Revolution spreads, it delays the possibility of widespread famine with loss of life to untold numbers. But at the same time, the more prevalent problems of hunger and malnutrition above the famine level may not be significantly alleviated. One yardstick which may be used to measure progress towards a solution to these problems is the per capita availability of essential nutrients, especially calories and protein. Current measurements of progress place too much emphasis upon production advances and are too often concerned only with economic self-sufficiency in foods, which means as much food is produced and available domestically as people can afford to buy. Measured against economic self-sufficiency, the Green Revolution has moved several countries, such as the Philippines, India, and West Pakistan, forward at a dramatic rate of speed in recent years. Present indications are that this advance has already slowed somewhat in India, and this may be occurring in other countries as well. The yardstick of nutrient per capita availability forces us to raise our focus of attention from economic self-sufficiency to nutritional self-sufficiency. Nutritional self-sufficiency costs more to reach than economic self-sufficiency, and the impact of the Green Revolution has been much less dramatic when measured against this yardstick. The Green Revolution has occurred mainly in cereal production, and available data seem to indicate that in some places this increased cereal production has occurred at the cost of decreased production of higher protein commodities such as pulses. It can be concluded that without increases in the level of income in developing countries, the Green Revolution will have little effect on alleviating poverty-caused malnutrition.

INCREASED INCOMES

Another possibility for overcoming poverty-caused malnutrition would obviously be increased incomes. Even the most optimistic projections of income

TABLE 2.5

INCOME AND NUTRITIONAL STATUS

Per Capita GNP ($)	Plant/Animal Ratio[1]	Nutritional Status
Over 1500	1:1	Adequate food
1000–1500	2:1	Malnutrition rare or seasonal
400–1000	3:1	Some malnutrition; no famine
125–400	5:1	Widespread malnutrition; occasional famine
Under 125	6:1	Extensive malnutrition; greater famine potential

Source: Altschul (1971).
[1] Ratio of plant to animal protein in the daily diet.

increases in the developing countries force us to conclude that this cannot provide the solution. Using a hypothetical rice- and livestock-producing country with a per capita income of $52 in 1962, Francois (1969) has projected an increase in protein consumption of only 7.9% by 1975 with an annual increase in per capita income of 3%. This is considered an unreasonably high rate of income gain for a developing country. While animal protein consumption increased 38%, this was from 7.9 gm to only 11 gm per capita per day. Altschul has estimated that a per capita GNP (Gross National Product) of at least $1000 is necessary to ensure the elimination of malnutrition through the income approach, as illustrated in Table 2.5. In the book, *The Year 2000* by H. Kahn and A. J. Weiner, as quoted by Altschul (1969), the per capita GNP in that year for Brazil and India were projected at $506 and $270, respectively. These figures represent substantial growth for both countries.

CEREAL FORTIFICATION

In order to provide lower income countries with alternative systems for obtaining protein, not primarily dependent upon socio-economic development, advantage must be taken of the explosive increase in our knowledge about foods and nutrition (Altschul 1965; Altschul and Rosenfield 1970). One possibility which we want to mention but not discuss is to increase the availability of new protein foods which utilize technological and marketing innovations. These protein foods include protein mixtures for infants, soft drink protein beverages, and textured protein products (Altschul 1969; Rosenfield and Stare 1970).

A complementary approach to nutrition improvement is food fortification in general and cereal fortification in particular (Rosenfield 1970, 1971). It is possible to alleviate malnutrition without increasing the variety of foods in the diet by improving the nutritive value of existing foods and commodities via the addition of vitamins, minerals, amino acids and/or protein concentrates. This

TABLE 2.6

EFFECT OF VARIOUS SUPPLEMENTS ON THE NUTRITIVE VALUE OF A CORN-BEAN DIET

Diet	PER
Basal[1]	1.09 ± 0.07
Basal + lysine + trytophan (AA)	1.10 ± 0.08
Basal + AA + vitamins	1.73 ± 0.08
Basal + AA + minerals	2.37 ± 0.06
Basal + AA + minerals + vitamins	2.55 ± 0.06

Source: Bressani (1970).
[1]Corn-bean basal: corn, 72.4%; beans, 8.1%; sugar, 13.8%; roots, 5.7% (9.1% protein, 374 cal/100 gm).

nutrient addition is called fortification. A cereal fortification scheme to increase protein impact would usually involve the addition of appropriate vitamins and minerals as well as amino acids or protein concentrates. Data in Table 2.6 clearly indicate the need for adding vitamins and minerals when fortifying with amino acids. However, the thrust of the following remarks will be devoted to fortification with amino acids to increase biologically utilizable protein of the major cereals: wheat, rice, and corn.

Economics of Fortification and Role of Amino Acids

The economics of amino acid fortification must be considered concurrently with its physiological action. The role of amino acids is illustrated by the work of Hegsted (1969) and Jansen (1969) as presented in Table 2.7. According to Hegsted, the utilizable protein ranged from 3% in corn meal to 3.2% in white flour and 4.5% in rice. In rat feeding experiments he showed that the utilizable protein in wheat flour in 1 sample increased from 3.2% to 5.3% upon the addition of 0.2% L-lysine-HC1. Jansen found the increase in the utilizable protein in water bread to be from 7.3% to 10.0% upon the addition of 0.3% L-lysine-HC1. While they differ on the levels of biologically available protein in unfortified wheat protein, they agree that each 0.1 gm of lysine added per 100 gm of wheat-based food increases the utilizable protein content by 1 gm. Similarly, addition of 0.1% tryptophan and 0.3% lysine increases the utilizable

TABLE 2.7

EFFECT OF AMINO ACIDS ON THE NUTRITIVE
VALUE OF CEREAL PROTEIN[1]

Food	Reference	Protein (%)	Utilizable Protein (%)	Increase in Utilizable Protein per 100 Gm Food (Gm)
White wheat flour		13.75	3.20	
White wheat flour + 0.2%	Hegsted (1969)			2.14
lysine · HC1		13.94	5.34	
Water bread		15.0	7.3	
Water bread + 0.3%	Jansen (1969)			2.7
lysine · HC1		15.0	10.0	
Yellow corn meal		7.95	3.0	
Yellow corn meal + 0.1%	Hegsted (1969)			2.1
tryptophan + 0.3% lysine		8.37	5.1	
Rice (Thailand variety)		7.12	4.55	
Rice + 0.3% lysine + 0.1%	Hegsted (1970)			3.02
threonine		7.52	7.57	

[1] Prepared with the assistance of A. M. Altschul (1971).

protein in corn from 3% to 5.1%. Addition of 0.3% lysine and 0.1% threonine to rice increases utilizable protein from 4.5% to 7.6%.

We would like to emphasize the importance of the concept that amino acids added to cereals increase their utilizable protein. This is a more informative concept than considering amino acid addition as a technique to increase protein quality. The term protein quality, while appropriate in a technical sense, has little meaning to people without a background in nutrition. Quality is an esoteric term to which it is difficult to apply numbers in a cost-benefit or cost-effectiveness analysis. The economist and political scientist in government can readily understand the concept of biologically utilizable protein. He can relate the increase in available protein to the cost of amino acids in order to determine the cost of the increased protein supply. And this cost can be compared to the cost of supplying extra protein by techniques other than amino acid addition.

For example, we have seen that Hegsted and Jansen agree that the utilizable protein in 100 gm of wheat can be increased approximately 1 gm by the addition of 0.1 gm of lysine. If human-grade lysine can be purchased for $1 per lb, the cost of the additional 1 gm of utilizable protein is 0.022¢ which is equivalent to 10¢ per lb of protein. Let us compare this cost to costs of protein given in the data of Kermit Bird (1971) (Table 2.4). He calculated that the least cost source of utilizable protein is low fat soybean flour at 31¢ per lb. For wheat, the cost is 41¢ per lb and so on up to beef at $3.26 per lb utilizable protein. We are not suggesting that wheat fortification with lysine to increase available protein in the diet is the answer in every country under all circumstances. Obviously, if a country has to import wheat while having other sources of protein available, it is hard to see how wheat fortification would make economic sense in view of the need to conserve scarce hard currency.

In considering the economic feasibility of fortification, one must consider fortification costs in terms of the cost of the processed grain itself. Examples of such considerations are given in Table 2.8. The ingredient cost of wheat fortification is less than 5% of the flour cost. The ingredient cost of rice fortification will vary depending on the need for and cost of L-threonine. There is no doubt that rat feeding studies have shown the benefits of adding threonine as well as lysine to rice. But some human feeding studies have indicated threonine may not be necessary. This point needs to be cleared up, for without threonine, rice fortification expenses would drop to approximately 27¢ per 100 lb. of rice or approximately 2.7% of the cost of rice. This makes rice fortification economically reasonable. Tryptophan is the critical factor in the ingredient costs of corn fortification; even at projected prices the relative expense is above 10% of the cost of the grain. It may be necessary to develop techniques of fortifying corn with lysine and soy protein as a source of tryptophan, as workers at the Institute of Nutrition of Central America and Panama (INCAP) and Rutgers Food Science Department are doing in order to reduce the expense of corn fortification to a manageable level. Finally, if the experience with synthetic vitamins

TABLE 2.8

RELATIVE COST OF FORTIFICATION

Cereal	Cost of 100 Lb (U.S. $)	Cost of Fortifying 100 Lb (U.S. $)	Relative Cost (%)
Wheat flour (Paraguay)	6.10	0.24[1]	4.0
(Tunisia)	6.80	0.24	3.5
Yellow corn meal			
(Guatemala)	6.00	0.91[2]	15.2
Rice[3] (Pakistan)	8.00	1.02[4]	13.0
		0.58[4]	7.2
(Panama)	10.00–15.00	1.02[4]	10.0–7.0
		0.58[4]	5.8–4.0
	(10.00)	0.27[5]	2.7

[1] U.S. Title II premix (includes vitamin A) + 0.2% L-lysine.
[2] Standard degerminated corn meal premix (high iron) + 0.3% L-lysine + 0.1% DL-tryophan.
[3] Vitamins A, B_1, B_2, iron, 0.2% l-lysine, and 0.1% L-threonine.
[4] Price differential based on current and projected costs of threonine. The assumed prices per pound for amino acids are as follows: L-lysine, $1.00; L-threonine, $7.50 and $3.00; DL-tryptophan, $5.90.
[5] Threonine not included.

is at all valid, we could expect significant reductions in costs of all amino acids as the size of their market increases.

In determining the economic feasibility of fortification, one should also compare the cost of protein gained through fortification against that of protein from unfortified foods. For example, under the Commodity Distribution Program of USDA, 150 million pounds of wheat flour are annually supplied to needy families. At a fortification level of 0.2% lysine, an additional 3 million pounds of utilizable protein would be supplied to these families. The cost of this additional protein (as well as vitamins, iron, and vitamin A) would range from $360,000 to $600,000 depending upon the price of lysine. The cost per pound of additional protein would be from 12 to 20¢. Contrast this with the cost per pound of utilizable protein in meat which is $3.26; in skim milk powder, $0.79; and in unfortified wheat, $0.41.

Fortification Technology

Cereals are milled in two major ways prior to final preparation for consumption (Senti and Pence 1971). The first type, which is used for wheat, corn, grain sorghum, and millets, involves milling the grain into a flour, meal, or grit. In the second type, the outer layer of the kernel is all or partially removed but the whole kernel form of the grain is retained (Mitsuda 1969). This type of milling is used for rice.

In the first type, amino acids, vitamins and minerals in powdered form can be mixed directly into flours or meals in so-called dry chemical feeders. Feeders used in the flour milling industry are capable of metering approximately 0.01–1.0% addition of nutrients into a product being moved by mechanical means such as a screw conveyor or by forced air or gravity flow. The conveyor must be of adequate length to allow uniform mixing without segregation to take place. Costs of chemical feeders are low relative to the capital cost of commercial roller mills. Approximate cost is about $1000. Simple, modified feeders have been developed for use in low-income countries at a cost of less than $100.

It is possible to fortify at the point where the flour is to be prepared into a food product. For example, wafer-thin tablets containing measured quantities of nutrients are commonly added in bakeries per 100 lb of flour in a dough batch.

Fortification of grains in whole kernel form is more difficult than fortifying flours. One way this can be done is to add the nutrients as a powder or spray them as a concentrated solution onto a moving stream of the grain. During subsequent movement, the nutrients are uniformly mixed throughout the grain kernels. A second method is to prepare a premix by adding fortification nutrients in high concentration to grain kernels and then adding these heavily-fortified kernels to other kernels at levels of 1 part per 100 or 200 parts, i.e., at 0.5–1%. The fortification agents may be added to the premix as a surface coating or by infusing into the interior of the kernel. Another possibility for the premix approach is to prepare a simulated grain kernel via extrusion processing. These extruded granules contain the fortification ingredients and can be added at the 0.5–1% levels to cereal kernels.

Powder, sprays, and natural rice premix kernels have been used to commercially enrich rice. The Agency for International Development is currently supporting a fortification study in Thailand in which synthetic rice fortification granules containing lysine, threonine, vitamins, and iron are being added to rice brought to village mills in the study area. The study is being carried out by Harvard's Department of Nutrition and the Thai Ministry of Public Health.

The only large-scale fortification of wheat kernels that we are aware of has been carried out with bulgur, which is parboiled wheat kernels. Lysine at the level of 0.1% is added to bulgur via the infusion procedure. In 1970, the U.S. Government shipped to developing countries over 5 million pounds of bulgur fortified with 0.1% lysine. Shipments of this product are continuing. In February 1971, 3.4 million pounds were purchased for use in the Indian weaning food Bal-Amul.

As noted previously, INCAP and the Rutgers Food Science Department are currently investigating the possibilities of using simulated corn kernels prepared from soy and containing lysine and other nutrients for addition to corn kernels being milled into flour. The soy serves as a source of tryptophan and protein. The project is noted here because of the similarity in concept to the rice fortification project being studied in Thailand. However, this is not an example of fortifying whole kernels; the fortified corn kernels leave the mill in flour form.

There is one problem which must be noted in fortifying grain kernels such as rice. If the kernels are presoaked and cooked in excess water, there is a possibility that the fortifying nutrients may leach out into the water and be discarded. Therefore, the home preparation process is a most important factor in determining the manner of fortifying grain kernels.

SUMMARY

We would like to conclude this presentation on amino acid fortification with a discussion of protein versus calorie shortages and the food versus diet approach to the alleviation of malnutrition. The only reason for amino acid fortification of foods is to increase the amount of biologically available protein in these foods. This approach presupposes there is a prime need for extra protein in human diets. There are some who argue the following: highest priority should be given to furnishing calories; if the calorie problem is solved, the protein problem will automatically be taken care of; and when both calories and protein are deficient, added protein will be used as energy and not as protein. Certainly, wherever people are consuming a near-starvation diet, it makes sense to first do all that is possible to increase calorie supply, be it from cereal grains or non-protein foods such as cassava. But there is questionable value in devoting scarce resources to increasing the intake of carbohydrate foods in diets which are merely suboptimal in calorie content. For these diets, it is clear that added protein would be beneficial. That is, for diets containing somewhat less than the optimum level of calories, most of the added protein would be physiologically utilized as protein; only a small percentage would be used as energy. Nutritionists know there is a continuum in the way protein is used depending upon both protein and calorie intakes. The simple aphorism that protein is used as energy when calories and protein are deficient is misleading, for it implies that all protein is "wasted" regardless of the level of caloric sufficiency. In short, it is not an all or nothing situation.

Finally, our concept of fortification is that it is a "food approach" concerned with adding needed nutrients directly to food staples. One of the basic assumptions of the food approach is that there is daily or seasonal uneven distribution of foods amongst poor people; at any point in time the variety of foods in the diet is limited. Therefore, by nutritionally upgrading the major food staples, the chances of low income people consuming their needed nutrients each day are greatly enhanced. The provision of adequate nutrition is not left to chance.

Let us now look at the "diet approach" which at first appears more efficient than the food approach. In the diet approach, average diets are so determined that they presume to guarantee a specific diet more or less every day and everytime for every individual. Any deficiencies can be overcome by schemes designed to change diet makeup or consumption patterns. The diet approach is disarming in its simplicity for it is like the animal feeding model. Animals in pens are fed a uniform diet; deficiencies can be easily recognized and corrected. The point need not be belabored that humans not in institutions do have at least some degree of freedom in their food choices. Therefore, unless there is complete control over the entire diet, the diet approach is of limited value.

Sometimes we feel that the benefits as well as the real problems of fortification with amino acids have become lost in emotional issues. The emotionalism surrounding amino acids seems to be practically identical to that surrounding fluoridation of our water supplies. If this is the case, the real losers are the poor people suffering from protein, vitamin, and mineral deficiencies, not the proponents and opponents of fortification.

BIBLIOGRAPHY

ALTSCHUL, A. M. 1965. Proteins: Their Chemistry and Politics. Basic Books, New York.

ALTSCHUL, A. M. 1969. Food: proteins for humans. Chem. Eng. News 47, Nov. 24, 68-81.

ALTSCHUL, A. M. 1971. Private communication. Professor, Georgetown University Medical School, Washington, D.C.

ALTSCHUL, A. M. and ROSENFIELD, D. 1970. Protein supplementation: satisfying man's food needs. Progress (Unilever Quart., London) 54, No. 305, 76-84.

BIRD, K. 1971. Private communication. Food and Nutrition Service, USDA, Washington, D.C.

BRESSANI, R. 1970. Private communication.Institute of Nutrition of Central America and Panama, Guatemala City, Guatemala.

BROWN, L. R. 1969. Seeds of Change. Praeger Publishers, New York.

FRANCOIS, P. 1969. Effects of income projection on the protein structure of the diet. Nutr. Newsletter 7, No. 4, FAO, Rome.

HEGSTED, D. M. 1969. Nutritional value of cereal proteins in relation to human needs. In Protein-Enriched Cereal Foods for World Needs, M. Milner (Editor). Am. Assoc. Cereal Chemists, St. Paul.

HEGSTED, D. M. 1970. Private communication. Nutrition Dept., Harvard University, Boston.

JANSEN, G. R. 1969. Total protein value of protein- and amino-acid-supplemented bread. Am. J. Clin. Nutr. 22, 38-43.

LADEJINSKY, W. 1970. Ironies of India's Green Revolution. Foreign Affairs July, 759-768.

MITSUDA, H. 1969. New approaches to amino acid and vitamin enrichment in Japan. In Protein-Enriched Cereal Foods for World Needs, M. Milner (Editor). Am. Assoc. Cereal Chemists, St. Paul.

ROSENFIELD, D. 1970. Enrichment of plant protein. In Proteins as Human Foods, R. A. Lawrie (Editor). In United States and Canada available from Avi Publishing Co., Westport, Conn.; otherwise available from Butterworths, London.

ROSENFIELD, D. 1971. Current amino acid fortification program. In Amino Acid Fortification of Protein Foods, N. Scrimshaw and A. M. Altschul (Editors). MIT Press, Cambridge, Mass.

ROSENFIELD, D., and STARE, F. J. 1970. The short-term solution of protein malnutrition. Mod. Govt. 11, No. 5, 47-61.

SENTI, F. R., and PENCE, J. W. 1971. Technological aspects of adding amino acids to foods. In Amino Acid Fortification of Protein Food, N. Scrimshaw and A. M. Altschul (Editors). MIT Press, Cambridge, Mass.

WHARTON, C. J., Jr. 1969. The Green Revolution: cornucopia or Pandora's box. Foreign Affairs April 464-476.

CHAPTER 3

G. Richard Jansen

Seeds as a Source
of Protein for Humans

In this country we have been accustomed to thinking about protein in terms of animal products and have relegated cereals and other plant foods to a role of energy supply. This is not realistic for all population groups in the United States, and for most of the world the converse is true with dietary protein being supplied primarily by vegetable sources. This point is illustrated for 12 representative developing countries by the data in Table 3.1 which were reported by the USDA (1962). Protein supplied an average 10.4% of dietary calories compared to 12.2% for the United States as a whole. However, the similarity ends here because in these developing countries vegetable sources accounted for over 82% of the total protein supply whereas for the United States the corresponding figure was only 34%. The relative constancy in the proportion of dietary calories supplied by protein in spite of large differences in the proportion of protein supplied from animal sources was observed and commented on by Dole (1957) based on FAO statistics for 32 countries.

TABLE 3.1

CONTRIBUTION OF ANIMAL AND VEGETABLE SOURCES
TO PROTEIN SUPPLY

Country	Protein (% of calories)	Animal	Vegetable	Total	% Vegetable
		(Grams Protein/Day)			
Iran	12.2	13	49	62	79
Iraq	13.1	15	59	74	80
Syria	11.5	11	54	65	83
Egypt	12.0	7	63	70	90
Congo	7.4	7	42	49	86
Nigeria	9.0	6	54	60	90
Brazil	9.1	20	44	64	69
Columbia	9.2	20	31	51	61
Mexico	10.1	18	51	69	74
India	11.1	6	51	57	90
Indonesia	9.0	4	44	48	92
Pakistan	10.6	10	44	54	82
Average	10.4	11	49	60	82

Source: USDA (1962).

The estimation of 82% for the proportion of protein from vegetable sources is an average and there are many individuals and families who consume significantly more and many significantly less than this average. The incidence of protein malnutrition is higher in the latter group than the former and our primary concern, in this paper, is with the diets of individuals who consume little, if any, animal protein. Even for such individuals who consume essentially all their protein from vegetable sources, the quality and, hence, adequacy of their protein supply will depend greatly on the particular protein sources consumed. Cereals are the most important sources of protein as well as energy for people in most of the developing countries of the world (Jansen and Howe 1964). However, generally speaking, cereal protein is low in concentration and poor in quality. The other major sources of seed protein in the diet are pulses such as peas and beans, and oilseeds. These foods generally contain higher amounts of protein than the cereals, and the protein is of better quality. In this paper the ability of combinations of cereals with pulse, oilseed, and animal protein supplements to satisfy the protein requirement of growing infants will be discussed, and the implications considered for the proposed fortification of cereals with amino acids.

DEFINITION OF THE PROTEIN REQUIREMENT

Before considering how effectively seed protein can meet the protein requirement of humans, it is necessary to consider briefly protein requirements in general. Much controversy and uncertainty surrounds this subject; but several excellent reviews have recently been published (Swaminathan 1969; WHO 1965). The amount of protein required per unit body weight depends on the

TABLE 3.2

PROTEIN INTAKE IN CHILDREN AND
ADULTS THEORETICALLY MEETING
CALORIE NEEDS ON GRAIN

Food	Protein Intake (GM/Kg Body Weight)			
	1–12 mo	1–3 yr	4–6 yr	Adults
Rice	2.3	2.2	1.9	0.9
Wheat	3.3	3.2	2.8	1.3
Corn	2.6	2.6	2.2	1.0
Human milk	2.3			
		Protein Standard		
FAO requirement	1.7	1.1	1.0	0.7

quality of the protein, i.e., on the efficiency with which it can be utilized. For example, twice as much of a protein would be required that was used only half as efficiently as the reference standard, which is usually considered to be whole egg protein. For protein 100% utilized, i.e., whole egg, the requirement has been estimated by the FAO to range from 2.3 per kg body weight for very young infants to 0.7 per kg for adults. Protein intakes for children and adults consuming enough cereal to meet their energy requirements have been calculated and are listed in Table 3.2. These data indicate that, even for infants, cereals contain enough protein to meet the gross requirement, not considering quality, if consumed as the sole item of food in the diet. The primary problem with cereals is the poor quality of the protein they contain, not the amounts. Available nitrogen balance data suggest that cereals can supply the protein requirements of adults, who require protein primarily for maintenance and not for growth (Bolourchi *et al.* 1968; Hegsted *et al.* 1956). Calculations of amino acid intakes would confirm this. For example, adults consuming a diet of 90% white rice and 10% sugar at 40 kcal per kg would consume 1.6 gm of the limiting amino acid lysine (Jansen 1962). This equals Rose's "safe" intake of lysine and is double the maximum amount of lysine required to maintain nitrogen balance in adult subjects which ranged from 0.4 to 0.8 gm daily (Rose *et al.* 1955). In contrast, there is no doubt that cereals by themselves are not an adequate protein source for infants (Jansen and Howe 1964), and it is to the protein needs of this vulnerable group that this paper is directed.

Protein quality considerations are crucial to this discussion. Protein is needed for growth as well as for maintenance and our concern is primarily with growth. As is well known the efficiency of utilization of dietary protein depends on its amino acid composition. The pattern required for growth is quite different from that required for maintenance. The growth pattern is greatly influenced by the amino acid composition of the new tissue formed, primarily muscle, while the maintenance pattern depends on the turnover rates of the individual essential amino acids. From a practical standpoint, the most critical difference between the two patterns concerns lysine. Tissue protein being formed contains a high proportion of lysine and hence the lysine requirement for growth is quite high. In contrast, of all the essential amino acids, lysine turnover is quite low and hence the amount of lysine required for maintenance is quite low. This difference in lysine requirement is particularly crucial since lysine is the first limiting amino acid in all cereals. The difference is also very pertinent to a discussion of the methods used to evaluate protein quality. An excellent review on the methodology of protein quality evaluation has been published recently (McLaughlin and Campbell 1969). The most widely used methods are based on rat growth and include PER, NPU or NPR measurement. Protein Efficiency Ratio (PER) considers growth only, whereas Net Protein Utilization (NPU) and Net Protein Ratio (NPR) consider maintenance as well. The NPU-NPR methods overestimate the value of poor quality lysine-deficient protein sources such as the cereals. The reason for this overestimation relates to the great difference in the lysine requirement for maintenance as compared to growth. It is of practical significance because the NPU-NPR methods underestimate the improvement of cereal protein resulting from fortification with proteins or amino acids.

Because the ability of proteins to support growth is well correlated with their essential amino acid composition, the latter is a reasonably good guide to the former although problems of amino acid availability complicate the relationship. The essential amino acid pattern of whole egg supports maximum growth in the rat at the lowest level in the diet and is generally considered to represent the ideal amino acid pattern for growth. Block and Mitchell (1946) introduced the concept of chemical scoring of proteins 25 yr ago. In this method the essential amino acid in a protein that is in shortest supply as compared to egg protein is considered to be the limiting amino acid and the chemical score is the percentage deficit in this amino acid subtracted from 100. In considering the amino acid pattern required for the growing infant we have much less available data than for the rat. Because the infant grows at a slower rate than the rat, one might expect that a lower percentage of essential amino acids would be needed than is required by the rat and present in egg protein. Data obtained by Graham and colleagues (1969) would support this view. These workers have reported that lysine-supplemented wheat flour supported nearly as good growth in infants as did casein. White flour has approximately 1/3 of its protein in the form of essential amino acids compared to over 1/2 for egg. The joint FAO/WHO Expert Group on Protein Requirements (WHO 1965) has suggested that proteins be rated in terms of their essential amino acid pattern, as compared to the essential amino acid pattern for whole egg. In the calculations that follow, this is the procedure that has been used.

PROTEIN VALUE OF CEREAL DIETS

It has been pointed out that cereals are the most important source of dietary protein in the developing countries and all cereals are first limiting in lysine (Jansen and Howe 1964). The major sources of supplemental plant proteins are the pulses, and these are all primarily deficient in the sulfur amino acids methionine and cystine. Therefore, there has been much controversy concerning the evaluation of diets as a whole versus individual food staples such as the cereals, and whether methionine or lysine is a more crucial deficiency in practical human nutrition. It is to these questions that the major portion of this paper is directed. Based on all available biological data, the only amino acids that appear to be of any real concern in practical feeding are lysine, threonine, tryptophan, and the sulfur amino acids methionine and cystine. Therefore, in the following calculations these are the amino acids that have been considered. The chemical score for these amino acids has been calculated individually for 6 major cereal products supplemented with up to 32 parts of protein supplements per 100 parts cereal. Chemical score was calculated by dividing the essential amino acid content in milligrams per gram of essential amino acid nitrogen (N) by the corresponding value for that amino acid in whole egg protein and multiplying by 100. The cereals evaluated were ragi millet, white wheat flour, whole wheat bulgur, cornmeal, sorghum, and polished rice. The protein supplements were soy flour (8.1% N), fish meal (12.0% N), groundnut flour (9.4% N), and cottonseed flour (8.0% N). The basic data on nitrogen content and amino acid composition used were obtained from published sources (FAO 1970; Orr and Watt 1957). Using a computer, the protein and amino acid content of the mixtures were calculated as were the individual chemical scores for the amino acids of concern. The chemical score for the limiting amino acid in each mixture was considered to be a numerical measure of quality and was multiplied by the protein content, using the appropriate nitrogen-to-protein conversion factors, to give a measure of overall protein value.

Chemical Scores of Mixtures

Ragi Millet, White Flour and Bulgur.—Figure 3.1 shows the calculated chemical scores for ragi millet supplemented with the four protein supplements.

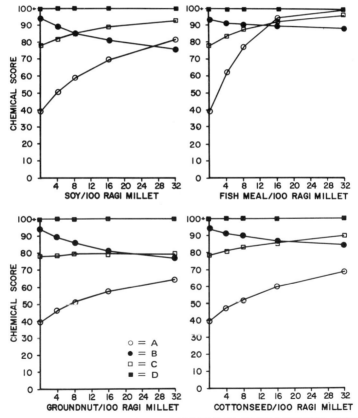

FIG. 3.1. INDIVIDUAL CHEMICAL SCORES FOR FOUR LIMITING AMINO ACIDS IN RAGI MILLET SUPPLEMENTED WITH SOY FLOUR, FISH MEAL, GROUNDNUT FLOUR, OR COTTONSEED FLOUR

A—lysine. B—sulfur amino acids. C—threonine. D—tryptophan. See Table 3.3 for nitrogen content of supplements.

Ragi millet is decidedly first limiting in lysine with a chemical score of 40 compared to 80 for second limiting threonine. Fish meal is the most effective supplement because of its high protein content and the high proportion of lysine in fish protein. Soy flour at a level of 16 parts per 100 parts millet raises the chemical score from 40 to 70 and is still limiting in lysine. Groundnut and cottonseed, because of their low lysine content are both poor supplements to millet, although they could be improved considerably by lysine addition. The situation for white flour and bulgur are both seen to be very similar to that for ragi millet, as illustrated by the data shown in Fig. 3.2 and 3.3. Both these wheat

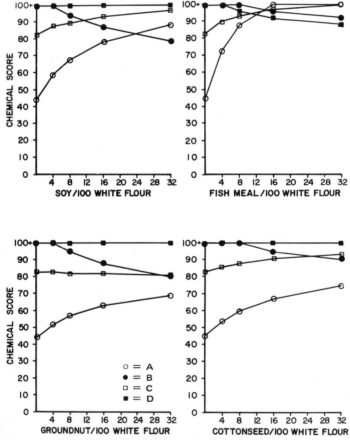

FIG. 3.2. INDIVIDUAL CHEMICAL SCORES FOR FOUR LIMITING AMINO ACIDS IN WHITE WHEAT FLOUR SUPPLEMENTED WITH SOY FLOUR, FISH MEAL, GROUNDNUT FLOUR, OR COTTONSEED FLOUR

A—lysine. B—sulfur amino acids. C—threonine. D—tryptophan. See Table 3.3 for nitrogen content of supplements.

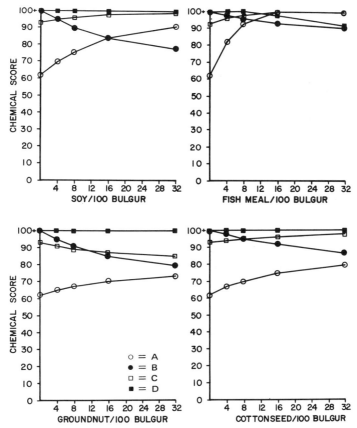

FIG. 3.3. INDIVIDUAL CHEMICAL SCORES FOR FOUR LIMITING AMINO ACIDS IN WHOLE WHEAT BULGUR SUPPLEMENTED WITH SOY FLOUR, FISH MEAL, GROUNDNUT FLOUR, OR COTTONSEED FLOUR

A—lysine. B—sulfur amino acids. C—threonine. D—tryptophan. See Table 3.3 for nitrogen content of supplements.

products are primarily limiting in lysine and second limiting in threonine, with wheat bulgur considerably higher in chemical score than 70% extraction white flour. Groundnut is an especially poor supplement to bulgur with 32 parts per 100 parts only raising the chemical score from 62 to 73. The reason, of course, is that groundnut protein contains very little more lysine than the wheat protein it is supplementing.

Corn, Sorghum and Rice.—Similar calculations have been made for corn, sorghum, and rice and the results are presented in Fig. 3.4, 3.5, and 3.6. Corn is first limiting in lysine, but is almost as limiting in tryptophan. Soy and fish are reasonably good supplements to corn with mixtures of the former becoming limiting in sulfur amino acids and mixtures of the latter in tryptophan. Ground-

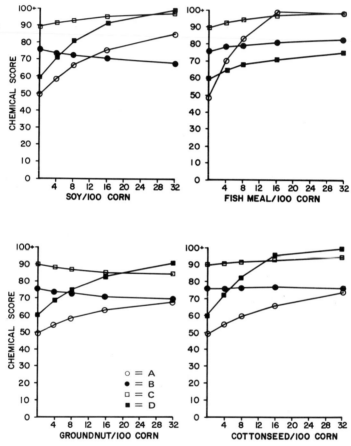

FIG. 3.4. INDIVIDUAL CHEMICAL SCORES FOR FOUR LIMITING AMINO ACIDS IN CORN MEAL SUPPLEMENTED WITH SOY FLOUR, FISH MEAL, GROUNDNUT FLOUR, OR COTTONSEED FLOUR

A—lysine. B—sulfur amino acids. C—threonine. D—tryptophan. See Table 3.3 for nitrogen content of supplements.

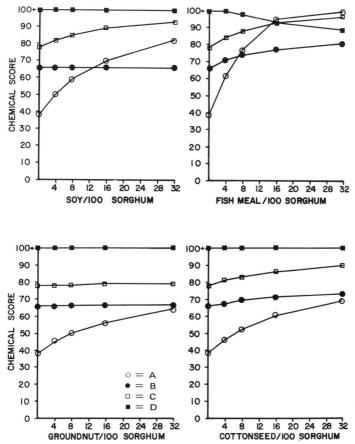

FIG. 3.5. INDIVIDUAL CHEMICAL SCORES FOR FOUR LIMITING AMINO ACIDS IN SORGHUM SUPPLEMENTED WITH SOY FLOUR, FISH MEAL, GROUNDNUT FLOUR, OR COTTONSEED FLOUR

A—lysine. B—sulfur amino acids. C—threonine. D—tryptophan. See Table 3.3 for nitrogen content of supplements.

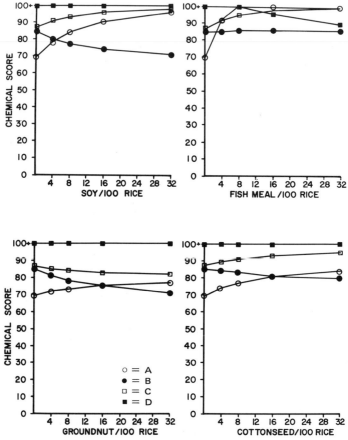

FIG. 3.6. INDIVIDUAL CHEMICAL SCORES FOR FOUR LIMITING AMINO
ACIDS IN POLISHED RICE SUPPLEMENTED WITH SOY FLOUR, FISH
MEAL, GROUNDNUT FLOUR, OR COTTONSEED FLOUR

A—lysine. B—sulfur amino acids. C—threonine. D—tryptophan. See Table 3.3 for
nitrogen content of supplements.

nut and cottonseed, again, are poor supplements because of their low content of lysine. They are more effective in raising the tryptophan level, however, thus making lysine supplementation of the mixtures more attractive. Sorghum, in contrast to corn, has ample tryptophan, but is second limiting in sulfur amino acids, as shown by Fig. 3.5. Soy and fish are effective in raising the lysine level but have very little effect on the concentration of sulfur amino acids. With groundnut and cottonseed added to sorghum, lysine remains the first limiting amino acid in all levels of supplementation up to and including 32 parts per 100 parts sorghum. The chemical score data shown for rice in Fig. 3.6 are not completely satisfactory because of the known problem of threonine un-availability in rice. Nevertheless, the data are still useful if one makes allowance for this. The intercept of the threonine lines should probably be around 80 rather than the 88 calculated from chemical analysis since threonine is definitely second limiting in rice. Another point to consider in interpreting the data is that, because of the low protein level of rice, the proportion of protein in the mixture derived from the protein supplement rises more rapidly than for the other cereals. Soy and fish raise the lysine and threonine levels effectively and thus are good supplements to rice. Fish protein does not contain an excess of sulfur amino acids, and hence the proportion of sulfur amino acids in the mixture does not rise as the proportion of soy protein increases. Because of the sulfur amino acid deficiency in soy, as the proportion of soy protein increases, sulfur amino acids become markedly first limiting. Neither groundnut nor cottonseed are effective supplements to rice, not only because of the low level of lysine, but also because of the low concentrations of both threonine and sulfur amino acids. Of the two, groundnut is a particularly poor supplement to rice.

Protein Value of Cereal-based Diets

As was stated earlier, the protein value for each theoretical diet has been estimated by multiplying the chemical score, or quality factor, by the quantity factor which is the percentage of protein in the diet. Although the estimates are based on chemical analysis with the known difficulties, they have the advantage that all estimates were made in the same way from the same basic data and hence should be reasonably comparable. In evaluating the estimates of protein value obtained by this method, the following point of reference will be used in judging adequacy of diets. White flour supplemented with lysine alone is calculated to have a protein value of 7.6 (Table 3.8). Graham *et al.* (1969) have reported that lysine-supplemented white flour gives growth in the human infant comparable to that obtained with casein. In the discussion to follow, a protein value of 7.5 is arbitrarily considered to be the dividing point between diets adequate and inadequate in dietary protein.

In Table 3.3, the limiting amino acids, chemical scores, and protein values are listed for all 6 cereals each supplemented with 8 parts per 100 parts cereal of

TABLE 3.3

SUPPLEMENTATION WITH CONCENTRATES AT 8% LEVEL[1]

Supplement	% Protein	Limiting Amino Acid	Chemical[2] Score	Protein[3] Value
		Ragi Millet		
None	7.4	Lysine	39	2.9
Groundnut	10.6	Lysine	51	5.4
Soy	10.3	Lysine	59	6.1
Cottonseed	10.0	Lysine	52	5.2
Fish meal	12.4	Lysine	78	9.7
		White Flour		
None	9.2	Lysine	45	4.1
Groundnut	12.3	Lysine	57	7.0
Soy	11.9	Lysine	68	8.1
Cottonseed	11.7	Lysine	60	7.0
Fish meal	14.1	Lysine	88	12.4
		Whole Wheat (Bulgur)		
None	11.2	Lysine	62	6.9
Groundnut	14.2	Lysine	67	9.5
Soy	13.8	Lysine	76	10.5
Cottonseed	13.5	Lysine	70	9.5
Fish meal	15.9	Lysine	93	14.8
		Corn		
None	9.5	Lysine	49	4.7
Groundnut	12.6	Lysine	58	7.3
Soy	12.2	Lysine	67	8.2
Cottonseed	11.9	Lysine	60	7.2
Fish meal	14.4	Tryptophan	69	9.9
		Sorghum		
None	10.1	Lysine	38	3.8
Groundnut	13.1	Lysine	50	6.6
Soy	12.8	Lysine	59	7.5
Cottonseed	12.5	Lysine	52	6.5
Fish meal	14.9	Methionine	74	11.0
		Rice		
None	6.7	Lysine	69	4.7
Groundnut	10.0	Lysine	73	7.3
Soy	9.6	Methionine	77	7.4
Cottonseed	9.3	Lysine	77	7.2
Fish meal	11.8	Methionine	86	11.2

[1] Supplements: Groundnut flour, 9.4%; N; soy flour, 8.1% N; cottonseed flour 8.0% N; fish meal, 12.0% N.
[2] Based on mg amino acid/gm essential amino acid nitrogen, compared to whole egg.
[3] Chemical score x % protein/100.

groundnut, soy, cottonseed, and fish meal. The level of supplementation (8%) on which these figures are based provides a greater proportion of protein from noncereal sources than is characteristic of national diets throughout the developing regions of the world (see Table 3.7). In almost every case the limiting amino acid of the mixture is lysine. A question to be answered is: What is the limiting amino acid the supply of which is most critical in cereal-based diets containing varying amounts of pulse, oilseed, or animal protein? In Table 3.4, limiting amino acids have been listed for all diets with protein values 7.5 or less. In every case but one the limiting amino acid is lysine. The only exception is rice supplemented with 8 parts of soy flour in which case methionine (or total sulfur amino acids) is calculated to be limiting. This diet is calculated to have a border-line protein value of 7.4. Another way to answer the same question is to con-

TABLE 3.4

CEREAL-BASED DIETS WITH PROTEIN VALUES 7.5 OR LESS[1]

Supplement to 100 Parts Cereal	Limiting Amino Acid	Protein Value
Ragi Millet		
None	Lysine	2.9
Groundnut, 8 parts	Lysine	5.4
Soy, 8 parts	Lysine	6.1
Cottonseed, 8 parts	Lysine	5.2
White Flour		
None	Lysine	4.1
Groundnut, 8 parts	Lysine	7.0
Soy, 4 parts	Lysine	6.3
Cottonseed, 8 parts	Lysine	7.0
Bulgur		
None	Lysine	6.9
Corn		
None	Lysine	4.7
Groundnut, 8 parts	Lysine	7.3
Soy, 4 parts	Lysine	6.3
Cottonseed, 8 parts	Lysine	7.2
Sorghum		
None	Lysine	3.8
Groundnut, 8 parts	Lysine	6.6
Soy, 8 parts	Lysine	7.5
Cottonseed, 8 parts	Lysine	6.5
Rice		
None	Lysine	4.7
Groundnut, 8 parts	Lysine	7.3
Soy, 8 parts	Methionine	7.4
Cottonseed, 8 parts	Lysine	7.2

[1] See Table 3.3 for nitrogen content of supplements and definition of protein value.

sider what the protein values are for diets calculated to be limiting in sulfur amino acids. As shown in Table 3.5, in all cases but one such diets have protein values considerably higher than 7.5. Supplementation of cereals with soy is an especially important situation to consider since soy, like most pulses, is deficient in sulfur amino acids. In Table 3.6, calculations are presented for the supplementation of each of the 6 cereal products with up to 32 parts of soy flour per 100 parts of cereal. In no case does a sulfur amino acid deficiency (methionine) show up with less than 36% of the protein of the mixture supplied by soy protein. With ragi millet and white flour it is necessary for over 1/2 the protein to be derived from soy for sulfur amino acids to limit the quality of the mixture.

TABLE 3.5

CEREAL-BASED DIETS LIMITING
IN SULFUR AMINO ACIDS[1]

Cereal	Supplement to 100 Parts Cereal	Protein Value
Ragi millet	32 parts soy	12.7
	16 parts fish meal	15.0
White flour	32 parts soy	14.3
Bulgur	16 parts soy	13.5[2]
	16 parts fish meal	18.6
Corn	16 parts soy	10.3
Sorghum	16 parts soy	10.0
	8 parts fish meal	11.0
Rice	16 parts groundnut	9.6[2]
	8 parts soy	7.4
	16 parts cottonseed	9.4[2]
	4 parts fish meal	7.9

[1] See Table 3.3 for nitrogen content of supplements and definition of protein value.
[2] Equally limiting in lysine.

NATURE OF DIETS IN DEVELOPING COUNTRIES

We have considered the characteristics of cereal diets supplemented with protein obtained from pulse, oilseed, or animal sources. A general conclusion has been made that poor diets will be limiting in lysine, and, hence, should benefit from lysine fortification, whereas diets limiting in sulfur amino acids will be adequate in protein value and do not need methionine fortification.

How do these calculations relate to the characteristics of diets consumed in the developing countries? Table 3.1 indicated that for 12 developing countries 82% of the dietary protein was supplied from vegetable sources. As has been discussed, the vulnerable inhabitants of these countries are those individuals consuming less than this proportion of animal protein. In Table 3.7, calculations have been made for 12 developing countries on the amount of plant protein

TABLE 3.6

SUPPLEMENTATION OF CEREALS WITH
SOY FLOUR CONTAINING 8.1% N[1]

Parts Soy/100 Parts Cereal	Soy Protein % of Total	Chemical Score	Protein Value	Limiting Amino Acid
		Ragi Millet		
0	0	39	2.9	Lysine
8	33	59	6.1	Lysine
16	50	70	8.9	Lysine
32	67	76	12.7	Methionine
		White Flour		
0	0	45	4.1	Lysine
8	28	68	8.1	Lysine
16	45	78	11.2	Lysine
32	62	79	14.3	Methionine
		Bulgur		
0	0	62	6.9	Lysine
8	25	76	10.5	Lysine
16	40	84	13.5	Lysine + Methionine
32	57	78	15.3	Methionine
		Corn		
0	0	49	4.7	Lysine
8	28	67	8.2	Lysine
16	43	71	10.3	Methionine
32	61	69	12.6	Methionine
		Sorghum		
0	0	38	3.8	Lysine
8	27	59	7.5	Lysine
16	42	66	10.0	Methionine
32	59	66	12.4	Methionine
		Rice		
0	0	69	4.7	Lysine
8	36	77	7.4	Methionine
16	52	74	8.9	Methionine
32	69	71	11.5	Methionine

[1] See Table 3.3 for Definition of Terms

TABLE 3.7

CONTRIBUTION OF CEREALS AND VEGETABLES TO DAILY PROTEIN SUPPLY[1]

Country	Cereals	Pulses and Nuts	Vegetables	% Cereal Protein
		Grams Protein/Day/Person		
Brazil	26.4	17.3	0.2	60
Taiwan	31.4	7.2	1.8	78
Columbia	17.8	2.9	0.5	84
Ethiopia	44.3	12.0	0.4	78
Guatemala	38.6	5.3	1.7	85
India	31.0	13.2	0.1	70
Iran	42.6	2.2	0.3	95
Iraq	37.2	3.4	1.9	88
Jordan	39.8	5.6	3.8	81
Mexico	33.0	13.6	0.5	70
Pakistan	32.8	3.3	0.7	89
Philippines	25.5	1.6	1.4	89
United Arab Republic	54.3	6.9	3.5	84
Average	34.9	7.3	1.3	81
Range	17.8–54.3	1.6–17.3	0.1–3.8	60–95

[1] Based on FAO Food Balance Sheets, 1960–1962.

supplied by cereals, pulses and nuts, and other vegetables based on FAO food balance sheets for 1960–1962. Cereals supplied 60–95% of the vegetable protein with the average 81%. These data confirm that the bulk of dietary protein in the developing countries is supplied by cereals and that the estimates made in the preceeding calculations are realistic in terms of the proportion of noncereal protein actually consumed in these countries.

PROTEIN IMPROVEMENT BY FORTIFICATION

How do the data presented in this report relate to the question that we are considering this morning, namely the improvement of practical diets in the developing countries by fortification with protein concentrates and/or amino acids? First, it is not advantageous to consider the average diets by country, but, instead, it is better to analyze what are the significant characteristics of critical or vulnerable diets and then determine what can be done to improve these diets. The greater proportion of protein in a diet that is supplied by cereals, the poorer will be the protein value of that diet and the more chance that the limiting amino acid will be lysine. In contrast, if enough animal, pulse, or oilseed protein is consumed along with cereals to make sulfur amino acids limiting, the overall diet that results will be adequate in protein value, even for the growing infant. The major exception to this general rule is for mixtures of rice with pulse protein deficient in methionine, when sulfur amino acids become limiting and where, because of the low protein level of rice, the protein value of the mixture is borderline but not seriously so. These data suggest that lysine fortification will prove to be more important than methionine addition in improving those poor diets throughout the world that are a major cause of protein malnutrition in children. The improvements in protein value made possible by lysine addition to a variety of diets with low protein values have been listed in Table 3.8. With few exceptions, addition of L-lysine monohydrochloride would improve the protein value to a level which would be out of the critical range. As has been emphasized elsewhere (Howe *et al.* 1967), all necessary vitamins and minerals should be added in combination with the lysine in order to ensure that full benefit of the amino acid supplement is realized. Another important consideration is that amino acids and protein concentrates should not always be considered as competitors but rather as partners in practical supplementation situations. For example, groundnut and cottonseed are seen to be poor supplements to most cereals because of the low level of lysine they contain. However, if lysine is added in combination with the groundnut or cottonseed these protein concentrates become effective supplements.

TABLE 3.8

LYSINE FORTIFICATION OF CEREAL DIETS

Diet[2]	Protein Value[1]	
	No LMH[3]	LMH Added[3]
Ragi millet	2.9	5.8
Millet + groundnut	5.4	8.4
Millet + cottonseed	5.2	8.3
White flour	4.1	7.6
Flour + groundnut	7.0	10.1
Flour + cottonseed	7.0	10.3
Bulgur	6.9	10.4
Bulgur + groundnut	9.5	12.6
Bulgur + cottonseed	9.5	12.8
Corn	4.7	5.7
Corn + groundnut	7.3	9.2
Corn + cottonseed	7.2	9.3
Sorghum	3.8	6.7
Sorghum + groundnut	6.6	8.6
Sorghum + cottonseed	6.5	8.6
Rice	4.7	5.8
Rice + groundnut	7.3	7.8
Rice + cottonseed	7.2	7.7

[1] Protein value defined in Table 3.3.
[2] 8 parts protein supplement/100 parts cereal.
[3] L-lysine monohydrochloride.

The conclusions that have been made in this report, based on consideration of the amino acid composition of diets, are supported by available rat growth data. Jansen (1969) has reported data on the net protein value of white bread supplemented with fish flour or soy, singly, or in combination with synthetic lysine. Even with as high a supplementation level of 16 parts of toasted soy flour or 6 parts fish flour per 100 parts white flour, lysine remained the first limiting amino acid.

Lysine addition alone to white bread converts a poor protein source into one adequate to support good growth in the rat and in the human infant as well, as was demonstrated by Graham *et al.* (1969). Daniel *et al.* (1970) studied the amino acid fortification of wheat diets consisting of a mixture of 75 parts Indian hard wheat and 5 parts red gram dhal. Addition of lysine raised the PER from 2.28 to 2.82. These workers concluded that the protein quality of poor wheat diets as actually consumed in India can be effectively improved by fortification with lysine. A similar conclusion was also reached in regard to the practical improvement of poor sorghum diets by lysine addition (Narayanaswamy *et al.* 1970.) A poor rice diet containing 8.2% protein of which 80% was supplied by rice and the rest by skim milk and red gram dhal was first limiting in methionine (Desai *et al.* 1970).

The above considerations apply to countries and diets where the food staples are cereals. There are countries in the world in which crops such as cassava and plantain are quantitatively important food items. Diets based on such crops are very low in the quantity of dietary protein and would benefit greatly from the addition of protein supplements. Amino acids would not be adequate. If such crops are eaten in combination with pulses the diet as a whole would be limiting in sulfur amino acids and would likely benefit from methionine addition. A rough estimate has been made that such diets would be characteristic of 5–10% of the total of undernourished people in the world today (Jansen and Howe 1954), and are quantitatively of much less importance than the cereal diets considered in this discussion.

SUMMARY

Vegetable protein typically supplies over 80% of the dietary protein in the developing countries of the world. The major sources of vegetable protein are cereal grains, pulses, and oilseeds. Based on published data, the limiting amino acids, chemical scores, and protein values have been calculated for 6 cereal foodstuffs each supplemented with varying levels of soy flour, cottonseed flour, groundnut flour or fish meal. The cereals considered were ragi millet, white wheat flour, bulgur, corn, sorghum, or rice. Based on the analyses made, the following conclusions were made. The greater the proportion of protein in a diet supplied by cereals the poorer in protein that diet is and the greater the proba-bility that the diet as a whole would benefit from lysine addition. If enough animal, oilseed, or pulse protein is consumed along with the cereal to make sulfur amino acids first limiting, the protein value of such a diet would be adequate for a growing infant and would not require .methionine addition. Lysine addition to most cereals should be of practical value in improving poor diets in which most of the protein is supplied by cereals. This is especially true for wheat, millet, and sorghum. Methionine addition should be beneficial when the diet is primarily noncereal, but is based on root crops and pulses. Poor rice diets in which small amounts of pulses are eaten together with the rice could conceivably also benefit from methionine addition. It is further suggested that protein and amino acid supplements be viewed as partners rather than as competitors in practical supplementation situations. For example, groundnut and cottonseed flours by themselves are poor supplements to lysine-poor cereals because of the low concentrations of lysine in the supplements. However, in combination with lysine both these widely available protein sources become very effective in improving the protein value of cereal diets.

ACKNOWLEDGEMENT

The author would like to gratefully acknowledge the invaluable assistance of Dr. Judson Harper, Head, Department of Agricultural Engineering, Colorado State University, for writing the computer program needed to make the calculations on multiple chemical scores that have been used in this analysis.

BIBLIOGRAPHY

BLOCK, R. J., and MITCHELL, H. H. 1946. The correlation of the amino acid composition of proteins with their nutritive value. Nutr. Abst. Rev. *16*, 249-278.

BOLOURCHI, S., FRIEDEMANN, C. M., and MICKELSEN, O. 1968. Wheat flour as a source of protein for adult human subjects. Am. J. Clin. Nutrition *21*, 827-832.

DANIEL, V. A. *et al.* 1970. Improvement of protein value of poor wheat diet by supplementation with limiting amino acids. Nutr. Repts. Intern. *1*, 169-174.

DESAI, B. L. M. *et al.* 1970. Improvement of protein value in poor rice diet by supplementation with limiting essential amino acids. Nutr. Repts. Intern. *1*, 113-118.

DOLE, V. P. 1957. A relation between dietary protein and total caloric intake of large population groups. Am. J. Clin. Nutr. *5*, 72-75.

FO. 1970. The Amino Acid Content of Foods. Food Agr. Organ. U.N., Rome.

GRAHAM, G. G. *et al.* 1969. Lysine enrichment of wheat flour: evaluation in infants. Am. J. Clin. Nutr. *22*, 1459-1468.

HEGSTED, D. M., TRULSON, M. F., and WHITE, H. S. 1956. Lysine and methionine supplementation of all vegetable diets for human subjects. J. Nutr. *59*, 555-576.

HOWE, E. E., JANSEN, G. R., and ANSON, M. L. 1967. An approach toward the solution of the world food problem with special emphasis on protein supply. Am. J. Clin. Nutr. *20*, 1134-1147.

JANSEN, G. R. 1962. Lysine in human nutrition. J. Nutr. *76*, Suppl. 1, Part II, 1-35.

JANSEN, G. R. 1969. Total protein value of protein- and amino acid-supplemented bread. Am. J. Clin. Nutr. *22*, 38-43.

JANSEN, G. R., and HOWE, E. E. 1964. World problems in protein nutrition. Am. J. Clin. Nutr. *15*, 262-274.

McLAUGHLIN, J. M., and CAMPBELL, J. A. 1969. Methodology of protein evaluation. *IN* Mammalian Protein Metabolism, Vol. III, H. N. Munro and J. B. Allison (Editors). Academic Press, New York.

NARAYANASWAMY, D. *et al.* 1970. Improvement of protein value of poor kaffin forn (sorghum vulgare) diet by supplementation with limiting amino acids. Nutr. Repts. Intern. *1*, 293-303.

ORR, M. L., and WATT, B. K. 1957. The amino acid content of foods. Home Econ. Res. Rept *4*, USDA, Washington, D.C.

ROSE, W. C., BORMAN, A., COON, M. J., and LAMBERT, G. F., 1955. The amino acid requirements of man. X. The lysine requirement. J. Biol. Chem. *214*, 579.

SWAMINATHAN, M. 1969. Amino Acid and protein requirements of infants, children and adults. J. Nutr. Dietetics *6*, 356-377.

USDA. 1962. The world food budget, 1962 and 1966. Foreign Agr. Econ. Rept. *4*. USDA, Washington, D.C.

WHO. 1965. Protein requirements. WHO Tech. Rept. Ser. *301*. World Health Organ., Geneva.

Arthur Dalby
Gloria B. Cagampang
Iolo ab I. Davies
and
John J. Murphy

Biosynthesis of Proteins in Cereals

Despite the breadth implied by the above title it seemed that perhaps an appropriate and useful purpose would be served by confining our remarks to maize, and, more especially, to bringing together various data on the effects of the mutant gene, *opaque-2*, on protein synthesis in order to discuss the regulatory aspects of this gene. In this way it is hoped a more comprehensive picture will result which may be of benefit both to maize researchers and to workers studying other cereals. To this end the subject will be treated somewhat speculatively.

The rational for this approach is that the discovery of the biochemical effects of *opaque-2* and *floury-2* in maize (Mertz *et al.* 1964; Nelson *et al.* 1965) has provided opportunities for investigating storage protein synthesis not yet matched in other cereals, with the possible exception of barley (Munck *et al.* 1969). Indeed, judging from the low concentrations of prolamines in rice and oats, perhaps the equivalents of the maize mutant genes are already present, and what is required is a good "normal" for comparison. Furthermore, in view of the unity which exists amongst the cereal grains, e.g., the occurrence in common of a unique group of proteins (the prolamines) or the fact that when bred for higher protein content cereals tend to behave in a similar fashion and accumulate more prolamine, we are encouraged to think that conclusions drawn regarding storage protein synthesis in maize may well be of value in understanding the same process in other cereals.

For purposes of discussion we may divide the proteins of maize endosperm into three groups. (1) Water- and saline-soluble proteins which during fractionations are generally accompanied by the free amino acids. (2) Alcohol-soluble prolamines of the zein complex. (3) Glutelins, generally soluble only under the more extreme conditions of higher pH and/or reducing conditions.

In terms of absolute amount, the first group—the saline- and water-soluble proteins—is of limited significance, contributing less than 10% to the total endosperm nitrogen. Functionally, it is of major importance because of its enzyme content. The converse is true of the other two protein groups, their primary functional role apparently being to act as a repository of nitrogen for the subsequent benefit of the embryo at the time of germination. It is interesting, therefore, to examine what effect the presence of the *opaque-2* gene has on these various protein groups during endosperm development.

In terms of total nitrogen the water- and saline-soluble fraction of *opaque-2* endosperm is higher than in normal endosperm throughout most of development (Dalby 1966). The difference can be largely accounted for on the basis of elevated levels of glutamine, asparagine and alanine (Sodek and Wilson 1970; Murphy and Dalby 1971). There are also quite specific changes in the concentration of certain enzyme and protein components, as demonstrated by polyacrylamide gel electrophoresis (Fig. 4.1).

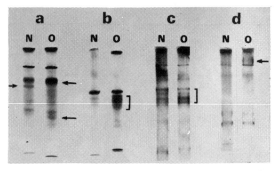

Courtesy of I. I. Davies and A. Dalby

FIG. 4.1. POLYACRYLAMIDE GEL ELECTROPHORETOGRAMS OF 22-DAY W64A NORMAL (N) AND *opaque-2* (O) CRUSHED ENDOSPERM (1000 GM SUPERNATE)

Standard 7% gel, stacking at pH 8.9 and running at pH 9.5. Overage acrylamide was used in preparing gels of b. a and b—proteins stained with 0.5% aniline black in 7% acetic acid. c—phosphatase substrate: α-naphthyl phosphate in 0.05 M acetate buffer, pH 5.5; coupling dye: Diazo Garnet GBC (Allen and Gockerman 1964). d—esterase substrate: α-naphthyl acetate in 0.2 M tris-HC1 buffer, pH 7.5; coupling dye: Fast Blue RR (Markert and Hunter 1959).

In Fig. 4.1a and 4.1b are shown the electrophoretic patterns of the proteins in the juice from 22-day normal and *opaque-2* endosperm (W64A) stained with aniline black. Equivalent amounts of each extract, based on protein determinations (Lowry *et al.* 1951), were applied. The normal pattern in Fig. 4.1a shows the presence of a band (arrowed) which is barely detectable in *opaque-2*. In the *opaque-2* pattern there are two bands of increased intensity relative to normal (arrowed). No other differences could be detected in the original gels. The corresponding embryo patterns showed no differences. Figure 4.1b shows the same endosperm extracts run in gels prepared from "aged" acrylamide. The *opaque-2* pattern now shows a series of 6 bands (bracketed) compared with only 2 in normal. It seems possible that these six bands are derived from the heavy band shown in the *opaque-2* pattern of Fig. 4.1a (upper arrow); other differences are also visible. Figure 4.1c shows bands of phosphatase activity. The four bands bracketed in the *opaque-2* pattern are also present in normal. However, in the *opaque-2* pattern the 2 lower bands are much more intense than the 2 upper bands; differences which are not seen in the normal pattern. In Fig. 4.1d are shown bands of esterase activity. The distinct band arrowed in the *opaque-2* pattern is not detectable in normal.

In addition to these changes it is known that the ribonuclease activity of *opaque-2* endosperm is markedly increased over that of normal (Wilson and Alexander 1967; Dalby and Davies 1967). However, there are at least 4 enzymes which are not changed in 22-day material (Tsai 1971). These are sucrose synthetase, glucose-6-phosphate isomerase, ADP-glucose pyrophosphorylase and UDP-glucose pyrophosphorylase. Finally, we may note the possible absence from immature *opaque-2* of an enzyme or enzyme system catalyzing the conversion of lysine to glutamic acid in normal (Sodek and Wilson 1970).

Considering next the zein proteins, we may simply remark that, of all the changes produced by the mutant gene, the lowering of the zein content of *opaque-2* endosperm is perhaps the most consistent and well known (Mertz *et al.* 1964; Mossé *et al.* 1966; Concon 1966; Jiménez 1966; Dalby 1966; Wolf *et al.* 1967; Murphy and Dalby 1971). So far as the zein of mature *opaque-2* seed is concerned, the reduction is accompanied by some variation from normal in the relative amounts of the various electrophoretic components according to Mossé (1966), while Jiménez (1968) found some components completely absent—there is some disagreement here. Both are agreed that there appears to be no new zein species produced.

Turning to the glutelin fraction, early results showed an absolute increase in the glutelins compensating for the decline in zein content in *opaque-2* (Jiménez 1966; Dalby 1966; Murphy and Dalby 1971). At the time, it was suggested that the attenuated synthesis of zein in *opaque-2* possibly permitted the plant's resources to be directed towards nonzein protein synthesis leading to, for example, more glutelin (Dalby 1966). However, it now appears that an increase in glutelin is not a necessary accompaniment to a lowering of zein content, but is probably a function of the particular inbred line (Sodek and Wilson 1970).

Taken as a whole, the data suggest that the effects of *opaque-2* involve the regulation of quite specific protein synthesis, and not protein synthesis *in toto* as at one time appeared possible. It should be emphasized that there is no intention of suggesting that the mutant gene is *directly* responsible for all these changes.

The *opaque-2* gene must be considered as a regulator and not a structural gene, since it is recessive and shows no dosage effect with respect to the various modifications it produces, although there is some indication of "leakiness" (Bates 1966; Dalby and Cagampang 1970). The only dosage effect to be anticipated would be with respect to the immediate gene product. The next point, then, is whether *opaque-2* regulates one key protein (enzyme) from which all else follows, or acts as a master regulator for a number of proteins in some way. At the present time there is insufficient data to decide. However, perhaps it is worth looking at one possibility of the first alternative which has received some attention.

At the height of their enthusiasm on finding an elevated level of ribonuclease in *opaque-2*, Dalby and Davies (1967) and Wilson and Alexander (1967) speculated that ribonuclease might be the key, possibly by destroying messenger RNA more rapidly in the mutant than in normal. Although the bulk of the ribonuclease must be held in some inactive state, it is possible to conceive of equilibria which would result in a higher free ribonuclease concentration in *opaque-2* than in normal. Denić (1970) has suggested the possibility of a balance between ribonuclease and ribonuclease inhibitors. The appeal of such a role for ribonuclease is, of course, that it attacks substances of immediate importance in protein synthesis. However, any such system would be nonselective, thereby modifying the synthesis of all proteins. From the examples already given this is patently not true. The alternative is to propose some form of compartmentation of protein synthesis with selective attack on, say, zein messenger RNA within its compartment. We are reminded here of the dual system of protein synthesis proposed by Morton and his colleagues for wheat endosperm (Graham *et al.* 1964; Morton and Raison 1964).

A further possibility is the selective destruction by ribonuclease of specific transfer RNA species. Wilson (1963, 1968) demonstrated that maize endosperm ribonuclease (RNase I) has a base specificity for purines exceeding that for pyrimidines. Thus, the possible preferential destruction of purine-rich anticodon sites in tRNA over corresponding pyrimidine-rich sites could lead to a limiting situation in which the relative rates of synthesis of various proteins were affected. The discovery (Murphy and Dalby 1971) of a grossly atypical "zein" in 15-day *opaque-2* prompted an examination of this possibility. Table 4.1 shows the amino acid composition of ethanol extracts from 15-day normal and *opaque-2* endosperm together with their corresponding anticodon triplet compositions based on the number of purine or pyrimidine bases in each triplet. The first group (I) of amino acids were all present at higher levels in *opaque-2* than in normal, and were represented by anticodon triplets having 3 or 2 pyrimidine bases. The third group (III) of amino acids were all present at lower levels in *opaque-2* than in normal, and had anticodon triplets containing either 3 or 2 purine bases. Alanine and histidine fall into an indeterminate group (II). If, as outlined above, preferential destruction by ribonuclease occurs, then it might be anticipated that in *opaque-2* the tRNA fraction would contain higher relative concentrations of active tRNA species for group I than group III, when compared to normal. This possibility was examined to the extent of measuring the amino acid acceptor activity of tRNA fractions isolated from 22-day normal and *opaque-2* endosperm. The activations were catalyzed by a ribonuclease-free, amino acyl-tRNA synthetase preparation from the high speed supernatant fraction of 22-day normal endosperm. A summary of the results is shown in Table 4.2.

TABLE 4.1

AMINO ACID COMPOSITION OF 70% ETHANOL EXTRACTS
OF 15-DAY NORMAL AND *OPAQUE-2* ENDOSPERM FROM W64A

Group	Amino Acid	Amino Acid Composition (Mole %)		Anticodon Triplet Composition[1] (Pu = purine; Py = pyrimidine)			
		opaque-2	Normal	Py_3	Py_2Pu	$PyPu_2$	Pu_3
I	Lysine	1.6	0.25	+			
	Glutamic acid	30.4	23.9	+			
	(GluN)				+		
	Glycine	6.1	2.5	+	+		
	Arginine	1.3	0.89	+	+	+	
	Aspartic acid	11.1	4.5		+		
	(AspN)				+		
	(Tryptophan)				+		
	Threonine	2.5	1.7		+	+	
	Serine	6.0	2.3		+	+	+
	Methionine	1.9	1.1		+		
II	Alanine	13.4	15.7	+	+		
	Histidine	0.79	0.76		+		
III	Valine	2.0	4.1	+	+		
	Isoleucine	2.8	4.1	+	+		
	Tyrosine	1.4	2.7		+		
	Cystine	—	—		+		
	Proline	5.8	10.0		+		+
	Leucine	10.4	20.2		+		+
	Phenylalanine	2.6	5.2				+

Source: Murphy and Dalby (1971).
[1] Based on Crick (1966).

TABLE 4.2

RELATIVE AMINO ACID ACCEPTOR ACTIVITIES OF
TRANSFER RNA FROM 22-DAY NORMAL AND *OPAQUE-2* W64A ENDOSPERM

Amino Acid (AA)	$tRNA^{AA}/tRNA^{Leu}$		Normal $tRNA^{AA}$ / *opaque-2* $tRNA^{AA}$	
	opaque-2 (%)	Normal (%)	per Mg RNA Basis	per Endosperm Basis
Leucine	100	100	1.4	1.1
Valine	74	86	1.7	1.2
Tyrosine	23	30	1.8	1.3
Arginine	60	82	2.0	1.5
Glycine	56	84	2.1	1.6

Source: Dalby, unpublished results.

Relative to leucine tRNA the other amino acid specific tRNAs were lower in *opaque-2* than in normal (columns 2 and 3). Further, the tRNA for any particular amino acid was higher in normal endosperm than in *opaque-2* endosperm, whether the results are expressed on a per milligram RNA basis or on a per endosperm basis (columns 4 and 5). It may be noted that 34% more tRNA fraction was recovered from *opaque-2* than from normal on a per endosperm basis; 26% on a per gram fresh endosperm basis. This is reflected in the lower normal:*opaque-2* of column 5 compared with column 4, the larger total amounts of *opaque-2* tRNA compensating to some extent for the lower specific amino acid acceptor activity of the *opaque-2* tRNA. Thus, there appeared to be more inactive RNA in the *opaque-2* tRNA preparation, a finding consistent with the higher ribonuclease content of *opaque-2*. However, the fact that the arginine and glycine tRNA acceptor activities in *opaque-2* were even lower than the acceptor activities for leucine, valine and tyrosine when compared to normal, seemed at the time to negate the hypothesis that ribonuclease is attacking tRNA species specifically and affecting protein synthesis thereby. As a result this line of research was abandoned. Upon reflection perhaps this decision was premature, since there is evidence that some tRNA species when split at the anticodon site are still capable of accepting amino acids, but not transferring them at the ribosomal level (Nishimura and Novelli 1965; Bayev *et al.* 1968). Thus, relative amino acid acceptor activity is not an adequate criterion, and the question remains open. Further, what was said previously with respect to compartmentation of protein synthesis is equally applicable here.

There also exists the possibility that the synthetase preparation from normal endosperm used in the preceding study was inadequate for *opaque-2* tRNA. That it was inadequate for some amino acids, including lysine, in the sense that the specific synthetases were not present in sufficient concentration, was established for both tRNA preparations. However, for the five amino acids listed in Table 4.2 the preparation was satisfactory, based on the test that the amount of labeled amino acid bound was proportional to the concentration of the tRNA preparation. Another type of inadequacy would arise if *opaque-2* endosperm contains synthetases differing in either specificity or activity from those found in normal. In this connection it is interesting that, for lysine and phenylalanine tRNAs, Denić (1969) has reported differences between normal and *opaque-2* synthetase preparations from immature endosperm towards a particular endosperm tRNA preparation, as well as differences between normal and *opaque-2* endosperm tRNA preparations similar to those reported here.

As an interesting aside, Denić (1969) appears to have found a difference between the embryos of the mature seed of normal and *opaque-2*. Amino acyl-tRNA synthetase preparations from normal embryo inhibit the activation of lysine, but not phenylalanine, by mutant embryo synthetase preparations. This, together with the findings of Jolivet *et al.* (1970) that normal and *opaque-2* seedlings show differing distributions of intermediary metabolites after pulse-labeling experiments with $^{14}CO_2$, suggests that the mutant gene may also be active in diploid tissue.

A second line of evidence tending to reduce any suggested role for ribonuclease arises from an examination of the ribonuclease development curves of the normal and *opaque-2* versions of nine inbred lines of maize. These are shown in Fig. 4.2. Seven of the lines behave similarly insofar as they exhibit an

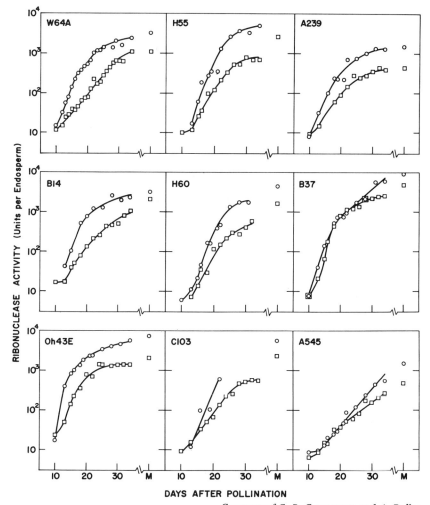

DAYS AFTER POLLINATION

Courtesy of G. B. Cagampang and A. Dalby

FIG. 4.2. RIBONUCLEASE DEVELOPMENT CURVES OF *opaque-2* (O_2) AND NORMAL (N) VERSIONS OF NINE INBRED LINES OF MAIZE

Opaque-2 is shown by circles. Normal is shown by squares.

early rapid increase in ribonuclease activity in the *opaque-2* version, contrasted with their corresponding normal, and maintain the resulting difference through to maturity. The two remaining inbreds, A545 and B37, do not follow this pattern. Here, differences in ribonuclease activity between the normal and mutant versions only arise later in development, but then continue through to maturity. Thus, it would appear that the pattern shown by the seven inbreds, which we may term the W64A pattern, is not a constant feature of *opaque-2*. It could be argued that, thereby, ribonuclease loses its position as a potential key in the effect of *opaque-2* on protein synthesis. It is interesting to note, however, that A545 and B37 were also anomalous inasmuch as they did not exhibit the homogenous floury type of endosperm customarily associated with *opaque-2*, but also contained regions of horny endosperm. The genetists postulate modifier genes to account for the changed phenotype of A545 and B37. It seems reasonable to suggest, therefore, that perhaps the same modifier genes are responsible for the changed developmental pattern of ribonuclease activity. Again then, a role for ribonuclease cannot be completely discarded.

Ribonuclease is of interest in another connection concerning the timing of *opaque-2* activity, as shown in Fig. 4.3. In this semilogarithmic plot of the accumulation of ribonuclease activity during development there are 2 linear phases in *opaque-2* but only 1 in normal. Phase I in *opaque-2* shows a faster rate

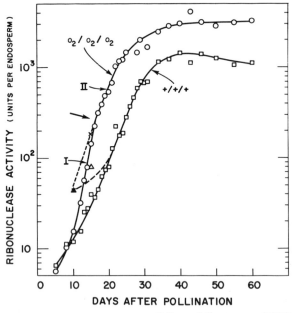

Dalby and Cagampang (1970)

FIG. 4.3. RIBONUCLEASE DEVELOPMENT CURVES OF *opaque-2* (O_2) AND NORMAL (N) VERSIONS OF W64A

of accumulation of ribonuclease than in normal. After the break at 16–17 days, the rate during phase II is the same as that shown by normal. These data have been interpreted as indicating that prior to 16 days the *opaque-2* gene is active, whereas after 16 days it becomes inactive (Dalby and Cagampang 1970). The last statement should be qualified by adding "so far as ribonuclease is concerned," since there is no evidence that ribonuclease accumulation is under the direct control of the *opaque-2* gene. A point of fundamental importance for the above interpretation is that the accumulation of ribonuclease, at least in its early stages, is an exponential process. This is difficult to understand as we are accustomed to thinking of protein synthesis as a linear process. If this were a bacterial culture growing exponentially then there would be no problem, but it is difficult to explain how the amount by which an enzyme increases is related to its concentration. Either it is necessary to invoke continuous gene replication after the fashion of a bacterial system or to look for a solution in the kinetics of the process. In the latter case the observations of Schimke (1970) on the kinetics of changes in enzyme concentration from one steady state to another may have relevance.

In the light of what has been said regarding ribonuclease accumulation during development, it is of interest now to examine the detailed pattern of zein accumulation in the same material, as shown in Fig. 4.4. The results confirm the

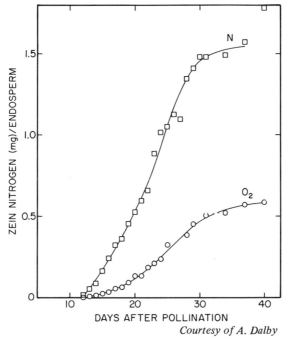

DAYS AFTER POLLINATION

Courtesy of A. Dalby

FIG. 4.4. ZEIN ACCUMULATION DURING DEVELOPMENT OF *opaque-2* (O_2) AND NORMAL (N) VERSIONS OF W64A

general pattern of zein accumulation in normal and *opaque-2* endosperm observed in our previous studies (Dalby 1966; Murphy and Dalby 1971). In absolute terms, the rate of zein accumulation is always greater in normal than in *opaque-2*. However, in neither case is the process linear, but rather sigmoid, and resembles the ribonuclease development curves when the latter are plotted using linear coordinates. In Fig. 4.5 the zein data have been transposed onto a semilogarithmic plot. It now becomes clear that even at very early ages some zein synthesis is occuring in *opaque-2*, but at a lower rate than in normal. Presumably, if it were possible to make the necessary zein determinations these curves would approach each other asymptotically. At no stage do the curves appear to be linear. However, it is interesting that by a simple shift in the position of the axes, the *opaque-2* curve from about 17–18 days can be superimposed upon the normal curve from about 15 days. It is difficult to accept that this ability to superimpose the whole of the later parts of the curves in this way is solely due to chance. Rather we would suggest that zein accumulation follows a decaying exponential curve which is the same in *opaque-2* after 18 days as in normal after 15 days; before these ages the curves, and hence the rates of zein accumulation, are different. If this is accepted, then the pattern for zein accumulation falls into line with the conclusions drawn concerning ribonuclease accumulation. Namely, that zein accumulation is only affected by the mutant gene prior to 17–18 days. After that time the process is the same as in normal

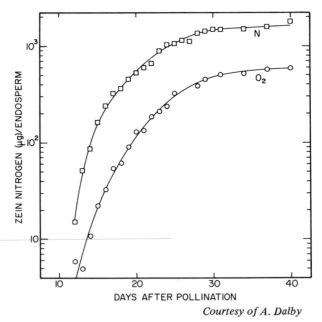

Courtesy of A. Dalby

FIG. 4.5. ZEIN ACCUMULATION DURING DEVELOPMENT OF *opaque-2* (O_2) AND NORMAL (N) VERSIONS OF W64A

endosperm. As with ribonuclease, it is only necessary to have the initial difference in rates, and hence in levels of zein attained, in order to account for the large differences in zein content seen in mature seed. (This would perhaps be the strongest argument against an important role for ribonuclease since, although the difference in ribonuclease concentrations still exists, zein accumulation in *opaque-2* is now following the normal pattern). In summary, it may be said that the ribonuclease and zein data are consistent with the notion that the *opaque-2* gene in W64A only affects their synthesis during the first 2½ weeks after pollination. Thereafter, these processes appear to be regulated as in normal endosperm. The "zein" of unusual amino acid composition isolated from 15-day *opaque-2*, and the more normal zeins found subsequently (Murphy and Dalby 1971), point in the same direction. On this basis then, it might be postulated that in the two inbreds not showing the W64A pattern of ribonuclease accumulation (Fig. 4.2) the timing of the effective phase for ribonuclease of the *opaque-2* gene is altered. It will be interesting to establish the pattern of zein accumulation in these two inbreds. Although it might be argued that the continuous action of the *opaque-2* gene throughout development would explain the results shown in Fig. 4.4, it is difficult to explain the semilogarithmic plots of either zein or ribonuclease accumulation on this basis.

The very interesting results of Sodek and Wilson (1970) on the fate of isotopically-labeled leucine and lysine in the developing maize seed suggests possibly supporting evidence for the above hypothesis. These workers found that whereas normal endosperm is capable of converting lysine into glutamic acid and proline, the endosperm of *opaque-2* is not, except during later development. Indeed, Sodek and Wilson suggest the possibility of a correlation between lysine conversion and zein synthesis. Since the inability to convert lysine to glutamic acid is related, directly or indirectly, to the activity of the mutant gene, it seems reasonable to suggest, on the basis of the above hypothesis, that the ability of *opaque-2* endosperm to make the conversion may start to develop at the same time that the normal mechanism of ribonuclease and zein accumulation takes over. This should be capable of being tested experimentally.

SUMMARY

Although the *opaque-2* gene produces many changes in the proteins of maize endosperm its effect appears to be directed towards specific protein synthesis rather than towards protein synthesis in general. The possibility that the higher ribonuclease activity found in *opaque-2* is responsible for the modified protein and amino acid composition seems, on balance, unlikely. A comparison of the detailed development curves of ribonuclease and zein accumulation could be interpreted as indicating that the *opaque-2* gene only affects their rates of increase during the first 2½ weeks after pollination in the inbred W64A.

BIBLIOGRAPHY

ALLEN, J. M., and GOCKERMAN, J. 1964. Electrophoretic separation of multiple forms of particle associated acid phosphatases. Ann. N.Y. Acad. Sci. *121*, 616-633.

BATES, L. S. 1966. Amino acid analysis. *In* Proceedings of the High Lysine Corn Conference, E. T. Mertz, and O. E. Nelson (Editors). Corn Ind. Res. Found., Washington, D. C.

BAYEV, A. A. *et al.* 1968. Studies of the primary structure and function of yeast valine transfer RNA 1. *In* Structure and Function of Transfer RNA and 5S-RNA, L. O. Fröholm, and S. G. Laland (Editors). Academic Press, London and New York.

CONCON, J. M. 1966. The protein of *opaque-2* maize. *In* Proceedings of the High Lysine Corn Conference, E. T. Mertz, and O. E. Nelson (Editors). Corn Ind. Res. Found., Washington, D.C.

CRICK, F. H. C. 1966. The genetic code. Cold Spring Harbor Symp. Quant. Biol. *31*, 1.

DALBY, A. 1966. Protein synthesis in maize endosperm. *In* Proceedings of the High Lysine Corn Conference, E. T. Mertz, and O. E. Nelson (Editors). Corn Ind. Res. Found., Washington, D.C.

DALBY, A., and CAGAMPANG, G. B. 1970. Ribonuclease activity in normal, *opaque-2*, and *floury-2* maize endosperm during development. Plant Physiol. *46*, 142-144.

DALBY, A., and DAVIES, I. ab I. 1967. Ribonuclease activity in the developing seeds of normal and *opaque-2* maize. Science *155*, 1573-1575.

DENIC, M. 1969. On the role of some components involved in the synthesis of storage protein in maize. Agrochimica *13*, 143-150.

DENIC, M. 1970. Role of the microsomal fraction in the regulation of protein synthesis in maize endosperm. *In* Improving Plant Protein by Nuclear Techniques. International Atomic Energy Agency, Vienna. (French)

GRAHAM, J. S. D., MORTON, R. K., and RAISON, J. K. 1964. The *in vivo* uptake and incorporation of radioisotopes into proteins of wheat endosperm. Australian J. Biol. Sci. *17*, 102-114.

JIMENEZ, J. R. 1966. Protein fractionation studies of high lysine corn. *In* Proceedings of the High Lysine Corn Conference, E. T. Mertz, and O. E. Nelson (Editors). Corn Ind. Res. Found., Washington, D.C.

JIMENEZ, J. R. 1968. The effect of the *opaque-2* and *floury-2* genes on the production of protein in maize endosperm. Ph.D. Thesis, Purdue Univ.

JOLIVET, E., NICOL, M., BAUDET, J., and MOSSE, J. 1970. Differences in $^{14}CO_2$ incorporation by normal and *opaque-2* maize seedlings exposed to light. *In* Improving Plant Protein by Nuclear Techniques. International Atomic Energy Agency, Vienna. (French)

LOWRY, O. H., ROSEBROUGH, N. J., FARR, A. L., and RANDALL, R. J. 1951. Protein measurement with the Folin phenol reagent. J. Biol. Chem. *193*, 265-275.

MARKERT, C. L., and HUNTER, R. L. 1959. The distribution of esterases in mouse tissues. J. Histochem. Cytochem. *7*, 42-49.

MERTZ, E. T., BATES, L. S., and NELSON, O. E. 1964. Mutant gene that changes protein composition and increases lysine content of maize endosperm. Science *145*, 279-280.

MORTON, R. K., and RAISON, J. K. 1964. The separate incorporation of amino acids into storage and soluble proteins catalyzed by two independent systems isolated from developing wheat endosperm. Biochem. J. *91*, 528-539.

MOSSE, J. 1966. Alcohol soluble proteins of cereal grains. Federation Proc. *25*, 1663-1669.

MOSSE, J., BAUDET, J., LANDRY, J., and MOUREAUX, T. 1966. Study on the proteins of maize. II. Comparison between the amino acid compositions, and the mutual proportions of the protein fractions, of normal and mutant grains. Ann. Physiol. Végétale *8*, 331-344. (French)

MUNCK, L., KARLSSON, K-E., and HAGBERG, A. 1969. Selection and characterization of a high-protein, high-lysine variety from the World Barley Collection. 2nd Intern. Barley Genetics Symp., Pullman, Wash. (in press).

MURPHY, J. J., and DALBY, A. 1971. Changes in the protein fractions of developing normal and *opaque-2* maize endosperm. Cereal Chem. *48*, 336-349.

NELSON, O. E., MERTZ, E. T., and BATES, L. S. 1965. Second mutant gene affecting the amino acid pattern of maize endosperm proteins. Science *150*, 1469-1470.

NISHIMURA, S., and NOVELLI, G. D. 1965. Dissociation of amino acid acceptor function of sRNA from its transfer function. Proc. Natl. Acad. Sci. U.S. *53*, 178-184.

SCHIMKE, R. T. 1970. Protein turnover and control of enzyme levels in animal tissues. Accounts Chem. Res. *3*, 113-120.

SODEK, L., and WILSON, C. M. 1970. Incorporation of leucine-^{14}C and lysine —[14]C into protein in the developing endosperm of normal and *opaque-2* corn. Arch. Biochem. Biophys. *140*, 29-38.

TSAI, C. Y. 1971. Personal communication. Purdue University, Lafayette, Ind.

WILSON, C. M. 1963. Substrate and product specificity of two plant ribonucleases. Biochim. Biophys. Acta *76*, 324-326.

WILSON, C. M. 1968. Plant nucleases. II. Properties of corn ribonucleases I and II and corn nuclease I. Plant Physiol. *43*, 1339-1346.

WILSON, C. M., and ALEXANDER, D. E. 1967. Ribonuclease activity in normal and *opaque-2* mutant endosperm of maize. Science *155*, 1575-1576.

WOLF, M. J., KHOO, U., and SECKINGER, H. L. 1967. Subcellular structure of endosperm protein in high-lysine and normal corn. Science *157*, 556-557.

CHAPTER 5

Julius W. Dieckert
and
Marilyne C. Dieckert

The Deposition of Vacuolar
Proteins in Oilseeds

INTRODUCTION

As early as 1872 Pfeffer noted that dormant seeds commonly contain numerous intracellular protein granules. Even earlier, Hartig had isolated these granules from several oilseeds by a nonaqueous technique, demonstrating that they contained protein (1855). Hartig (1856) named these granules "aleuron grains."

Pfeffer (1872) also reported that aleurone grains contain metals and organic phosphorus compounds. Recently, Dieckert et al. (1962) demonstrated that the aleurone grains of peanut seeds contain virtually all of the phytic acid of the seed and a large fraction of the cellular potassium, magnesium, manganese, and copper.

In 1964, Altschul et al. proposed that the proteins found in the aleurone grains be called "aleurins." Under this definition, the following seed proteins may be classed as aleurins: arachin, from peanut seed (Altschul et al. 1964; Daussant et al. 1969); edestin, from hempseed (St. Angelo et al. 1968); the reserve globulins of pea seeds (Varner and Schidlovsky 1963); vicilin and legumin, from Vicia faba (Graham and Gunning 1970).

The protein granules of plant cells may originate from a variety of subcellular organelles. For example, Bonnet and Newcomb (1965) reported that proteinaceous material accumulates in dilations of the granular endoplasmic reticulum in the root cells of radish. In a more recent paper, Newcomb (1967) demonstrated the presence of protein crystals in the plastids of bean root tips. Mottier (1921) concluded that the protein bodies of corn, castor bean and Conopholis arise from plastids. The conclusion was based on observations with the light microscope that plastids and protein granules seemed to derive from spherical or rod-shaped precursors. In the case of corn and castor bean, the protein bodies were typical aleurone grains.

The hypothesis that the aleurone grains of seeds develop inside of plastids is not currently accepted. A more likely view is that they develop in a vacuole. The observations supporting the vacuolar hypothesis are as follows: anthocyanins and calcium oxalate, which are known to occur in plant vacuoles, are found in certain aleurone grains [See Guilliermond (1941) for a general review of the older literature]; aleurone grains derive from structures that can accumulate neutral red, another characteristic of plant vacuoles; the forming aleurone grains of wheat (Graham et al. 1962; Buttrose 1963) and of cotton (Engleman 1966) are surrounded by a lipoprotein membrane; a single unit membrane, the structure of which is similar to the tonoplast, was observed around the mature aleurone grains of cotton seed (Yatsu 1965). If the vacuolar origin of the aleurone grain is correct, the question still remains open as to where the aleurins are synthesized and how they get into the vacuole.

Some insight into the problem can be gained from a consideration of an analogous process found in animal cells, zymogen granule formation. For example, cells of the exocrine pancreas elaborate zymogens of some of the digestive enzymes. These are excreted from the cell into the pancreatic duct, which, in turn, empties into the duodenum. A brief annotated review of the process is given by Dieckert (1971). Studies based on light and electron microscopy show that the cytoplasm in the neighborhood of the acinus contains numerous protein granules, each bounded by a lipophilic membrane. The proteins of the zymogen granule are synthesized by the polyribosomes of granular endoplasmic reticulum and concentrated in elements of the Golgi apparatus to form the intracellular protein granules. In response to certain stimuli, the bounding membrane of the zymogen granule fuses with the plasma membrane facing the acinus. The inside of the vesicle, containing the protein granule, is now continuous with the outside of the cell, leaving the protein in the acinus.

The data summarized in the present report support the hypothesis that the process of aleurone grain formation is analogous to that of zymogen granule formation in animal cells. Brief descriptions of the work on shepherd's purse and cotton appear elsewhere (Dieckert 1969, 1971).

MATERIALS AND METHODS[1]

Embryos at different stages of development were obtained from shepherd's purse (*Capsella bursa-pastoris, Medic.*), peanut (*Arachis hypogaea, L.,* large-seeded variety), and cotton (*Gossypium hirsutum, L.,* variety Delta Pineland, No. 14). Shepherd's purse embryos were taken from wild plants growing on the Texas A & M campus or from wild plants grown from seed in a greenhouse. Peanut and cotton embryos were obtained from plants cultivated in a greenhouse.

For a given experiment, a developmental series of shepherd's purse embryos, beginning with the "torpedo" stage, were obtained from a single raceme. The flowers of shepherd's purse are at the tip of the inflorescence; and, the siliques, which contain the embryos of increasing physiological age, are arranged in a spiral series toward the base. The stages of development of the embryo *in ovulo* are reviewed by Rijven (1952). Some of the later stages of particular interest here are shown in Fig. 5.1. Capsella embryos were excised from the ovules and fixed *in toto*.

[1] The authors are grateful for the technical assistance of Annamaria Wood, Netty Perez, Gertrude Adam, Anne Sandel, P. S. Baur and K. S. Derrick, who assisted in the work supporting this portion of the manuscript.

The research was supported in part by NSF Grant No. GB29, ARS-USDA Grant No. 12-14-100-7988 (34), and Cotton Producers Institute Project 69-140.

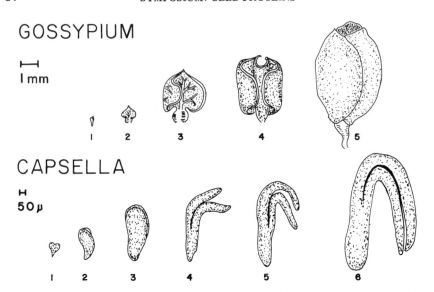

FIG. 5.1. STAGES IN THE DEVELOPMENT OF THE EMBRYOS OF COTTON AND SHEPHERD'S PURSE

Gossypium: 3—barely-rolled; 4—loose scroll; 5—mature seed. Capsella: 4 and 5—walking stick; 6—early upturned-U.

Cotton flowers were tagged at anthesis, and bolls were collected daily from 19 days post-anthesis onward. The cotton embryo exhibits characteristic changes in morphology with development. An approximate timetable for some of the important stages, and the gross morphology of each, are given in Fig. 5.1. Transverse strips about 1 mm wide were excised from the mid-region of the cotton embryos at key stages, beginning with the "barely-rolled" embryo (Fig. 5.1 item 3). Such strips were usually fixed as such, but occasionally they were cut into smaller pieces beforehand.

Unlike shepherd's purse and cotton, the peanut embryo develops underground at the end of a specialized stalk, called a gynophore, sometime after anthesis (Smith 1950). Soon after the tip of the gynophore penetrates the soil, the embryo starts to develop. Since a variable length of time is required for the tip of the gynophore to enter the soil, the age of the embryo was estimated from the time elapsed after entrance. The thickness and length of the cotyledons were also used as a measure of physiological age. Cubes of peanut tissue 1 mm long were cut from a transverse slice taken about midway along the length of the cotyledon and fixed.

All tissues were fixed in glutaraldehyde (Fisher biological grade) to immobilize proteins in the cells. Some tissue samples were also postfixed in osmium tetroxide to stabilize the lipids of the cytomembranes and cellular fat droplets. The precise conditions of fixation varied, depending upon the source of the tissue, the point of the experiment, and the state of the art at the time the experiments were carried out. The variables were as follows: concentration of glutaraldehyde or osmium tetroxide, the nature and strength of the buffer, the duration and temperature of the fixation.

Initially, shepherd's purse embryos were fixed in 6% glutaraldehyde in 0.1 M phosphate buffer, pH 6.8, for 1 ½ hr at 0°–4°C and postfixed in 1% osmium tetroxide in the same buffer for 1 hr at room temperature. An alternate procedure consisted of fixation in 2% glutaraldehyde in 0.1 M collidine buffer, pH 6.8, for 1 hr and postfixation in 2% osmium tetroxide in the same buffer for 2 hr at room temperature. In later experiments the tissue was fixed in 3% glutaraldehyde in 0.05 M phosphate buffer, pH 6.7–7.0, for 1 hr at 0°–4°C and postfixed for 1 or 4 hr in 1% osmium tetroxide in the same buffer at 0°–4°C.

All of the cotton embryos were fixed in 3% glutaraldehyde in 0.05 M sodium cacodylate buffer, pH 7.35, for 2 hr at room temperature. Some of these were subsequently postfixed in 1% osmium tetroxide in 0.05 M cacodylate buffer, pH 6.9, for 4 hr at 0°–4°C. After each fixation, residual fixative was removed from the tissue by suitable wash media (usually the buffer) before the next step was taken. For the purposes of the present paper, all of the fixatives were adequate; therefore, the only information recorded will be whether glutaraldehyde alone, or glutaraldehyde followed by osmium tetroxide, served as the fixative. The choice of fixatives and buffers was inspired by the pioneering work of Sabatini *et al.* (1963) on the use of aldehydes for cytochemistry and electron microscopy, and Wood and Luft (1965) on the influence of buffer systems on fixation with osmium tetroxide. The variations in concentration of buffer and fixative components and the duration and temperature of fixation were part of our effort to find optimal conditions for preserving the tissue at hand.

The fixed tissue was prepared for examination with the electron microscope as follows: tissues of shepherd's purse and peanut, as well as some of the cotton tissues, were embedded in epoxy resins according to the method of Winborn (1965). After dehydration in an alcohol series, most of the cotton was embedded in Maraglas according to the method of Spurlock *et al.* (1963). The embedded tissue was sectioned with an Ivan Sorvall MT-2 Ultramicrotome or an LKB Ultratome III, using glass knives. The tissue embedded according to the Winborn procedure sectioned easily, while that in Maraglas was considerably more difficult. Both embedding media gave useful material for ultrastructural analysis. Sections of silver to gray interference colors were routinely obtained. The thin sections were mounted on naked 400 mesh athene-type copper grids or on 0.2% formvar-coated copper slot grids (0.2 X 1 mm slot). The sections on naked grids were stabilized with a thin film of evaporated carbon, as were the formvar films. The sections were stained for electron microscopy first with saturated aqueous uranyl acetate (Watson 1958), methanolic uranyl acetate (Stempak and Ward 1964) or magnesium uranyl acetate (Frasca and Parks 1965), followed by staining with lead citrate (Venable and Coggeshall 1965). The stained thin sections were observed with a Hitachi HU11A electron microscope at 50 or 75 kV accelerating voltage, with objective lens apertures of 30 or 50 μ diameter.

Several experiments were designed to determine the intracellular distribution of protein deposits. Thick sections (0.25–0.5 μ) of shepherd's purse tissue, fixed only in glutaraldehyde and embedded in epoxy resin, were mounted on microscope slides, stained with mercuric bromphenol blue by the method of Mazia *et al.* (1953) and examined by light microscopy. Subcellular structures containing relatively large deposits of protein stained blue. The subcellular distribution of protein deposits in cells of the cotton embryo was determined on thin sections of glutaraldehyde-fixed tissue. Two or more serial sections were separately treated. To make the plastic porous to enzymes, each section was treated with 5% hydrogen peroxide in water. Each section was rinsed free of hydrogen peroxide; then, alternate sections of the series were digested with pronase solution for 2–10 min at 37°C. In the experiments reported here, pronase (Nomoto and Narahashi 1958) was suspended in 0.05 M sodium phosphate, pH 7.4, at a rate of 5.1 mg per ml. The suspended solids were removed by centrifugation to give a clear pronase solution. This pronase solution was used to digest the protein from the sections (Anderson and Andre 1968). The remaining sections in the series, which served as controls, were treated with 0.05 M sodium phosphate buffer without the enzyme for the same length of time and at the same temperature, 37°C. All sections were washed with millipore-filtered water at room temperature and stained with magnesium uranyl acetate (7.5 mg per ml). Each section was mounted on a separate carbon-stabilized 0.2% formvar-coated slot grid. The results of the experiment were observed and recorded with the electron microscope. The pronase experiment was similar to that described by Anderson and Andre (1968) in research with animal cells.

Rabbit antisera against the total antigens of peanut seed aleurone grains and α-conarachin were prepared for use in the analysis of extracts of developing peanut embryos. Aleurone grains were prepared from mature peanut seeds by the nonaqueous method of Dieckert *et al.* (1962). Aleurone grains were extracted with 7 volumes of 0.5 M sucrose solution at room temperature. The extract was clarified by centrifugation in the cold. The clarified extract was quick-frozen in a Dry Ice-acetone bath and stored in the deep-freeze until needed. Just before injection, the frozen extract was thawed and dialyzed against 0.109 M sodium phosphate buffer, pH 7.4, containing 0.5 M sucrose. The dialyzed extract was used to immunize white rabbits. α-Conarachin was prepared from the above extract as follows: the extract was brought to 40% saturation with ammonium sulfate at 4°C and the arachin removed by centrifugation. The supernatant was then made 85% saturated with ammonium sulfate at 4°C, and the precipitate of crude α-conarachin was collected by centrifugation at 2°C. α-Conarachin was isolated from the crude preparation by DEAE cellulose column chromatography with a linear sodium chloride gradient for elution (Dechary *et al.* 1961). The fraction eluting between 0.1 and 0.2 M sodium chloride was taken as α-conarachin. The buffer for the α-conarachin was changed to 0.109 M sodium phosphate, pH 7.4, containing 0.5 M sucrose by dialysis. This solution was used to immunize rabbits.

Antisera against the total antigens of the peanut aleurone grains were prepared from white rabbits, which were housed and cared for by the staff of the College of Veterinary Medicine at Texas A & M University. Each rabbit was injected subcutaneously at the back of the neck with the following emulsions: each of 2 rabbits received 0.5 ml of the phosphate buffer emulsified with 0.5 ml of complete Freund's adjuvant (DIFCO). These animals provided a nonimmune control serum. Each of 4 animals was injected with 0.5 ml of the buffered sucrose solution containing 45 mg of the total aleurone grain protein emulsified with 0.5 ml of complete Freund's adjuvant. Each of 4 additional rabbits received injections of 1.5 ml of buffered sucrose containing 6.5 mg of α-conarachin isolate emulsified with 1.5 ml of complete Freund's adjuvant. After 18 days, each animal received a booster shot as above. All protein solutions were sterilized by filtration through $0.45\ \mu$ millipore filters before injection. Fifteen days after administration of the booster shot, 5–15 ml of blood was taken from the marginal vein of the ear or a leg vein. Fifty-one days after the booster shot, 20–50 ml of blood was taken from each rabbit by heart puncture without anesthetic. The whole blood was allowed to stand at room temperature for 2 hr., then chilled to $4°C$. The serum was clarified by centrifugation in the cold. To prevent microbial action, each serum was brought to 0.25% phenol or 0.01% merthiolate and stored at $-10°C$.

The proteins of an extract of peanut seed aleurone grains were fractionated on a DEAE cellulose column with a linear sodium chloride gradient by a modification of the procedure of Dechary *et al.* (1961). Aleurone grains prepared according to the method of Dieckert *et al.* (1962) were extracted with about 13 volumes of 0.05 M Tris-HC1, pH 7.8, containing 0.5 M sucrose (Mann enzyme grade) and 0.015% sodium azide at $0°–4°C$ for 1 hr. The extract was clarified by centrifugation in the cold and dialyzed against the same buffer in the cold to remove small molecules coming from the aleurone grains. Approximately 75% of the protein of the aleurone grain sample was recovered in the dialyzed solution. A sample of the dialyzed extract was applied to the column in the dialysis buffer at a concentration of 47 mg per ml. The DEAE column was eluted with a linear sodium chloride gradient of 0–0.5 M sodium chloride in the dialysis buffer at $0°–4°C$. Fractions were collected with an automatic fraction collector and the progress of the fractionation followed by measuring the optical density of each cut at $278\ m\mu$.

Peanut embryos were classified into the following 4 groups according to the length of the embryo: Group 1, 2–5 mm; Group 2, 6–8 mm; Group 3, 10–13 mm; and, Group 4, 14–19 mm. The embryos of each class were collectively ground in the Tris-HC1 sucrose buffer mentioned above, and the insoluble debris was removed by centrifugation. The extracts were carefully separated from the pellet in the bottom of the centrifuge tube, and from the pellicle of fat at the top, and dialyzed against the same buffer overnight at $0°-4°C$. The protein concentration as determined by the Lowry method (Lowry *et al.* 1951) was adjusted to 625 μg per ml by dilution with buffer or by concentration against dry Sephadex. The latter was conducted by placing the sample to be concentrated in a dialysis tube and coating the tube with dry Sephadex. When the coating of Sephadex became wet, it was replaced with more dry Sephadex. The process was repeated until the proper concentration was reached.

The embryo extracts were analyzed by a micro double immunodiffusion test using templates (Mann Research) on agar films on microscope slides. The procedure is an adaptation from Crowle (1961). The procedure for cleaning the microscope slides and matrices, and for setting up the immunodiffusion plates, is described in detail elsewhere (Singh 1970). The agar was dissolved in a 0.05 M tris buffer, pH 7.8, containing 0.5 M sucrose, 0.15 M NaCl and 0.015% NaN_3. The matrix provides a hexagonal array of six wells, with a seventh in the center. Solutions to be analyzed, the immune serum, and the proper controls are placed in the wells according to the experimental design. Immunodiffusion experiments a and b, Fig. 5.2, were incubated in a moist chamber for 10 days at $2°-3°C$. The other experiments were incubated in a moist chamber at $2°-4°C$ for 7 days. After incubation the matrix was pried loose from the slide, the excess reactants were removed from the agar gel by extraction overnight with buffer after a brief rinse with distilled water and the precipitin bands fixed and stained with amidoschwarz according to a standard method (Crowle 1961). After the final wash with 1% acetic acid containing 0.1% glycerol, the slides were dried at $37°C$.

RESULTS

Ultrastructural Analysis of Shepherd's Purse

At the "upturned U" stage, the cotyledons of the shepherd's purse embryo begin to thicken but are still green in color. Electron micrographs of two thin sections of cells of the parenchyma at this stage are shown in Fig. 5.3. The cells are populated by a centrally located nucleus containing a large nucleolus. Regions of condensed chromatin are also present in the nucleus, which is bounded by the characteristic double membrane. There are numerous mitochondria, chloroplasts with typical stacks of thylakoids, and often a starch grain. The cytoplasm is populated by dictyosomes, rough endoplasmic reticulum, and vacuoles containing densely staining components. The cell wall is relatively thin and is penetrated at many points by plasmodesmata. Large numbers of spherosomes occupy the peripheral regions of the cell.

The cell shown in Fig. 5.3a was fixed in glutaraldehyde and postfixed in osmium tetroxide, while the one in Fig. 5.3b was fixed in glutaraldehyde, extracted with alcohol, and then postfixed with aqueous osmium tetroxide. The section of tissue that was extracted with alcohol before postfixation with

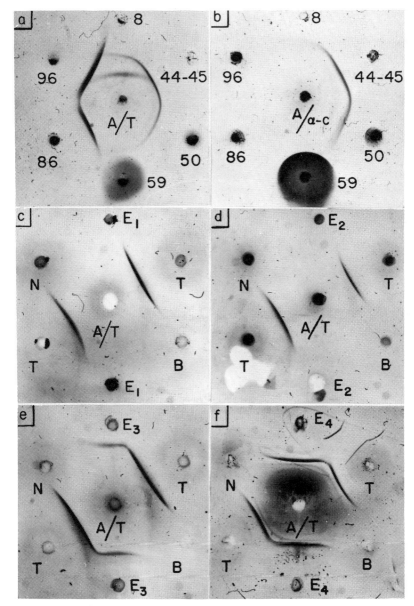

FIG. 5.2. MICRO DOUBLE DIFFUSION ANALYSIS OF MATURE ALEURONE GRAINS
AND DEVELOPING EMBRYOS OF THE PEANUT

a—Analysis of fractions 8, 44–45, 50, 59, 86, and 96 from DEAE cellulose column (Fig.
5.18) with rabbit antiserum against an extract of mature aleurone grains (A/T). b—Analysis
of the same fractions with rabbit antiserum against α-conarachin (A/α-c). c—Analysis of an
extract (E$_1$) of 2–5 mm embryos with A/T. The extract (T) of mature aleurone grains
served as reference; a nonimmune serum (N) and the buffer (B) served as controls. d—Analysis
of an extract (E$_2$) of 6–8 mm embryos with A/T. e—Analysis of an extract (E$_3$) of 10–13 mm
embryos with A/T. f—Analysis of an extract (E$_4$) of 14–19 mm embryos with A/T.

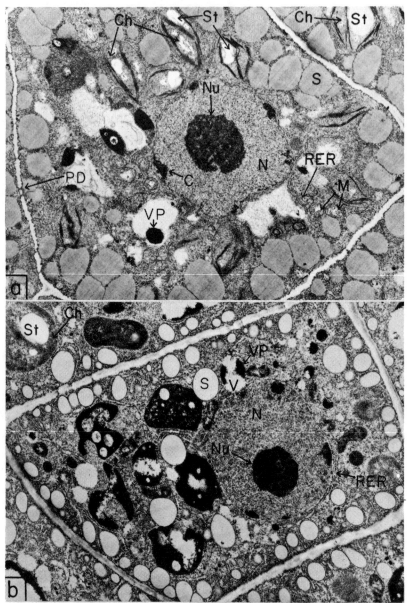

FIG. 5.3. THE EFFECT OF ALCOHOL EXTRACTION BEFORE OSMIUM TETROXIDE
FIXATION ON THE ULTRASTRUCTURE OF COTYLEDONARY CELLS OF THE SHEP-
HERD'S PURSE EMBRYO AT THE UPTURNED-U STAGE

a—Tissue fixed in glutaraldehyde and postfixed in osmium tetroxide. The following cell
organelles can be recognized: Nucleus (N) with nucleolus (Nu) and chromatin (C), chloroplasts (CH) with starch grains (St), mitochondria (M), rough endoplasmic reticulum (RER),
vacuoles (V) with vacuolar protein (VP), spherosomes (S), and cell walls with plasmodesmata (PD). X 4200. b—Tissue fixed in glutaraldehyde, extracted with alcohol and postfixed with osmium tetroxide. The same structures are visible, but the spherosomes and
cytomembranes do not stain. X 4200.

osmium tetroxide is well preserved except for the lipid components. The spherosomes and the various cytomembranes no longer appear stained. This verifies that the spherosomes contain lipids. The stainable vacuolar content remains after alcohol extraction; therefore, it must not be lipid. It is concluded that the aleurins of shepherd's purse are located in these vacuoles. Cytochemical evidence is consistent with this interpretation. Thick sections of a shepherd's purse embryo of the same stage, that was fixed in glutaraldehyde only, were stained with mercuric bromphenol blue, a stain that is specific for protein. The results are shown in Fig. 5.4a, and b. The vacuoles contain heavy deposits of densely-staining material along the cytoplasm-vacuole interface. Sections from a more mature shepherd's purse embryo showed increasing protein content by the same test. The end result of the maturation process is illustrated by the heavily stained mature aleurone grains, as seen in Fig. 5.4a. A similar sequence of events was observed with the electron microscope (Dieckert, unpublished results). For convenience, the vacuoles containing the presumptive aleurins will be called aleurone vacuoles.

The rough endoplasmic reticulum at the thickening upturned-U stage is highly developed, but usually there are no protein granules observed in the cisternae of this organelle. Sometimes, however, substances that electron-stain like proteins are found in the lumina of the rough endoplasmic reticulum (Fig. 5.4c). This phenomenon seems to occur when there are large masses of cytomembranous materials in the cytoplasm. The staining properties of these granules and of the protein of the aleurone vacuoles are indistinguishable.

Substances that stain like protein are found in dictyosomal vesicles (Golgi vesicles) of shepherd's purse embryos that have vacuolar deposits of protein. The staining properties of the material in these vesicles and in the aleurone vacuoles are indistinguishable by any of the three staining schedules used in the electron microscopy work (Fig. 5.5a,b,c, and d). The dictyosomal vesicles with the densely staining component are bounded by a single unit membrane. Similar vesicles in the cytoplasm have no obvious immediate connection with the dictyosome. Occasionally, sections are obtained showing one of these fused to the membrane bounding the aleurone vacuole (Fig. 5.5b, and d). In some instances the inside of the vacuole appears to be confluent with the inside of the aleurone vacuole, as seen in Fig. 5.5b.

The cells of mature shepherd's purse embryos contain a nucleus and nucleolus and are filled with spherosomes and aleurone grains (Fig. 5.6b). The aleurone grains are bounded by a single unit membrane (Fig. 5.6a).

The aleurins were not detected in the vacuoles of shepherd's purse embryos at the "walking-stick" stage (cf Fig. 5.7a and b). There are several other interesting differences. At the walking-stick stage the endoplasmic reticulum is not highly developed, and the dictyosomal vesicles have no densely-staining content.

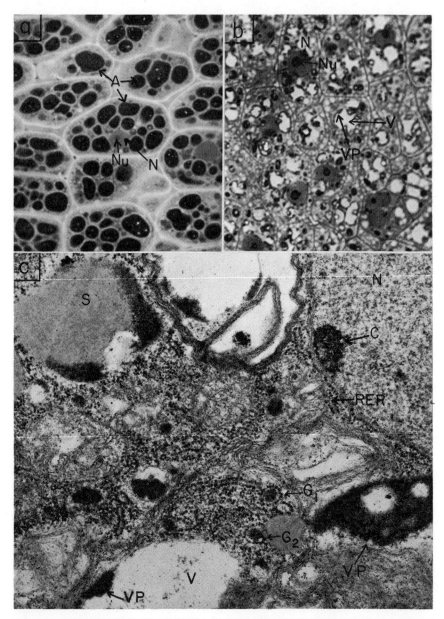

FIG. 5.4. DISTRIBUTION OF PROTEIN IN CELLS OF SHEPHERD'S PURSE EMBRYO

a—Mature embryo fixed only in glutaraldehyde. Aleurone grains (A) stain deeply with mercuric bromphenol blue. Light micrograph X 715. b—Upturned-U fixed only in glutaraldehyde. Peripheral content of vacuole (VP) stains deeply with bromphenol blue. Light micrograph X 715. c—Upturned-U stage fixed in glutaraldehyde and OsO_4. Sometimes, protein-like dense bodies (G_1 and G_2) are in the rough endoplasmic reticulum (RER). These stain like protein (VP) in the vacuole (V). Electron micrograph X 18,350.

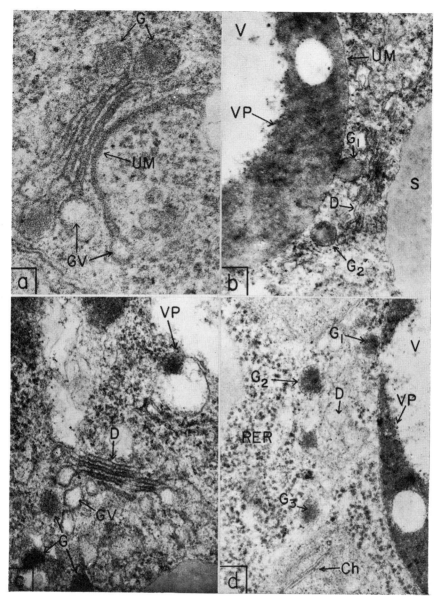

FIG. 5.5. DICTYOSOMES AND ALEURONE VACUOLES OF SHEPHERD'S PURSE AT
THE UPTURNED-U STAGE

a—Golgi vesicles (G) are filled with electron-dense material staining like protein. Stained
with aqueous uranyl acetate. Note unit membrane (UM). X 46,500. b—aleurone vacuole (V)
with protein (VP) and an osculating dense granule (G₁). Note also dictyosome (D), dense
granule (G₂) and spherosome (S). Stained with aqueous uranyl acetate. X 26,550. c—
Dictyosome with vesicles (GV) containing electron-dense substance like vacuolar protein
(VP). Stained with alcoholic uranyl acetate. X 26,550. d—Dictyosome (D) with associated
dense granules (G₁, G₂, G₃), vacuole (V) with protein (VP). Stained with magnesium uranyl
acetate. Sections also stained with lead citrate. X 26,550.

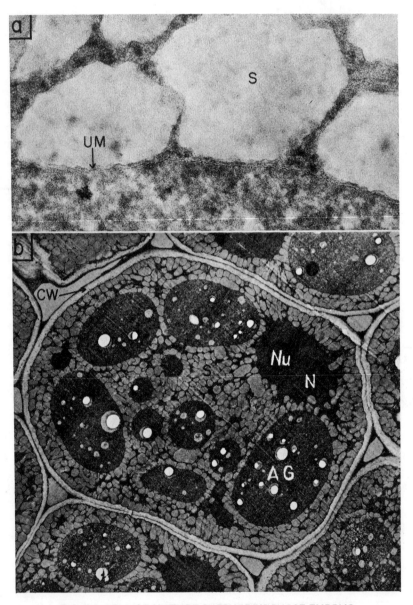

FIG. 5.6. CELL OF MATURE SHEPHERD'S PURSE EMBRYO

a—spherosomes (S) and aleurone grain bounded by a single unit membrane (UM). X 46,500.
b—Cell contains a nucleus (N) with a nucleolus (Nu), aleurone grains (AG), and numerous
spherosomes (S). X 2900.

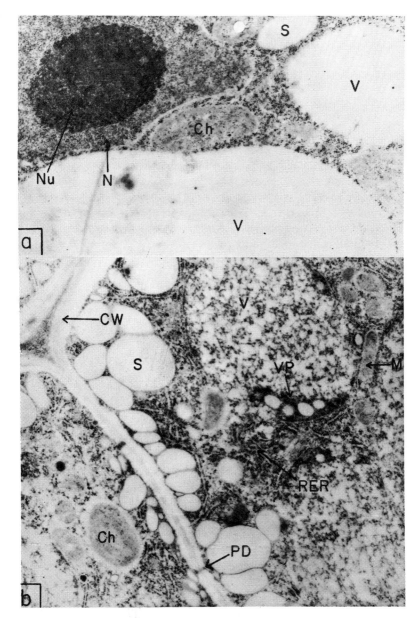

FIG. 5.7. TIME OF APPEARANCE OF ALEURINS IN THE SHEPERD'S PURSE EMBRYO

a—Walking stick embryo does not contain visible vacuolar protein. Rough endoplasmic reticulum is poorly organized. b—Upturned-U stage. Vacuole (V) contains protein (VP). The rough endoplasmic reticulum (RER) is highly developed. Both tissues fixed only in glutaraldehyde. X 6900.

Cotton Embryos

Much the same ultrastructural pattern may be observed in cotton embryos with developing aleurone vacuoles as in shepherd's purse embryos (cf Fig. 5.3a and Fig. 5.8b). All of the parenchyma cells contain a centrally located nucleus, a nucleolus (not shown in the figure) and patches of condensed chromatin, bounded by a double membrane. The cytoplasm contains mitochondria, leuco-plasts containing starch granules, rough endoplasmic reticulum, dictyosomes, aleurone vacuoles and spherosomes. The cell is bounded by a thin cell wall penetrated by plasmodesmata. No electron-dense granules are observed in the cisternae of the rough endoplasmic reticulum, but such granules are to be noted in the dictyosomal or Golgi vesicles (Fig. 5.9). Membrane-bound electron-dense granules having no obvious connection with the dictyosomes are observed in the cytoplasm (Fig. 5.8b and 5.9). Occasionally, one of these granules is found associated with the surface of the aleurone vacuole (Fig. 5.10b). In each case the staining properties of the substance in the Golgi vesicles proved to be indistin-guishable from those of the electron-dense content of the aleurone vacuoles.

The identification of the electron-dense deposits of the aleurone vacuoles of the cotton cells was verified by cytochemical analysis of individual vacuoles, as shown in Fig. 5.11a, b, and c. Sections a and c of three serial sections were digested with pronase, a mixture of serine-proteases. Section b was left un-treated. All three sections were stained by the same method. The staining prop-erty of the contents of each vacuole was abolished by pronase. Only the plastic remained, outlining the location of the protein in the original sections. This same type of experiment showed that the electron-dense material was removed from dictyosomal cisternae (cf D in Fig. 5.12a and c, with D in Fig. 5.12b). The Golgi vesicles illustrate the same result (cf G_2 in Fig. 5.12a and c with Fig. 5.12b; also cf G_1, G_2, G_3, G_4 and G_5 in Fig. 5.10a with the same in Fig. 5.10b). A similar analysis of an isolated cytoplasmic dense granule shows that it, too, consisted of protein (cf G_1 in Fig. 5.12a and c with G_1 in Fig. 5.12b). By the same argument, the electron-dense body G_6 fused to the aleurone vacuole, AV in Fig. 5.10b, is protein.

As in the case of shepherd's purse embryos at the walking-stick stage, cotton embryos in the barely-rolled stage (Fig. 5.8a) do not seem to have aleurins in the vacuoles. The rough endoplasmic reticulum is not highly developed, and the dictyosomal vesicles do not contain the electron-dense protein granules found in the older embryos.

Peanut Embryos

Peanut embryos 7 mm long have larger cells than cotton or shepherd's purse embryos, but the population of subcellular organelles is similar. Figure 5.13a and b show regions of the cytoplasm. The cell contains a large nucleus (not shown) with a centrally located nucleolus and condensed chromatin. The nucleus is

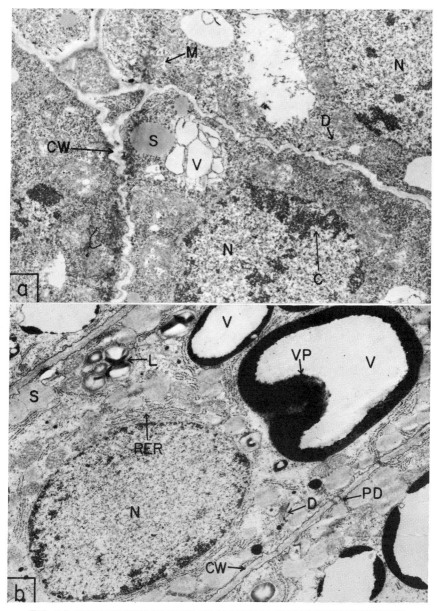

FIG. 5.8. TIME OF APPEARANCE OF ALEURINS IN THE COTTON EMBRYO

a—Barely-rolled stage. The vacuoles (V) do not contain aleurins (VP) and the dictyosomes (D) do not have electron-dense vesicles. b—Loose scroll stage. The vacuoles and dictyosomal vesicles contain protein. The rough endoplasmic reticulum (RER) is highly developed. X 4550.

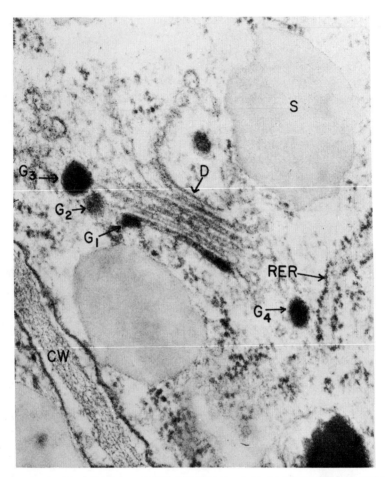

FIG. 5.9. DICTYOSOME OF A COTTON EMBRYO AT THE LOOSE SCROLL STAGE

Dictyosomal vesicles (G_1, G_2, G_3) and one sacule contain substance that stains like the vacuolar protein and not like the fat in the spherosomes (S). X 17,200.

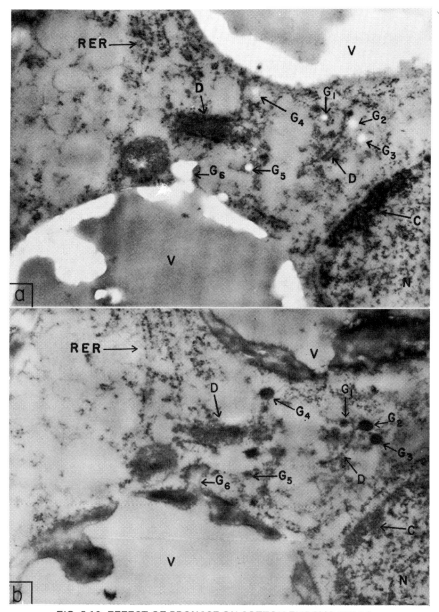

**FIG. 5.10. EFFECT OF PRONASE ON COTTON EMBRYO CELLS
AT THE LOOSE SCROLL STAGE**

The staining of dense granules (G_1, G_2, G_3, G_4, G_5, G_6) and the aleurone vacuole content is abolished by pronase. Other structures are relatively unaffected. a—Pronase-treated section. X 16,750. b—Untreated section. X 16,750.

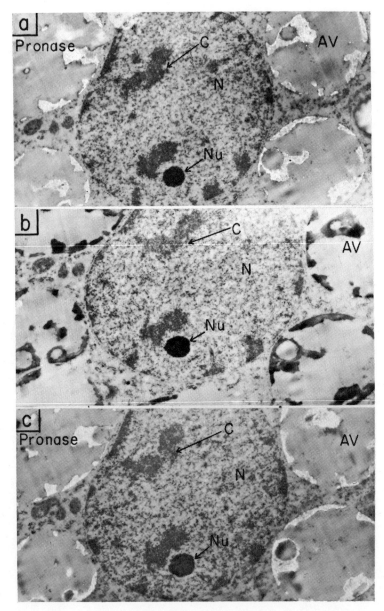

FIG. 5.11. THE EFFECT OF PRONASE ON THE ALEURONE
VACUOLES OF COTTON.

The capacity of the substance (AV) of the vacuoles to stain is abolished by pronase treatment. Serial sections a and c, but not b, were treated with pronase. X 4800.

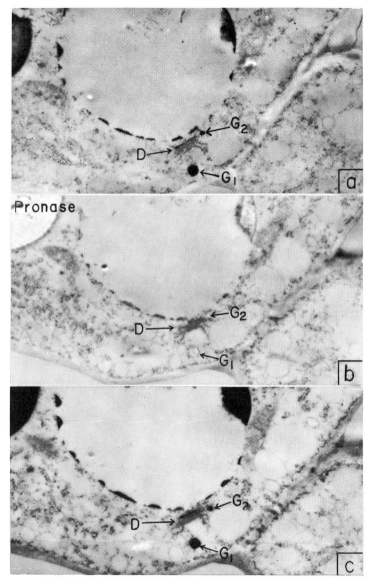

FIG. 5.12. EFFECT OF PRONASE ON THE CYTOPLASMIC DENSE BODIES
AND DICTYOSOMES

Staining is abolished for dense granules (G_1, G_2, G_3), the dictyosome (D), and the aleurone vacuole. Serial section b was treated with pronase, and sections a and c were untreated. X 7000.

bounded by the usual double membrane. In the cytoplasm are mitochondria, large leucoplasts with starch grains (not shown), dictyosomes (Fig. 5.14b and Fig. 5.15a and b), aleurone vacuoles with electron-dense inclusions of various sizes (Fig. 5.13a and b), and spherosomes (Fig. 5.12b). The dictyosomes appear slightly different under the three uranyl acetate stains. In cross-sectional views of the dictyosomes stained with saturated uranyl acetate, the unit membranes of the sacules and vesicles are indistinct or not visible (Fig. 5.15a), while the membranes of cross-sections of dictyosomes stained with alcoholic uranyl acetate or magnesium uranyl acetate are clearly visible (Fig. 5.14b and 5.15b). As might be expected, evidence of the stacked sacks of the dictyosome is absent in sections that are more or less coplanar with the individual sacules (Fig. 5.14a and 5.15b). The rough endoplasmic reticulum is highly developed in peanut cells that are depositing aleurins in vacuoles (Fig. 5.16d). Tangential sections through the cisternal membranes show the presence of numerous polyribosomes containing 18–19 ribosomes in spiral array (Fig. 5.16c).

Electron-dense granules that stain like vacuolar protein are found in cisternae of the rough endoplasmic reticulum (Fig. 5.14a, 5.16a, and b, and 5.17a). Occasionally, such granules are found in cisternae in close proximity to a dictyosome (Fig. 5.14a). Numerous electron-dense granules of the same kind are found in the peripheral regions of dictyosomes with cisternae coplanar with the section (Fig. 5.14a and 5.15c). In sections normal to the plane of the dictyosomal sacules the electron-dense granules are in vesicles at the edge of individual sacules (Fig. 5.15a and b). As was the case for shepherd's purse and cotton embryos, membrane-bound electron dense granules which stain like the aleurins but which are not clearly associated with dictyosomes are found in the cytoplasm in small vacuoles (Fig. 5.14b). Occasionally, electron-dense granules of the size found associated with the dictyosomes are found just inside the membrane of an aleurone vacuole (Fig. 5.15b). The partially filled aleurone vacuoles of the peanut embryo are bounded by a single unit membrane (Fig. 5.17b). The aleurins of partially filled vacuoles are present as clumps near the vacuolar membrane (Fig. 5.17b) or as globular masses of various sizes (Fig. 5.13a and b and 5.17a and b). In mature peanut aleurone vacuoles the unit membrane is closely appressed to the protein granule (Fig. 5.17a and inset).

Peanut embryos as small as 6 mm long contained a few vacuoles with aleurin-like deposits. Embryos less than 5 mm long seemed to lack such vacuoles. Similarly, dictyosomal vesicles of embryos less than 5 mm long did not contain substances staining like that in the aleurone vacuoles of older embryos. Apparently, the smaller peanut embryos were not yet making the aleurins.

Peanut Proteins

The chromatography of the extract of peanut seed aleurone grains on DEAE cellulose shows major components (Fig. 5.18) that eluted at 0.19 M and 0.27 M

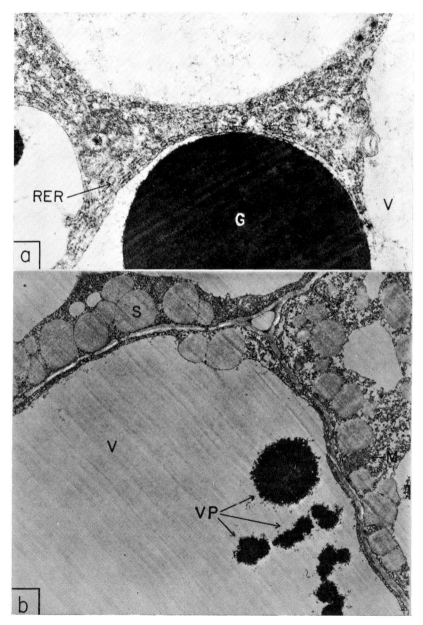

FIG. 5.13. CELLS OF PEANUT EMBRYOS

The rough endoplasmic reticulum (RER) is highly developed. a—10-day embryo with a large protein granule (G) in the vacuole. X 6900. b—7 mm embryo with a constellation of protein granules (VP) in the vacuole (V). X 2800.

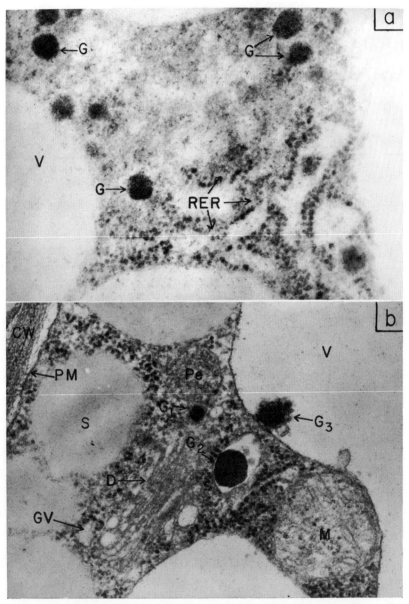

FIG. 5.14. DICTYOSOMES OF MATURING PEANUT EMBRYOS

a—Section coplanar with sacules of the dictyosome. Protein granules (G) in the peripheral regions of the dictyosome. A protein granule is in a sacule of rough endoplasmic reticulum (RER). X 24,400. b—Cross section of dictyosome (D). Dense granule (G_1) in dictyosomal vesicle. Same material in larger vesicle (G_2) and one (G_3) in the vacuole (V). X 24,400.

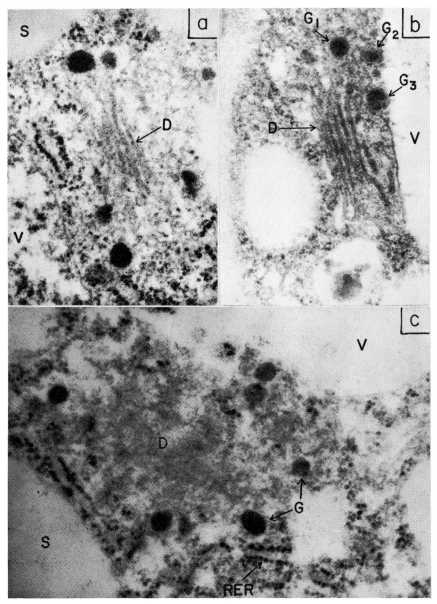

FIG. 5.15. DICTYOSOMES OF THE PEANUT EMBRYO

a—Dense granules in the peripheral region of the dictyosome (D) stains differently from spherosomes (S). Stained with aqueous uranyl acetate. X 28,300. b—Dictyosome (D) with dense granule (G_1). Also note dense granules (G_2 and G_3) are inside vacuole (V). Stained with alcoholic uranyl acetate. X 28,300. c—Dictyosome in section coplanar with sacule. Dense granules (G) in the peripheral region of the dictyosome. Rough endoplasmic reticulum in close proximity to dictyosome. Stained with aqueous uranyl acetate All sections stained with lead citrate. X 28,300.

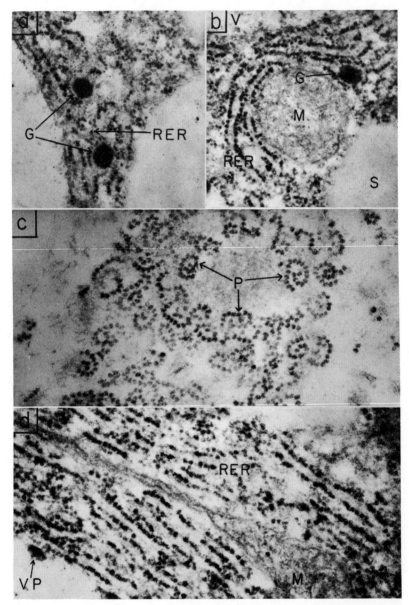

FIG. 5.16. ENDOPLASMIC RETICULUM OF THE CELLS OF A PEANUT EMBRYO
FILLED WITH ALEURONE VACUOLES

a—Dense granules (G) in sacules of the rough endoplasmic reticulum (RER). X 21,650.
b—Same, with mitochondrion (M) and spherosome (S). X 21,700. c—Section coplanar with
the membrane of the rough endoplasmic reticulum, showing the polyribosomes (P). X
27,450. d—Cross section of the rough endoplasmic reticulum (RER) showing stacks of
sacules. X 27,450.

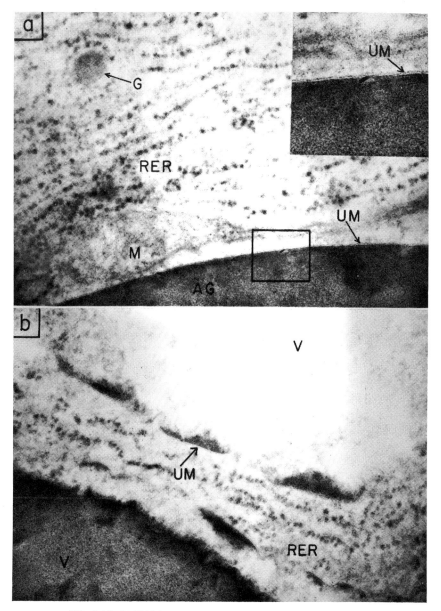

FIG. 5.17. ALEURONE VACUOLES OF THE PEANUT EMBRYO

a—Mature aleurone grain (AG) with closely appressed unit membrane (UM). Inset shows the unit membrane. X 25,650; inset X 46,900. b—Partially filled aleurone vacuoles (V) are bounded by a single unit membrane (UM). X 25,650.

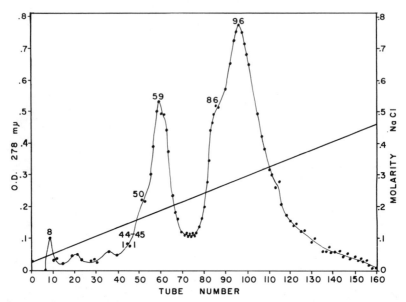

FIG. 5.18. CHROMATOGRAPHY OF AN EXTRACT OF ALEURONE GRAINS ON A DEAE CELLULOSE COLUMN

Solvent for the extraction and column chromatography was 0.05 M tris-HC1, pH 7.8, 0.5 M sucrose, 0.015% NaN$_3$. The proteins were eluted with a linear 0.0–0.5 M NaC1 gradient at 0°–4°C.

NaC1, respectively. The elution profile of each peak was asymmetric and exhibited shoulders. A minor component appeared at the break through volume, and several even less conspicuous ones eluted at salt concentrations less than 0.19 M. The fraction eluting at 0.19 M NaC1 is mostly α-conarachin, and the one eluting at 0.27 M NaC1 is arachin. The identity of the minor components is unknown.

The analysis of the rabbit antiserum against the total extract of peanut seed aleurone grains by micro double immunodiffusion analysis demonstrated that the immune serum contained antibodies against α-conarachin and arachin (Dieckert, unpublished data). This antiserum and a rabbit antiserum against an α-conarachin isolate was used to analyze key fractions from the DEAE cellulose column. Also, the important fractions from the column were used to test the antisera for additional antigens. The results of these analyses are given in Fig. 5.2a and b. Fraction 59 from the column gave a strong test for α-conarachin, but fractions 86 and 96 did not. These results confirm the conclusion that the first major component issuing from the column is predominantly α-conarachin. The combined fraction 44-45 and fraction 50, on analysis with anti-α-conarachin, exhibited a different pattern of precipitin bands from that of fraction 59, which contains α-conarachin. The results show that the α-conarachin isolate which was used to immunize the rabbits was not homogeneous, that the contaminating components are eluted from the column in advance of fraction 59 but mostly after fraction 8. From the results obtained with the antiserum against the total aleurone grain extract (Fig. 5.2a), it is again apparent that the antigens eluting from the DEAE cellulose column before α-conarachin are distinct from α-conarachin and arachin. In addition, these data verify that the antiserum against the total extract contains a broad spectrum of antibodies against the components of the isolate of mature aleurone grains used to induce them.

Extracts of developing peanut embryos were analyzed by double immuno-diffusion analysis with the broad spectrum antiserum against the extract of mature peanut seed aleurone grains. The results (Fig. 5.2c, d, e, and f) show that one or more of the aleurone antigens (including arachin) are present in peanut embryos 10–13 mm long or longer. These observations confirm a similar conclusion drawn from the ultrastructural analysis of peanut embryos. No aleurone grain antigens were detected in embryos 6–8 mm long or in those 2–5 mm long. It may be recalled that protein granules were observed in cells of embryos as small as 7 mm long. Vacuolar protein granules were not found by ultrastructural analysis in the embryos with blade-like cotyledons. These were usually less than 5 mm long. Presumably, the reason for the discrepancy between the ultra-structural and immunochemical analyses is that the aleurins in 6–8 mm embryos were below the detection limit of our double diffusion test. Clarification of this point remains for future experiments. The immunodiffusion tests were repeated with different sets of peanut embryos, with the same results. In each case, however, the protein concentration of the extract of the 6–8 mm embryo class never exceeded 650 μg per ml. At this protein concentration, the extracts of 10–13 mm embryos always gave a significantly weaker precipitin test than did similar extracts of larger peanut embryos.

DISCUSSION

The results of the present work are summarized in Fig. 5.19. It shows that the aleurone grains of mature seeds of cotton, shepherd's purse, and peanut develop in vacuoles bounded by a single unit membrane. The mature aleurone grain of shepherd's purse and peanut embryos retains the single unit membrane at maturity, as shown here. The same is true for mature cotton aleurone grains (Yatsu 1965). Aleurone grains with a single unit membrane were also reported for the seeds of *Vicia faba* (Briarty *et al.* 1969). These observations confirm the long-held opinion that aleurone grains develop in vacuoles (cf Guilliermond 1941).

There are variations in the ultrastructure of embryos from different species that are depositing proteins in the aleurone vacuoles. For example, shepherd's purse embryos have chloroplasts, while those of cotton and peanut plants have leucoplasts. In all three species studied here the cytoplasm contains voluminous

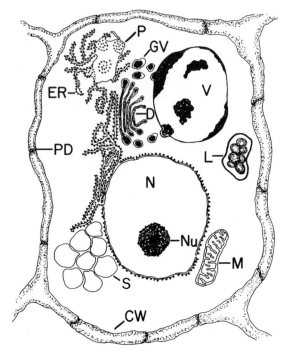

FIG. 5.19. CELL OF A HYPOTHETICAL OIL SEED PRODUCING
ALEURINS AND SEQUESTERING THEM IN A VACUOLE

The aleurins are synthesized on the polyribosomes (P) of the rough endoplasmic
reticulum (RER), concentrated by the dictyosome (D) in membrane-bound vesicles
(GV). The aleurins are transferred from the vesicles to the vacuole (V) by membrane fusion. Such cells also contain plastids (L), mitochondria (M), and spherosomes (S), and are encased in a cell wall (CW) penetrated by plasmodesmata (PD).

spherosomes, but in the case of pea, variety Alaska, and bush bean, variety Top
Crop, these lipid droplets are less conspicuous (Mollenhauer and Totten 1971).

Certain ultrastructural features are common to embryos from all three species
that are actively depositing proteins in the aleurone vacuoles. The rough endoplasmic reticulum is highly developed, the dictyosomal vesicles contain material
that stains like protein, the cytoplasm contains many isolated relatively small
vesicles bounded by a single unit membrane and containing similar electron-
dense material inside; and occasionally, such vesicles are seen fused to the membrane of the aleurone vacuole. Some of the same patterns are reported for
developing seeds of *Vicia faba L.*, variety Sharp's Conqueror (Briarty *et al.* 1969)
and maize endosperm (Khoo and Wolf 1970). Some of the same features can be
seen in the section of a maturing pea seed shown in Fig. 5.3 (Mollenhauer and
Totten 1971). These interesting characteristics of the subcellular architecture are
probably common to all developing seeds that are depositing proteins in

aleurone vacuoles. As will be brought out later, this common pattern is absent from developing seeds that are not yet forming aleurone vacuoles.

If the space enclosed by the vacuolar membrane is considered to be topologically outside of the plant cell, then the deposition of aleurins into vacuoles might be thought of as internal secretion. These plant cells would then be analogous to animal cells that synthesize and secrete specialized proteins such as hormones or zymogens. Indeed, the zymogen-producing acinar cells of the pancreas have many of the ultrastructural features of the analogous plant cells. The rough endoplasmic reticulum is highly developed, often with stacks of more or less flattened cisternae (Fawcett 1966). In most mammals no protein granules are located in the sacules of the rough endoplasmic reticulum, but in the guinae pig such granules are observed (Palade 1959). At certain phases of the secretory process, the vesicles of the Golgi bodies are filled with protein, as are larger membrane-bound vesicles that show no obvious connection with the Golgi apparatus (see review by Dieckert 1971). The cytoplasm adjoining the acinus of the gland is filled with zymogen granules, each bounded by a single membrane. And, at the time of secretion, the cytomembrane of the zymogen granule is found fused with the cell membrane (Palade 1959; Fawcett 1966). Other glandular cells present similar ultrastructural patterns.

There is a large body of evidence indicating that secretory proteins are synthesized in conjunction with polyribosomes on the rough endoplasmic reticulum, transported to the Golgi apparatus through the lumina of the rough endoplasmic reticulum, concentrated into intracellular protein granules in the Golgi region, and secreted from the cell by a kind of membrane fusion [see Dieckert (1971) for a more detailed review of zymogen granule formation]. The many similarities in ultrastructure noted between oilseed cells forming aleurone-filled vacuoles and animal cells synthesizing zymogen granules suggest that the aleurins of shepherd's purse, peanut, and cotton seeds are synthesized in the rough endoplasmic reticulum, concentrated into protein droplets in the dictyosomes, and transported to the neighborhood of an aleurone vacuole in membrane-bound vesicles. There, it is emptied into the vacuole by a process of membrane fusion. Such a process was proposed earlier by Dieckert (1969) for shepherd's purse embryos.

Several additional pieces of evidence support this hypothesis for aleurone grain formation. Polyribosomes were observed on the surface of the rough endoplasmic reticulum of peanut embryos that were depositing protein in the vacuoles. The electron-dense material in the dictyosomal vesicles, in the related but isolated cytoplasmic dense granules, and in the aleurone vacuoles, were identified by cytochemical analysis of serial sections of cotton embryos. Pronase, but not hydrogen peroxide or warm buffer, abolished the staining power of these structures. One of the droplets found fusing to an aleurone vacuole was also determined to contain protein by the same procedure. The identification of the proteins in the above structures as the corresponding seed-specific proteins has not yet been established. Some work in this direction has been reported recently for developing embryos of *Vicia faba* (Graham and Gunning 1970). Results of experiments with fluorescein-labeled antibodies showed that vicilin and legumin are present in the same aleurone vacuoles, but a few such bodies were found that contained vicilin but no detectable legumin.

Immunochemical experiments show that peanut seed aleurins (arachin, at least) appear shortly after the characteristic ultrastructure of the cells forming aleurone vacuoles is present. Prior to this time a few dense protein granules are present in the cells. It is possible that the early protein granules do not contain arachin, but it is also possible that the concentration of arachin and the other peanut aleurins is below the detection limit of the immunodiffusion test. The embryos 5 mm long, or shorter, failed to give a precipitin band and also did not contain electron-dense protein granules. Presumably, at the 5-mm stage the aleurins were not being synthesized in appreciable amount.

Additional evidence indicates that the aleurins are not synthesized and sequestered in the aleurone vacuoles until a certain critical stage is reached. The shepherd's purse embryo at the walking-stick stage has neither a highly developed rough endoplasmic reticulum, dictyosomal vesicles filled with protein, nor protein in the vacuoles. Just after the upturned-U stage, these ultrastructural characteristics appear. In the cotton embryo the critical cytodifferentiation is apparent just after the barely-rolled stage.

SUMMARY

Cells of maturing cotton, peanut, and shepherd's purse embryos sequester proteins in vacuoles. At the time the process is underway the cells of the cotyledons contain mitochondria, nuclei with nucleoli and regions of condensed chromatin, and relatively thin cell walls. The protoplasts of adjoining cells are connected by plasmodesmata that penetrate the common cell wall. The cytoplasm of shepherd's purse embryos contain chloroplasts. These are bounded by a double membrane and contain stacks of thylakoids, isolated thylakoids, and starch granules. Cotton and peanut embryos contain leucoplasts with characteristic starch granules and bounding cytomembranes.

There are certain ultrastructural features that are common to the cells of embryos of each species that are actively depositing vacuolar proteins. The rough endoplasmic reticulum is highly developed and has spirally arranged polyribosomes attached to the surface interfacing with the cytoplasm. In the peanut, protein granules are found in sacules of the rough endoplasmic reticulum. Such is not normally the case in cotton and shepherd's purse. In the latter species, protein granules are occasionally observed in vesicles of the rough endoplasmic reticulum of cells showing an abnormal proliferation of smooth cytomembranes. In each case the protein granules of the rough endoplasmic reticulum stain like the protein in the aleurone vacuoles. Dictyosomes and dictyosomal vesicles filled with protein are frequent cellular components. Here also the protein of the dictyosomal vesicles and the aleurone vacuoles stain the same. In peanut embryos the intracisternal protein granules of the rough endoplasmic reticulum are often in close association with dictyosomes and appear to be in passage from one subcellular organelle to the other. In all three species, dictyosome-like vesicles packed with densely staining protein are observed free in the cytoplasm and not associated in any obvious way with the dictyosomes. Fat droplets (spherosomes), which are also common cellular components during this period of seed maturation, are easily distinguished from the protein droplets by their differential staining properties. The cells of the three species contain several to many vacuoles partially filled with protein. These proteins stain like the protein droplets in the sacules of the rough endoplasmic reticulum and the dictyosomal vesicles. The vacuoles are bounded by a single unit cytomembrane. Occasionally one of the small, smooth membrane-bounded droplets is observed fused to an aleurone vacuole, giving the impression that the protein of the small vesicle is being transferred to the vacuole. No sacule of the rough endoplasmic reticulum was ever observed to open directly into an aleurone vacuole in cotton, peanut, or shepherd's purse.

The protein concentration of the vacuoles increases with seed maturation until the classical aleurone grains are formed. The aleurone grains, or protein bodies, are enclosed at maturity by a single membrane and may contain one or more globular inclusions consisting mostly of potassium magnesium phytate.

The available ultrastructural evidence suggests that aleurone grain formation occurs by the following process. The aleurins, or seed-specific proteins, are synthesized on the polyribosomes of the rough endoplasmic reticulum and are passed through the cytomembrane to the lumen of the organelle. The protein is transported to the dictyosomes via the lumina of the rough endoplasmic reticulum. There the protein is concentrated into droplets. The membrane-bounded protein droplets then migrate through the cytoplasm to the aleurone vacuoles and pass through the vacuolar membrane by a process of membrane fusion. Finally, the mature aleurone grains form by dehydration of the aleurone vacuoles.

The hypothesis receives additional support from correlations noted between the character and timing of modulations of the endoplasmic reticulum-dictyosome system and the appearance of protein in the aleurone vacuoles. Until a certain critical stage of development is reached, neither cotton, peanut, nor shepherd's purse cotyledonary cells deposit appreciable quantities of proteins in the large vacuoles. Up to this time the rough endoplasmic reticulum is not highly developed and protein is not detected in the dictyosomal vesicles and large vacuoles. The other subcellular organelles seem ultrastructurally the same during the transition period. Immunochemical analysis of extracts of developing peanut embryos shows that the principal aleurins of mature peanut seed aleurone grains, the arachins, are not detectable until just after the aleurin deposits are first detectable by ultrastructural analysis.

Cytochemical analyses confirm the presence of fat in the spherosomes and protein in the dictyosomal vesicles and aleurone vacuoles. Staining of the spherosomal content, but not the content of the dictyosomal vesicles and aleurone vacuoles, is abolished by extracting the tissue with alcohol after fixation in glutaraldehyde but before postfixation in osmium tetroxide. A similar result is obtained with tissue fixed in glutaraldehyde alone. The organic solvents employed to dehydrate and embed the tissue are expected to extract fat preferentially. Mercuric bromphenol blue selectively stains protein. The vacuolar deposits of the maturing shepherd's purse embryo give a strong positive test, as expected. The contents of individual dictyosomal vesicles, the larger dictyosome-related sacules, and the aleurone vacuoles of cotton embryos give a strong positive test for protein. In this analysis, alternate serial sections of tissue fixed in glutaraldehyde were digested with pronase, a mixture of serine proteases. All sections were subsequently stained for electron microscopy. The protease treatment selectively removed the stainable component from each of the structures mentioned above.

BIBLIOGRAPHY

ALTSCHUL, A. M., NEUCERE, N. J., WOODHAM, A. A., and DECHARY, J. M. 1964. A new classification of seed proteins: application to the aleurins of *Arachis hypogaea*. Nature *203*, 501-504.

ANDERSON, W. A., and ANDRE, J. 1968. The extraction of some cell components with pronase and pepsin from thin sections of tissue embedded in an Epon-Araldite mixture. J. Microscopic (Paris) *7*, 343-354. (French)

BONNET, H. T., Jr., and NEWCOMB, E. H. 1965. Polyribosomes and cisternal accumulations in root cells of radish. J. Cell Biol. *27*, 423-432.

BRIARTY, L. G., COULT, D. A., and BOULTER, D. 1969. Protein bodies of developing seeds of *Vicia faba*. J. Exptl. Botany *20*, 358-372.

BUTTROSE, M. S. 1963. Ultrastructure of the developing wheat endosperm. Australian J. Biol. Sci. *16*, 305-317.

CROWLE, A. J. 1961. Immunodiffusion. Academic Press, New York.

DAUSSANT, J., NEUCERE, N. J., and YATSU, L. Y. 1969. Immunochemical studies on *Arachis hypogaea* proteins with particular reference to the reserve proteins. I. Characterization, distribution, and properties of α-arachin and α-conarachin. Plant Physiol. *44*, 471-479.

DECHARY, J. M. *et al.* 1961. α-Conarachin. Nature *190*, 1125-1126.

DIECKERT, J. W. 1969. Ultrastructural analysis of aleurone grain formation in embryos of *Capsella bursa-pastoris*. Abstr. XI Intern. Botan. Congr. Seattle, Wash.

DIECKERT, J. W. 1971. Cell particles. *In* Chemistry of the Cell Interface, H. D. Brown (Editor). Academic Press, New York.

DIECKERT, J. W. *et al.* 1962. Composition of some subcellular fractions from seeds of *Arachis hypogaea*. J. Food Sci. *27*, 321-325.

ENGLEMAN, E. M. 1966. Ontogeny of aleurone grains in cotton embryo. Am. J. Botany *53*, 231-237.

FAWCETT, D. W. 1966. The Cell. Its Organelles and Inclusions. W. B. Saunders Co., Philadelphia.

FRASCA, J. M., and PARKS, V. R. 1965. A routine technique for double-staining ultrathin sections using uranyl and lead salts. J. Cell Biol. *25*, 157-161.

GRAHAM, J. S. D. *et al*. 1962. Protein bodies and protein synthesis in developing wheat endosperm. Nature *196*, 967-969.

GRAHAM, T. A., and GUNNING, B. E. S. 1970. Localization of legumin and vicilin in bean cotyledon cells using fluorescent antibodies. Nature *228*, No. 5266, 81-82.

GUILLIERMOND, A. 1941. The Cytoplasm of the Plant Cell, Chronica Botanica Co., Waltham, Mass.

HARTIG, T. 1855. Concerning protein granules. Botan. Zeit. *13*, 881-882. (German)

HARTIG, T. 1856. Further investigations concerning protein granules. Botan. Zeit. *14*, 257-268. (German)

KHOO, U., and WOLF, M. J. 1970. Origin and development of protein granules in maize endosperm. Am. J. Botany *57*, 1042-1050.

LOWRY, O. H., ROSEBROUGH, N. J., FARR, A. L., and RANDALL, R. J. 1951. Protein measurement with the folin-phenol reagent. J. Biol. Chem. *193*, 265-275.

MAZIA, D., BREWER, P. A., and ALFERT, M. 1953. The cytochemical staining and measurement of protein with mercuric bromphenol blue. Biol. Bull. *104*, 57-67.

MOLLENHAUER, H. H., and TOTTEN, C. 1971. Studies on seeds. I. Fixation of seeds. J. Cell Biol. *48*, 387-394.

MOTTIER, D. M. 1921. On certain plastids, with special reference to the protein bodies of *Zea, Ricinus,* and *Conopholis*. Ann. Botany (London) *35*, 350-364.

NEWCOMB, E. H. 1967. Fine structure of protein-storing plastids in bean root tips. J. Cell Biol. *33*, 143-163.

NOMOTO, M., and NARAHASHI, Y. 1958. A proteolytic enzyme of *Streptomyces griseus*. I. Purification of a protease of *Streptomyces griseus*. J. Biochem. *47*, 653-667.

PALADE, G. E. 1959. Functional changes in the structure of cell components. *In* Subcellular Particles, T. Hayashi (Editor). Ronald Press Co., New York.

PFEFFER, W. 1872. Investigations about protein granules and the significance of asparagine during seed germination. Jahrb. Wiss. Botany *8*, 429-574. (German)

RIJVEN, A. H. G. C. 1952. *In Vitro* Studies on the Embryo of *Capsella bursa-pastoris*. North-Holland Publishing Co. Amsterdam.

SABATINI, D. D., BENSCH, K., and BARRNETT, R. J. 1963. Cytochemistry and electron microscopy. The preservation of cellular ultrastructure and enzymatic activity by aldehyde fixation. J. Cell Biol. *17*, 19-57.

SINGH, J. 1970 The isolation and partial characterization of arachin (P6) from the seeds of *Arachis hypogaea*. Ph.D. Dissertation, Texas A & M Univ.

SMITH, B. W. 1950. *Arachis hypogaea*. Aerial flower and subterranean fruit. Am. J. Botany *37*, 802-815.

SPURLOCK, B. O., KATTINE, V. C., and FREEMAN, J. A. 1963. Technical modifications in Maraglas embedding. J. Cell Biol. *17*, 203-207.

ST. ANGELO, A. J., YATSU, L. Y., and ALTSCHUL, A. M. 1968. Isolation of edestin from aleurone grains of *Cannabis sativa*. Arch. Biochem. Biophys. *124*, 199-205.

STEMPAK, J. G., and WARD, R. T. 1964. An improved staining method for electron microscopy. J. Cell Biol. *22*, 697-701.

VARNER, J. E., and SCHIDLOVSKY, G. 1963. Intracellular distribution of proteins in pea cotyledons. Plant Physiol. *38*, 139-144.

VENABLE, J. H., and COGGESHALL, R. 1965. A simplified lead citrate stain for use in electron microscopy. J. Cell Biol. *25*, 407-408.

WATSON, M. L. 1958. Staining of tissue sections for electron microscopy with heavy metals. J. Biophys. Biochem. Cytol. *4*, 475-478.

WINBORN, W. B. 1965. Dow epoxy resin with triallyl cyanurate, and similarly modified Araldite and Maraglas mixtures as embedding media for electron microscopy. Stain Technol. *40*, 227-231.

WOOD, R. L., and LUFT, J. H. 1965. The influence of buffer systems on fixation with osmium tetroxide. J. Ultrastructure Res. *12*, 22-45.

YATSU, L. Y. 1965. The ultrastructure of cotyledonary tissue from *Gossypium hirsutum L*. seeds. J. Cell Biol. *25*, 193-199.

CHAPTER 6

Robert L. Ory | **Enzyme Activities Associated with Protein Bodies of Seeds**

INTRODUCTION

In 1855, Hartig was the first to report the isolation of colorless, round bodies from oilseeds by cutting the seeds into thin slices and washing them in oil. These particles were extracted into the oil phase and behaved towards chemical reagents as protein substances contained within a skin or membrane. A century later, protein bodies of seeds are again attracting the interests of plant scientists. In the past decade, the number of reports on these organelles has increased considerably. Intact protein bodies, or aleurone grains, have been prepared from a variety of ungerminated seeds: peanuts (Altschul *et al.* 1964; Dieckert *et al.* 1962), cottonseed (Lui and Altschul 1967; Yatsu and Jacks 1968), soybean (Saio and Watanabe 1966; Tombs 1967; Wolf 1970), hempseed (St. Angelo *et al.* 1968, 1969A), lima beans (Morris *et al.* 1970), corn (Christianson *et al.* 1969), wheat (Graham *et al.* 1963, 1964; Morton *et al.* 1964), rice (Mitsuda *et al.* 1967A,B, 1969), barley (Ory and Henningsen 1969; Tronier and Ory 1970; Tronier *et al.* 1971). and from germinating peas (Matile 1968; Varner and Schidlovsky 1963).

Electron microscopic techniques are employed to determine the purity (lack of contamination by other subcellular elements) and integrity of these preparations and to examine them for further substructures. In general, protein bodies of oilseeds are rather similar in appearance. They may vary in size from 1 to 20 μ in diameter, they contain the storage proteins of the seed enclosed by a single membrane, they stain evenly with osmic acid or $KMnO_4$, some contain globoids, the sites of phytic acid storage (Dieckert *et al.* 1962; Lui and Altschul 1967), some contain crystalloids (Ory *et al.* 1968; St. Angelo *et al.* 1968) which appear to be pure crystals of the major storage protein within (St. Angelo *et al.* 1968), and they are surrounded by spherosomes (storage organelles containing the oil of the seed) which fill the remainder of the cell. In cereals, the pictures are not as similar. The protein bodies contain neither globoids nor crystalloids of storage protein as seen in oilseeds, they are generally circled by a single ring of spherosomes which also line the inner wall of the cell, and each of the cereals examined thus far has different types of associated substructures (Graham *et al.* 1963; Khoo and Wolf 1970; Mitsuda *et al.* 1969; Ory and Henningsen (1969).

In order to demonstrate enzyme activity associated with a particular organelle, the best way is to isolate the particles intact and uncontaminated by other elements. To demonstrate enzyme activity *in situ*, it it necessary to control the reaction conditions in such a way that the substrate for the enzyme is converted at the site of activity either to colored or fluorescent end products for light microscopic examination, or to electron-dense products for electron microscopy. In some of the forementioned papers hydrolytic activities were localized in the protein bodies by either or both of these techniques.

It is not possible to give a comprehensive review of all of these different studies in a brief report. Therefore, this paper will be confined to studies on protein bodies isolated from three representative ungerminated seeds: cottonseed, a typical oilseed; hempseed, a seed having both globoids and crystalloids; and barley, a cereal grain.

COTTONSEED PROTEIN BODIES AND ASSOCIATED ENZYMES

Intact protein bodies (aleurone grains) have been successfully isolated from cottonseed by differential centrifugation in cottonseed oil:carbon tetrachloride solutions (Lui and Altschul 1967), the nonaqueous medium described by Dieckert *et al.* (1962), for preparing protein bodies from peanuts. However, if the isolated organelles are to be assayed for enzyme activity, chlorinated hydrocarbons could have adverse effects on any enzymes present. In order to minimize changes in enzymes present in the particles, Yatsu and Jacks (1968) homogenized and fractionated dehulled cottonseed in water-free glycerol. The pellet of intact protein bodies obtained by this procedure was examined by electron microscopy and shown to be free of other subcellular particles. They were 2-5 microns in diameter and contained numerous globoids which appeared as clear inclusions in the permanganate-stained protein material. Both cytochemical and *in vitro* chemical techniques were utilized to confirm the presence of acid hydrolytic activities in these organelles.

For the cytochemical demonstration of acid phosphatase activity (E.C. 3.1.3.2) *in situ*, small pieces of cottonseed cotyledonary tissue were fixed in calcium-formal overnight, then incubated in pH 5 acetate buffer containing lead nitrate and β-glycerol phosphate. Pieces of tissue heated to 90°C and tests minus β-glycerol phosphate served as controls.

Figure 6.1-A shows a normal cell of dormant cottonseed fixed in glutaraldehyde and potassium permanganate. The protein bodies (P) are easily recognized next to the clear oil-containing spherosomes (S). The small inclusions (G) within the protein bodies are globoids, which Lui and Altschul (1967) isolated from cottonseed protein bodies and characterized. The results of the *in situ* tests for acid phosphatase activity by Yatsu and Jacks (1968) are shown for comparison in Fig. 6.1-B. The lead phosphate produced by enzymic action is deposited primarily in the protein bodies, most heavily in the globoids.

To confirm the association of acid hydrolytic activities with isolated organelles, they fractionated the cottonseed into aleurone grain and nonaleurone grain components. Seeds were homogenized in glycerol throughout to prevent bursting of the particle membranes and the two fractions were assayed for enzyme activity. Acid proteinase activity (E.C. 3.4.4...) was determined by measuring the rate of hydrolysis of hemoglobin (Anson 1938); acid phosphatase,

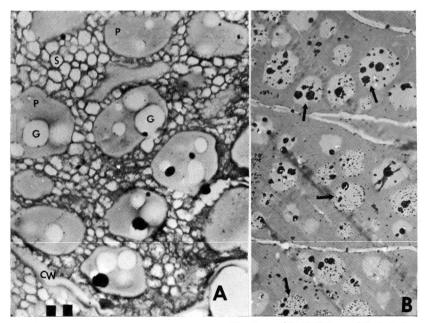

Courtesy of L. Y. Yatsu and T. J. Jacks

FIG. 6.1. A—ELECTRON MICROGRAPH OF PART OF A CELL IN COTTONSEED
COTYLEDON: P—PROTEIN BODIES; G—GLOBOIDS; S—SPHEROSOMES; CW—CELL
WALL. B—CYTOCHEMICAL STAIN FOR ACID PHOSPHATASE ACTIVITY IN
COTTONSEED COTYLEDON

by measuring the inorganic phosphate released from disodium phenylphosphate
spectrophotometrically (Salomon *et al.* 1964); and alcohol dehydrogenase (E.C.
1.1.1.1), by measuring the rate of reduction of NAD (Racker 1950).

All of the acid proteinase activity was located in the isolated aleurone grains;
none was found in the nonaleurone grain supernatant. The acid phosphatase was
distributed 77% in the aleurone grains and only 23% in the supernatant. This
latter amount probably represents activity in the nonaleurone grain cytoplasm.
In order to determine the amount of nonparticulate cytoplasm in the aleurone
grain fraction, alcohol dehydrogenase activity, considered a marker enzyme for
soluble cytoplasm, was measured in both fractions. All alcohol dehydrogenase
activity was confined to the supernatant, confirming the absence of cytoplasmic
contaminants in the isolated protein bodies (aleurone grains).

Recently, Morris *et al.* (1970) examined protein bodies isolated from ungerminated lima beans and determined the degree of contamination by other
organelles. By assay for marker enzymes of mitochondria and starch grains, they
showed that only 1.3% of the total starch and 7.4% of the total mitochondria
were present with the protein bodies. Further examination of the isolated particles showed the presence of acid protease and acid phenylphosphatase

activities. Phytase activity was absent in dormant seed preparations but was detected after four days germination.

HEMPSEED PROTEIN BODIES AND ASSOCIATED PROTEINASE ACTIVITY

The next example of an acid hydrolase in protein bodies is the work on hempseed by St. Angelo *et al.* (1968, 1969A,B, 1970). Acid proteinase activity was first localized in the particles, then the enzyme was subsequently purified and characterized.

Hempseed is one of those seeds having crystalloids within the protein bodies. Fig. 6.2-A shows a magnification of part of a cell in hempseed. The protein bodies are surrounded by spherosomes and contain globoids, as does cottonseed (Fig. 6.1-A). However, hempseed differs from other oilseeds studied in that the protein bodies contain distinct crystalloid-type inclusions. St. Angelo *et al.* (1968) isolated intact protein bodies from hempseed using the cottonseed oil:carbon tetrachloride method of Dieckert *et al.* (1962). The crystalloid-type structures were maintained in the isolated particles, as seen in Fig. 6.2-B. The crystalloids were then separated from the protein bodies by a combination of sonication, washing in NaCl containing sodium dodecylsulfate, and centrifugation. The resulting precipitate, when analyzed by various biochemical methods, was shown to be essentially pure crystals of edestin, the major storage protein.

Since the crystalloid material was pure edestin, attention then shifted to the

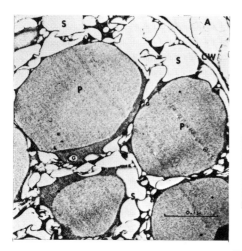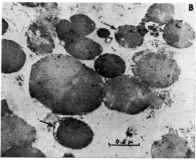

Courtesy of A. J. St. Angelo and L. Y. Yatsu

FIG. 6.2. A—ELECTRON MICROGRAPH OF PART OF A CELL IN HEMPSEED COTYLEDON: P—PROTEIN BODIES; G—GLOBOIDS; S—SPHEROSOMES; CW—CELL WALL. B—ELECTRON MICROGRAPH OF PROTEIN BODIES ISOLATED FROM HEMPSEED: ARROWS POINT TO GLOBOIDS RETAINED IN ISOLATED PARTICLES

protein surrounding the crystalloids in the protein bodies. In order to prevent bursting of the protein bodies during extraction and to preserve any enzyme activity present, the glycerol method of Yatsu and Jacks (1968) was employed to prepare the protein bodies from hempseed this time (St. Angelo *et al.* (1969A). The active enzyme was subsequently separated by the procedure described in Fig. 6.3. About 75% of the total proteolytic activity was recovered in this fraction, suggesting protein bodies as the major site of the enzyme. Activity was detected using hemoglobin as substrate (Anson 1938). With hemoglobin, optimum activity was observed at pH 3.2; no activity was found below pH 2 or above 5.

PREPARATION OF ACID PROTEINASE FROM HEMPSEED PROTEIN BODIES

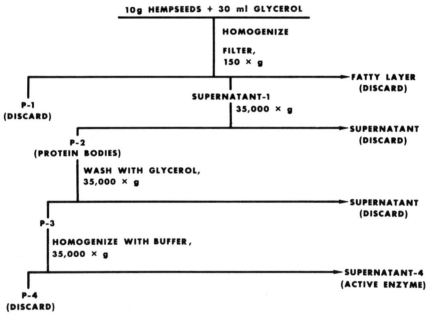

FIG. 6.3. FLOW DIAGRAM FOR PREPARATION OF ACID PROTEASE FROM ISOLATED HEMPSEED PROTEIN BODIES

However, since this enzyme is believed to be responsible for the *in vivo* hydrolysis of the reserve proteins upon initial germination of the seed, it was thought necessary to test its activity on edestin, the crystalloid protein. Fig. 6.4 shows that the enzyme does hydrolyze edestin but optimum activity is at pH 4.3, higher than the optimum observed with hemoglobin as substrate. It was also noted that the hydrolysis of edestin at 50° [the temperature of greater solubility

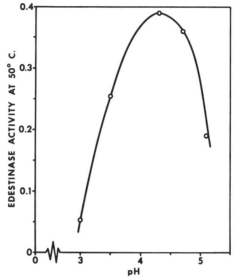

FIG. 6.4. ACTIVITY OF HEMPSEED ACID PROTEASE
ON EDESTIN AT 50°C AS A FUNCTION OF pH

of edestin as Stockwell *et al.* (1964) had shown] was twice that measured at 24°.

After confirming the presence of this protease in hempseed protein bodies, larger quantities of the supernatant fraction containing the protein bodies were prepared and the enzyme purified 25-fold from it (St. Angelo. *et al.* 1969B). Examination of this material in the ultracentrifuge showed a single component with an estimated molecular weight of 20,000. The purified enzyme, which was named edestinase by St. Angelo *et al.* (1969B) because of its location and its activity on edestin, was treated with a large number of compounds known to inhibit various types of proteolytic activity, to try and determine the type of functional group in the active site. None of these: mercuric acetate, N-ethyl maleimide, dithiothreitol, diisopropylfluorophosphate, phenylmethylsulfonyl-chloride, diphenylcarbamylchloride, and phenylpyruvic acid, nor EDTA and cysteine, generally considered activators, had any effect on edestinase activity (St. Angelo *et al.* 1970). The closest comparison possible was to cathepsin D, a mammalian lysosomal hydrolase, which has a pH optimum of 3.0 with hemo-globin as substrate and 4.2 with albumin. This would suggest a lysosomal-like role of protein bodies in intracellular digestion.

BARLEY PROTEIN BODIES AND ASSOCIATED ENZYMES

Of the investigations on protein bodies prepared from cereal grains, associated hydrolase activities were reported in barley only (Ory and Henningsen 1969;

Tronier and Ory, 1970; Tronier *et al* 1971). Morphologically, the picture one sees of protein bodies in cereals is different from that observed in oilseeds. Electron microscopic examination of immature wheat cells *in situ* showed some protein bodies to have layers of alternating dark and light lamellar structures appressed to the outer membranes (Jennings *et al*. 1963). The lamellar structures appear to be part of the protein bodies and the authors suggested that they are probably fine lipoprotein membranes, external to which are parallel layers derived from the endoplasmic reticulum. However, the isolated protein bodies did not seem to retain these lamellar structures (Morton *et al*. 1964). Rice protein bodies showed concentric dark and light lamellae within the particles when examined *in situ* by electron microscopy (Mitsuda *et al*. 1967A,B). This structure was retained in the isolated particles (Mitsuda *et al*. 1967A). Some of the protein bodies observed *in situ* in maize endosperm seem to have a dark speckled lumen and a wide lamellar ring inside the membrane (Khoo and Wolf

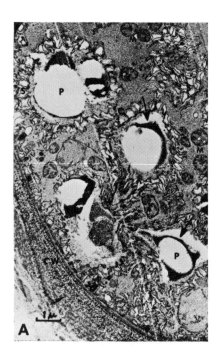 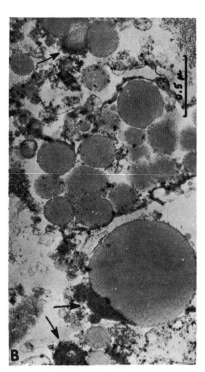

Courtesy of K. W. Henningsen

FIG. 6.5. A—ELECTRON MICROGRAPH OF PART OF A CELL ᴎ BARLEY ALEURONE TISSUE: P—PROTEIN BODIES; S—SPHEROSOMES; CW—CELL WALL; M—MITOCHONDRIA; ARROWS POINT TO ELECTRON-DENSE FINE STRUCTURE SURROUNDING PROTEIN BODIES. B—ELECTRON MICROGRAPH OF PROTEIN BODIES ISOLATED FROM BARLEY: ARROWS POINT TO FINE STRUCTURE MATERIAL ASSOCIATED WITH PROTEIN BODIES

1970), but these substructures were not evident in the organelles isolated from the tissue (Christianson *et al*. 1969).

Electron microscopic analysis of ungerminated barley aleurone tissue (Fig. 6.5-A) shows that spherosomes form a single layer around the protein bodies, rather than filling the cell space as in oilseeds. There are no globoids or crystalloids in these protein bodies but there appears to be a fine structure attached to many of the particles. This fine structure is retained by the organelles (Fig. 6.5-B) subsequently isolated by density gradient centrifugation over different sucrose concentrations (Ory and Henningsen 1969). Because of the low oil-high starch content of cereals, use of sucrose gradient centrifugation is more suitable for preparing intact protein bodies than are the glycerol or cottonseed oil:carbon tetrachloride methods.

Phosphatase Activity

Isolated barley protein bodies were examined for acid phosphatase and for proteolytic enzyme activities. However, because hydrolysis of nitrophenyl phosphate or synthetic substrates not native to the seed would not identify specific phosphatases which might be present, it was decided to employ natural substrates which might be found there: β-Glycerol phosphate fructose-1,6-phosphate, ATP, and sodium phytate. The results in Table 6.1 indicate that the acid phosphatase in barley protein bodies is a substrate-specific phytase. If whole barley seeds were extracted with buffer and the dialyzed, soluble extract tested on these same substrates for comparison to the protein bodies, all substrates tested were hydrolyzed by the total barley extract, confirming the presence of acid phosphatases active on other substrates in the whole seed.

TABLE 6.1

PHOSPHATASE ACTIVITY[1] IN BARLEY PROTEIN BODIES

Substrate	Activity in Barley	
	Protein Bodies	Whole Seeds
β-Glycerol phosphate	0	149.0
Fructose-1,6-phosphate	0	3.5
ATP	0	87.0
Sodium phytate	66.0	96.0

Source: Ory and Henningsen (1969) methods for preparing protein bodies and for assay of acid phosphatase activity.
[1] Activity measured as 10^{-3} micromole inorganic phosphate liberated per minute per 115 μ g enzyme nitrogen per test.

To localize the phytase cytochemically small pieces of barley aleurone tissue were incubated in lead nitrate at pH 5.0 for short periods, then examined in the electron microscope without further staining with osmium tetroxide or KMnO$_4$. The heavy deposits of lead phosphate were localized in the fine structure around the protein bodies (Tronier *et al.* 1971) suggesting this fine structure as a possible counterpart of the globoids in oilseed protein bodies. Poux (1963A,B) localized both the phosphate and the acid phosphatase activity in aleurone grains of wheat embryo cells by cytochemical methods.

Protease Activity

Assays for proteolytic activity in barley protein bodies were made by two methods (Ory and Henningsen 1969). Enari *et al.* (1963) have separated five different proteases from dormant barley seeds; one, called BAPA-ase because it hydrolyzes BAPA (α-N-benzoyl-DL-arginine *p*-nitroanilide), is a trypsin-like en-

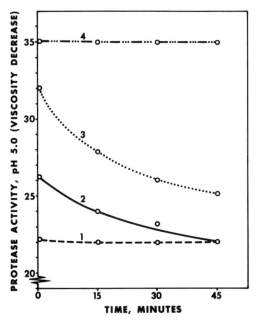

FIG. 6.6. ACID PROTEASE ACTIVITY IN ISOLATED
BARLEY PROTEIN BODIES

Activity at pH 5.0 is indicated by enzyme-catalyzed decrease in viscosities of gelatin solutions measured at 15 min intervals for 45 min. Curve 1, a control, 2 mg protein bodies boiled for 15 min before testing. Curve 2, same amount as Curve 1 but not boiled. Curve 3, 3 mg of protein bodies not boiled. Curve 4, 4 mg of protein bodies not boiled, but prepared without β-mercaptoethanol in extracting buffer.

zyme with a pH optimum at 8.6. The other four are acid proteases. When whole barley seed extracts and protein bodies were tested on BAPA, only the whole seed extracts were active indicating that only acid hydrolases may be associated with protein bodies of seeds. Acid protease activity was detected by measuring the decrease in viscosity of a 5% gelatin solution in pH 5.0 acetate buffer in viscometer tubes according to Enari *et al.* (1963). The results in Fig. 6.6 show that acid proteolytic activity is present in barley protein bodies, as was found in cottonseed (Yatsu and Jacks 1968), hempseed (St. Angelo *et al.* 1969A), and lima beans (Morris *et al.* (1970). The absence of activity when β-mercaptoethanol was not included in the homogenized protein bodies mixture (curve 4) indicates that this acid protease may be a sulfhydryl enzyme. Enari *et al.* (1963) reported that one of the four acid proteases is SH-sensitive. This provides further evidence that the initial hydrolysis of reserve proteins upon germination of seeds is catalyzed by acid proteases in the protein bodies.

β-Amylase Activity

Continuing the search for possible lysosomal-like enzymes associated with protein bodies, a third acid hydrolase was detected.

β-Amylase activity in barley is present in two forms; a free form found in germinating seeds of barley malt and a bound form present in dormant seeds. Nummi *et al.* (1965) showed that the bound β-amylase could be released with reducing agents and had the same molecular size as the free β-amylase. Daussant and Grabar (1965), using immunoelectrophoretic analysis (IEA), showed that the two forms had the same antigenic structure.

When isolated protein bodies were examined by the starch-iodine IEA technique of Daussant and Grabar (1965), some form of β-amylase activity was detected as a *halo* around the well containing the proteins (Tronier and Ory 1970). This is shown in Fig. 6.7-B. The failure to migrate in the electric field on the gels indicated that the amylase was bound to insoluble material of the protein bodies remaining at the point of application. If the homogenized protein bodies were treated with β-mercaptoethanol before IEA (Fig. 6.7-C, lower well), the β-amylase was released and migrated to the zone identical with free β-amylase of barley malt. Matile (1968) found acid protease, phosphatase, and β-amylase activities in aleurone grains isolated from germinating pea seedlings.

Since starch is the substrate for β-amylase, finding this enzyme bound to protein storage organelles was unexpected. However, Daussant (1968) reported that phytic acid can bind β-amylase as heterogeneous aggregates. Since phytase and phytic acid are associated with the protein bodies, the β-amylase may be bound to this phytic acid.

SUMMARY

The foregoing results suggest the following roles for these enzymes when seed

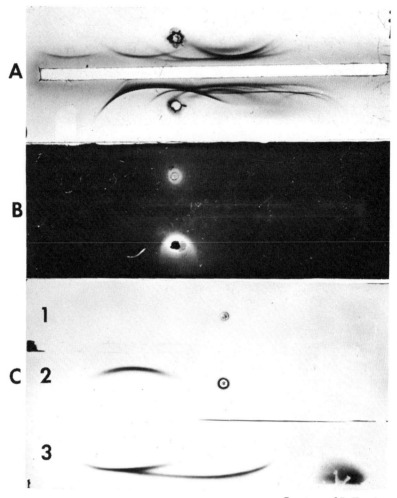

Courtesy of B. Tronier

FIG. 6.7. IMMUNOELECTROPHORETIC ANALYSIS OF BOUND β-AMYLASE
ACTIVITY ASSOCIATED WITH BARLEY PROTEIN BODIES

A—Total proteins of dormant barley stained with amido-black protein stain. B—Starch-iodine test for β-amylase activity after IEA of a protein bodies homogenate minus β-mercaptoethanol. C—Starch-iodine tests as in B, but using the photographed slide as a negative to increase contrast of the amylase activities: upper well (1) contains protein bodies homogenate; middle well (2) contains protein bodies homogenate treated with β-mercaptoethanol before IEA to release enzyme; lower well (3) contains proteins of germinating barley.

germination begins. Acid phosphatases release the inorganic phosphate from phytate which is required for phosphorylation of sugars in the energy reactions. The primary role of acid proteases is to release amino acids from the reserve proteins for new protein synthesis in the emerging seedlings. In barley, however, there appears to be an additional function.

β-Amylase which catalyzes the hydrolysis of starch is bound to protein bodies in dormant seeds. Nummi *et al.* (1965), in their studies on the free and bound β-amylases found that sulfhydryl proteases such as papain and ficin, produced the same effect on bound β-amylase as did sulfhydryl reducing agents or the malting process. They proposed a "splitting enzyme system" as the natural mechanism of release of the bound enzyme in germinating barley. The SH-sensitive protease associated with the protein bodies in barley (Ory and Henningsen 1969) may also function *in vivo* as the splitting enzyme system proposed by Nummi *et al.* (1965).

BIBLIOGRAPHY

ALTSCHUL, A. M., NEUCERE, N. J., WOODHAM, A. A., and DECHARY, J. M. 1964. A new classification of seed proteins: application to the aleurins of *Arachis hypogaea.* Nature *203,* 501-504.

ANSON, M. L. 1938. The estimation of pepsin, trypsin, papain, and cathepsin with hemoglobin. J. Gen. Physiol. *22,* 79-89.

CHRISTIANSON, D. D. *et al.* 1969. Isolation and chemical composition of protein bodies and matrix proteins in corn endosperm. Cereal Chem. *46,* 372-381.

DAUSSANT, J. 1968. Recent research on the proteins in brewery malt. Biotechnique *9,* 2-12. (French)

DAUSSANT, J., and GRABAR, P. 1965. β-amylase. II. Identification of the different β-amylases of barley and malt. European Brewery Conv. Proc., Stockholm 62-69. (French)

DIECKERT, J. W. *et al.* 1962 Composition of some subcellular fractions from seeds of *Arachis hypogaea.* J. Food Sci. *27,* 321-325.

ENARI, T.-M., PUPUTTI, E., and MIKOLA, J. 1963. Fractionation of the proteolytic enzymes of barley and malt. European Brewery Conv. Proc., Brussels 36-44.

GRAHAM, J. S. D., MORTON, R. K., and RAISON, J. K. 1963. Isolation and characterization of protein bodies from developing wheat endosperm. Australian J. Biol. Sci. *16,* 375-383.

GRAHAM, J. S. D., MORTON, R. K., and RAISON, J. K. 1964. The *in vivo* uptake and incorporation of radioisotopes into proteins of wheat endosperm. Australian J. Biol. Sci. *17,* 102-114.

HARTIG, T. 1855. Concerning the glutenflour. Botan. Zh. *13,* 881-882. (German)

JENNINGS, A. C., MORTON, R. K., and PALK, B. A. 1963. Cytological studies of protein bodies of developing wheat endosperm. Australian J. Biol. Sci. *16,* 366-374.

KHOO, U., and WOLF, W. J. 1970. Origin and development of protein granules in maize endosperm. Am. J. Botany *57,* 1042-1050.

LUI, N. S. T., and ALTSCHUL, A. M. 1967. Isolation of globoids from cottonseed aleurone grains. Arch. Biochem. Biophys. *121,* 678-684.

MATILE, P. 1968. Aleurone vacuoles as lysosomes. Z. Pflanzenphysiol. *58,* 365-368.

MITSUDA, H. *et al.* 1967A. Protein bodies in rice endosperm. Kyoto Univ. Coll. Agr. Mem. *92,* 17-28.

MITSUDA, H. *et al.* 1967B. Studies on the proteinaceous subcellular particles in rice endosperm: electron microscopy and isolation. Agr. Biol. Chem. (Tokyo) *31,* 293-300.

MITSUDA, H., MURAKAMI, K., KUSANO, T., and YASUMOTO, K. 1969. Fine structure of protein bodies isolated from rice endosperm. Arch. Biochem. Biophys. *130,* 678-680.

MORRIS, G. F. I., THURMAN, D. A., and BOULTER, D. 1970. The extraction and chemical composition of aleurone grains (protein bodies) isolated from seeds of *Vicia faba.* Phytochemistry *9,* 1707-1714.

MORTON, R. K., PALK, B. A., and RAISON, J. K. 1964. Intracellular components associated with protein synthesis in developing wheat endosperm. Biochem. J. *91,* 522-528.

NUMMI, M., VILHUNEN, R., and ENARI, T.-M. 1965. β-Amylase. I. β-Amylases of different molecular size in barley and malt. European Brewery Conv. Proc., Stockholm 51-61.

ORY, R. L., and HENNINGSEN, K. W. 1969. Enzymes associated with protein bodies isolated from ungerminated barley seeds. Plant Physiol. 44, 1488-1498.

ORY, R. L., YATSU, L. Y., and KIRCHER, H. W. 1968. Association of lipase activity with the spherosomes of Ricinus communis. Arch. Biochem. Biophys. 123, 255-264.

POUX, N. 1963A. Localization of phosphates and acid phosphatase in the cells of wheat embryo (Triticum vulgare Vill.). J. Microscopie 2, 557-568. (French)

POUX, N. 1963B. Localization of acid phosphatase activity and phosphates in the aleurone grains. I. Aleurone grains enclosing at the time globoids and crystalloids. J. Microscopie 4, 771-782. (French)

RACKER, E. 1950. Crystalline alcohol dehydrogenase from baker's yeast. J. Biol. Chem. 184, 313-319.

SAIO, K., and WATANABE, T. 1966. Preliminary investigation on protein bodies of soybean seeds. Agr. Biol. Chem. (Tokyo) 30, 1133-1138.

SALOMON, L. L., JAMES, J., and WEAVER, P. R. 1964. Assay of phosphatase activity by direct spectrophotometric determination of phenolate ion. Anal. Chem. 36, 1162-1164.

ST. ANGELO, A. J., YATSU, L. Y., and ALTSCHUL, A. M. 1968. Isolation of edestin from aleurone grains of Cannabis sativa. Arch. Biochem. Biophys. 124, 199-205.

ST. ANGELO, A. J., ORY, R. L., and HANSEN, H. J. 1969A. Localization of an acid proteinase in hempseed. Phytochemistry 8, 1135-1138.

ST. ANGELO, A. J., ORY, R. L., and HANSEN, H. J. 1969B. Purification of acid proteinase from Cannabis sativa L. Phytochemistry 8, 1873-1877.

ST. ANGELO, A. J., ORY, R. L., and HANSEN, H. J. 1970. Properties of a purified proteinase from hempseed. Phytochemistry 9, 1933-1938.

STOCKWELL, D. M., DECHARY, J. M., and ALTSCHUL, A. M. 1964. Chromatography of edestin at 50°C. Biochim. Biophys. Acta 82, 221-230.

TOMBS, M. P. 1967. Protein bodies of the soybean. Plant Physiol. 42, 797-813.

TRONIER, B., and ORY, R. L. 1970. Association of bound β-amylase with protein bodies in barley. Cereal Chem. 47, 464-471.

TRONIER, B., ORY, R. L., and HENNINGSEN, H. W. 1971. Separation and characterization of the fine structure and proteins from barley protein bodies. Phytochemistry 10, 1207-1211.

VARNER, J. E., and SCHIDLOVSKY, G. 1963. Intracellular distribution of proteins in pea cotyledons. Plant Physiol. 38, 139-144.

WOLF, W. J. 1970. Scanning electron microscopy of soybean protein bodies. J. Am. Oil Chemists' Soc. 47, 107-108.

YATSU, L. Y., and JACKS, T. J. 1968. Lysosomal activity in aleurone grains of plant seeds. Arch. Biochem. Biophys. 124, 466-471.

CHAPTER 7

Joe H. Cherry
Jan Kanabus
Witold Jachymczyk
and
Marianna B. Anderson

Protein Synthesis
In Germinating Seeds

INTRODUCTION

Protein synthesis in plants has been shown (Jachymczyk and Cherry 1968; Marcus and Feeley 1964; Marcus 1970; Williams and Novelli 1964) to involve the same reactions and general mechanisms as described for bacterial (Gilbert 1963; Matthaei and Nirenberg 1961; Henshaw *et al.* 1963) and for animal (Hardesty *et al.* 1963; Herriman and Hunter 1965; Wilkinson and Kirby 1966) systems. Usually in plants a quiescent or dormant state is associated with a low level of polyribosomes in comparison to monoribosomes. A shift upwards in metabolic activity, such as with seed germination, is associated with a large increase in polyribosomes. This increase in amount of polyribosomes is also correlated with an increase in protein synthesis. The "activation" of the protein synthesizing machinery may be attributed to one or more of the following components: an increase in synthesis or availability of mRNA, and changes in amounts of RNA species, aminoacyl-tRNA species aminoacyl-tRNA synthetase activities, and various factors (initiation, termination etc.).

This report describes some of our progress in an attempt to determine the role of mRNA synthesis, and changes in tRNA species and aminoacyl-tRNA synthetases during seed germination.

MATERIALS AND METHODS[1]

Plant Material

Peanut seed (*Arachis hypogaea* L. variety Virginia 56 R) or soybean seed (*Glycine max* L. variety Wayne) was germinated at 27°C on moist paper or in vermiculite as previously described (Jachymczyk and Cherry 1968). The cotyledons of unimbibed or germinated (2–4 days) seeds were used to study the general properties of *in vitro* amino acid incorporation into protein.

Cotyledon and hypocotyl tissue of soybean seedlings (Anderson and Cherry 1969) (four days after planting) were used to study changes in tRNA species and aminoacyl-tRNA synthetases.

Preparation and Characterization of Ribosomes

Peanut cotyledons were homogenized in a sucrose-tris buffer and the ribo-

---

[1] Abbreviations: mRNA, messenger RNA; tRNA, transfer RNA; lrRNA, light ribosomal RNA; hrRNA, heavy ribosomal RNA; MAK, methylated albumin-Kieselguhr; poly (U), polyuridylic acid; HA column, a column made of a mixture of hydroxylapatite and cellulose powder at the ratio of 10:1 (w/w).

somes were isolated and characterized as previously described (Jachymczyk and Cherry 1968). A ribosome-free supernatant containing various factors was obtained from cotyledons of dry peanut seed also as described.

Isolation of RNA

Cotyledons from 3-day old peanut seedlings were labeled with $^{32}PO_4^{3-}$ or [^{3}H] adenosine for 2.5 hr Subsequently, the RNA was extracted by a method of Kirby (1965). The RNA was passed through a Sephadex G-50 column and fractionated on a methylated albumin-Kieselguhr column to obtain an mRNA fraction (Jachymczyk and Cherry 1968).

Incorporation of Amino Acids into Protein

The incorporation of labeled leucine and phenylalanine into protein was determined using the techniques previously described (Jachymczyk and Cherry 1968) as adapted from Marcus and Feeley (1964). The details on the concentration of Mg^{2+}, ribosomes, supernatant proteins, and template RNA are given in the legends of the Figures and Tables. The amount of radioactive amino acid incorporated into protein was followed using the filter paper disk method of Mans and Novelli (1961).

Preparation of Transfer RNA and Aminoacyl-tRNA Synthetase

Transfer RNA was extracted from soybean cotyledon and hypocotyl tissues by a phenol method and then solubilized in 1 M NaCl. The methods are essentially the same as those described by Zubay (1962) and used in this laboratory (Anderson and Cherry 1969). Occasionally, 2 M potassium acetate was used for tRNA solubilization to reduce the amount of ribosomal RNA contamination (Cherry and Osborne 1970).

Fractionation of Leucyl-tRNA Species on the Freon Column

Transfer RNA preparations from soybean cotyledons and hypocotyls were charged with radioactive leucine (either ^{3}H or ^{14}C) using an aminoacyl-tRNA synthetase preparation from either tissue. Charged tRNA was then pumped onto a Freon column (RPC-2) and eluted with a linear gradient of NaCl as described by Weiss and Kelmers (1967) and adapted for plant systems by Anderson and Cherry (1969).

Purification of Leucyl-tRNA Synthetase

For the purpose of enzyme fractionation leucyl-tRNA synthetase was extracted from soybean cotyledons and hypocotyls as described by Kanabus and Cherry (1971). Briefly, the enzyme was purified by techniques employing (NH$_4$)$_2$SO$_4$ precipitation, followed by Sephadex G-150 and DEAE cellulose column fractionation. Finally, this purified enzyme preparation was fractionated on a hydroxylapatite (HA) column (Kanabus and Cherry 1971). Alternatively, the unfractionated enzyme was purified by the procedure of Anderson and Cherry (1969).

RESULTS

In Vitro Amino Acid Incorporation

Nucleic acids from plant tissues have been partially characterized by a number of workers (Cherry 1964; Cherry *et al*. 1965; Chroboczek and Cherry 1966; Ingle *et al*. 1965). Following the hrRNA fraction, a distinct peak of radioactive RNA is eluted from the MAK column. In a previous paper from this laboratory, Van Huystee and Cherry (1966) showed that DNA-RNA hybrid formation occurred to a greater extent with this than any other RNA fraction. Therefore, in order to determine the relative template activity of various nucleic acid fractions obtained from the MAK column (Fig. 7.1) each (except tRNA) was incubated with monoribosomes and the incorporation of $[^{14}C]$-leucine was determined. The results show that the addition of mRNA fraction stimulated the $[^{14}C]$-leucine incorporation by 30-fold, which is 3 times more than the stimulation observed when lrRNA was used. The hrRNA fraction had no stimulatory effect. The second best native nucleic acid fraction is DNA with its associated RNA. In agreement with Marcus and Feeley (1964, 1965), we find that during imbibition the capacity of ribosomes to support amino acid incorporation (counts per minute per milligram of RNA) increases 6-fold during 48 hr of imbibition. When ribosomes obtained from dry seed were incubated with a fraction of mRNA isolated from cotyledons of 3-day seedlings by MAK column chromatography, amino acid incorporation was stimulated 5-fold (Fig. 7.2). The rate of amino acid incorporation of the monoribosomes-mRNA system is essentially the same as that for polyribosomes of imbibed seed. It is also of interest that polyuridylic acid [Poly (U)] stimulates $[^{14}C]$-phenylalanine incorporation to the degree of that for native mRNA.

Ribosome particles isolated from dry seed in the presence of 10 mM Mg^{2+} are comprised mostly of monoribosomes with few polyribosomes. Conversely, a ribosomal preparation from imbibed cotyledons shows a large quantity of polyribosomes (cf Fig. 7.3-A,B). Further, the monoribosomes are nonfunctional in amino acid incorporation unless mRNA is added (Table 7.1). The data of Table 7.1 show that actinomycin D had essentially no effect on amino acid incorporation. Cycloheximide inhibited both systems but the polyribosomes of imbibed seed were more severely affected than the monoribosome-exogenous mRNA system.

A relatively high level of polyribosomes extracted from tissues indicates availability of mRNA for the formation of the complex and the occurrence of protein synthesis. On the other hand, an apparent increase in polyribosome

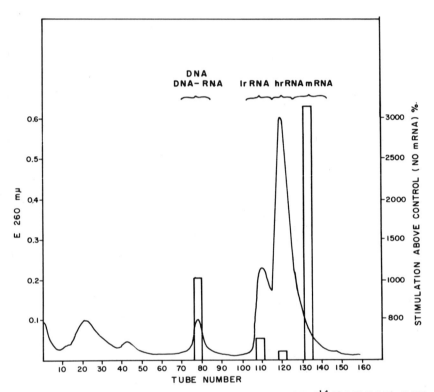

FIG. 7.1. STIMULATION OF THE INCORPORATION OF [^{14}C]-LEUCINE INTO PROTEIN BY DIFFERENT FRACTIONS OF RNA ISOLATED ON THE MAK COLUMN

Two milligrams of nucleic acids extracted from cotyledons of 3-day-old peanut seedlings were eluted from the column with a linear gradient of NaC1 from 0.4 to 1.2 M in 0.05 M sodium phosphate buffer (pH 6.7). Five-milliliter fractions were collected and analyzed for A260 mμ. The relative template activity of 4 different nucleic acid components was determined by incubating in an *in vitro* system containing monoribosomes (0.4 mg) from dry seed. Other additives (in 1 ml) were as follows: 0.73 mg of supernatant protein from dry seed, 15 μmoles of MgCl$_2$, nucleic acid (110 μg of DNA-RNA or 200 μg of hrRNA or 180 μg of lrRNA or 90 μg of mRNA) and 0.5 μC of [^{14}C]-leucine (240 μC/μmole). Tris buffer, KCl, ATP, GTP, and ATP regenerating system as described by Jachymczyk and Cherry (1968) were included. The reaction mixture was incubated at 37°C for 60 min. The amount of [^{14}C]-leucine incorporation for the control (no RNA added) was 2050 counts per minute per milligram ribosomal RNA.

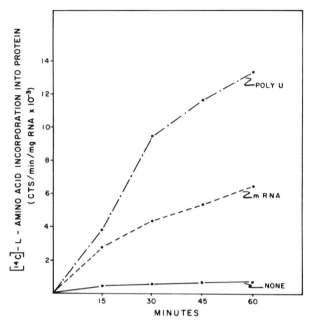

FIG. 7.2. STIMULATION OF L-[^{14}C]-LEUCINE INCORPORATION
INTO PROTEIN BY MESSENGER RNA

Ribosomes and 105,000 x g supernatant protein were from dry peanut
cotyledons. Incubation medium contained in 1 ml of the following: 0.1
mg of ribosomes (as RNA), 0.73 mg of supernatant protein, 15 μmoles
of MgCl$_2$, 150 μg of yeast tRNA, 0.5 μC of L-[^{14}C]-leucine (240
μC/μmole) or [^{14}C]-phenylalanine (302 μC/μmole), and 90 μg of
mRNA (from MAK column) or 100 μg of Poly (U).

content does not indicate a rapid synthesis of mRNA. In the case of dry peanut
seed, mRNA appears unavailable for the synthesis of certain proteins during the
early stage of seed germination (Gientka-Rychter and Cherry 1968). Thus, the
release of "masked" mRNA may be important in the initial stage of germination.
However, in the case of seeds, the demonstration of "masked" mRNA release
from some inactive state appears very difficult. Thus, because of the technical
difficulties we decided to study various parameters of mRNA translation, with
special reference to different tRNA species and aminoacyl-tRNA synthetases.

Differences in Leucyl-tRNA's in the Soybean Seedling

Aminoacylation of tRNA with leucine by synthetases from soybean
cotyledon and hypocotyl follows standard enzyme kinetics, the rate being linear
for 10 min and leveling off after 40 min. No loss in charged tRNA after maximal
acylation, no production of new acylated-tRNA peaks, and no broadening of

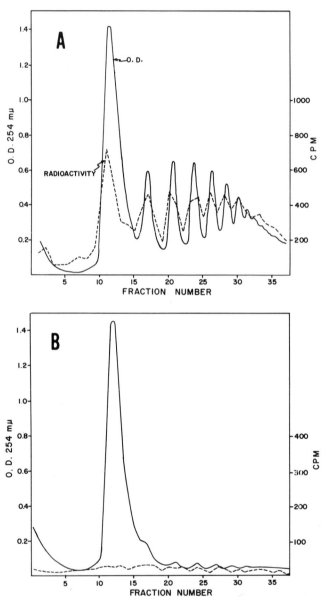

FIG. 7.3. SUCROSE GRADIENT PROFILES OF RIBOSOME PREPARATIONS FROM
COTYLEDONS OF DRY AND IMBIBED SEEDS

Four cotyledons from 48 hr imbibed (A) and dry (B) seed were incubated 3 hr at 30°C
with 10 μC of reconstituted protein hydrolysate containing 13 [^{14}C]-labeled amino
acids purchased from Schwarz Bio Research in a medium of 1% sucrose, 10^{-4} M ammonium citrate (pH 6.0) in a total volume of 5 ml. The ribosomes were isolated and
characterized by sucrose gradient centrifugation.

TABLE 7.1

CONDITIONS FOR THE *IN VITRO* INCORPORATION OF AMINO ACID INTO PROTEIN BY RIBOSOMES ISOLATED FROM DRY AND IMBIBED PEANUT SEEDS

Source of Ribosomes	Additions	Counts/Min per Mg Protein
Dry seed	Complete system[1]	2188
Dry seed	Complete system + actinomycin D	2080
Dry seed	Complete system + cycloheximide	312
Dry seed	ATP omitted	90
Dry seed	Supernatant omitted	411
Dry seed	mRNA omitted	413
48 h imbibed seed	Complete system	3982
48 h imbibed seed	Complete system + actinomycin D	3410
48 h imbibed seed	Complete system + cycloheximide	1422
48 h imbibed seed	ATP omitted	62
48 h imbibed seed	Supernatant omitted	945
48 h imbibed seed	mRNA omitted	2480

Source: Jachymczyk and Cherry (1969).
[1] The complete system contained in 1 ml of incubation medium the following: 0.30 mg of ribosomes from dry seeds or 0.32 of ribosomes from imbibed seeds, 0.8 mg supernatant protein, 125 μg of mRNA, 15μg actinomycin D, 15 μg of cycloheximide and 0.5 μC of [^{14}C] leucine (260 mC/mmole). Ten and 15 μmoles/ml of Mg^{2+} were added to the ribosomal systems from imbibed and dry seed, respectively. Incubation was at 37°C for 30 min.

peaks were observed with time; thus, the enzyme preparations seem to contain very little ribonuclease.

Leucyl-tRNA's from cotyledons can be fractionated in six discrete peaks by Freon column chromatography (Fig. 7.4). These tRNA peaks will be referred to as tRNA 1, tRNA 2, etc., in order of elution from the column. The first peak (not numbered; usually at tubes 40-45) is radioactive leucine charged to degraded tRNA and possibly some uncharged amino acid not completely removed from the DEAE-cellulose column. Leucyl-tRNA 5 and -tRNA 6 from the hypocotyl are only slightly acylated with hypocotyl enzyme (Fig. 7.4). The degree of acylation of tRNA 1, 2, 3, and 4 with the 2 enzymes is similar; in contrast, there is a 10-fold difference between tRNA 5 and 6 charged using a homologous system from cotyledons and hypocotyls (see insert of Fig. 7.4). The comparatively small amount of tRNA 5 and 6 being charged by the hypocotyl system indicates differences that could be based on either 1 of 2 rate-limiting factors, namely, the synthetases or the tRNA.

The enzyme from cotyledon tissue acylated tRNA 5 and 6 from both tissues. On the other hand, the enzyme from hypocotyl tissue did not charge hypocotyl or cotyledon tRNA 5 and 6. An attempt to acylate leucyl-tRNA 5 and -tRNA 6 using concentrated hypocotyl tRNA with an increased amount of hypocotyl enzyme preparation failed (Table 7.2). Thus, hypocotyl tissue has some leucyl-tRNA 5 and -tRNA 6, but the hypocotyl synthetase preparation contains insuffi-

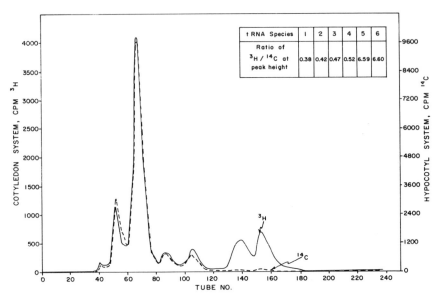

tRNA Species	1	2	3	4	5	6
Ratio of $^3H/^{14}C$ at peak height	0.38	0.42	0.47	0.52	6.59	6.60

FIG. 7.4. LEUCYL-tRNA PATTERNS FROM HOMOLOGOUS TISSUE SYSTEMS FRAC-
TIONATED ON A FREON COLUMN

Solid line represents cotyledon tRNA species acylated with L-[^{3}H]-leucine by cotyledon
protein for 10 min. Dashed line is hypocotyl tRNA acylated by hypocotyl protein with
L-[^{14}C]-leucine. In both preparations, 1 mg tRNA and 0.4 mg protein were used.

cient activity to acylate these two tRNA's. These results clearly indicate an
organ specificity in tRNA and aminoacyl-tRNA synthetase in the soybean
seedling. Furthermore, other results (Cherry and Osborne 1970) show species
specificity in the leucyl-tRNA's and leucyl-tRNA synthetases.

Differences in Leucyl-tRNA Synthetases

To explain the difference between the enzymes of the 2 tissues, we adopted a
hypothesis that the leucyl-tRNA synthetase activity of soybean cotyledons consists of more than 1 enzyme of which 1 or more are absent from the hypocotyls
Investigation of this hypothesis resulted in fractionation of the total, partially
purified, synthetase preparation from soybean cotyledons into three fractions on
a hydroxylapatite column. Under similar conditions, the hypocotyls appeared
deficient in 1 of the 3 fractions.

The results show that purification of the total enzyme does not produce any
changes in its range of specificity toward the tRNA^{Leu} species. Co-
chromatography on a Freon column of the cotyledon tRNA charged with
leucine using the cotyledon enzymes after Sephadex and DEAE-cellulose column
chromatography is shown in Fig. 7.5. The two enzymes are similar to those
obtained by Anderson and Cherry (1969) using total enzyme isolated from the

TABLE 7.2

CHANGES IN ACYLATION OF LEUCYL-tRNA's WITH VARIOUS COMBINATIONS OF
COTYLEDON AND HYPOCOTYL tRNA's AND ENZYMES

Source of Enzyme	Mixture (Mg)	Source of RNA	Mixture (Mg)	Relative Amount of Leucyl-tRNA Acylation of Each Peak (% of Total)[1]					
				1	2	3	4	5	6
Hypocotyl	0.4	Cotyledon	1.0	13.1	79.1	1.4	6.4	...	...
Hypocotyl	0.4	Hypocotyl	1.0	13.5	79.6	1.6	5.4	...	...
Cotyledon	0.4	Cotyledon	1.0	8.1	50.8	7.3	5.8	10.1	17.9
Cotyledon	0.4	Hypocotyl	1.0	14.4	67.9	6.0	4.2	2.3	5.2
Hypocotyl	1.9	Hypocotyl	5.9	6.5	77.4	7.9	7.0	0.7	1.2
Hypocotyl	1.9	Cotyledon	0.6	11.9	68.5	7.7	7.9	1.7	2.6

[1] tRNA was acylated in a 2-ml reaction mixture with either [^{3}H]- or [^{14}C]-leucine and fractionated on a Freon column as described in Methods. The amount of radioactivity in each peak was summed and expressed as a percentage of the sum total in the 6 peaks.

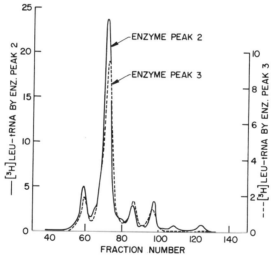

FIG. 7.5. FREON COLUMN CHROMATOGRAPHY OF LEU-
tRNA'S PRODUCED BY COTYLEDON ENZYME AT TWO
STAGES OF PURIFICATION

Cotyledon tRNA was acylated for 30 min using the following conditions: (1) Protein from the peak of activity from Sephadex G-150 column (0.40 A$_{280}$ units), 0.5 mg tRNA, and 50 μCi of L-4,5[^{3}H]-leucine per milliliter (solid line). (2) Protein from the peak of activity from DEAE-cellulose column (0.194 A$_{280}$ units), 0.5 mg tRNA, and 1.63 μCi of L- U-[^{14}C]-leucine per milliliter (broken line). Samples corresponding to 0.65 ml and 0.71 ml of reaction mixture 1 and 2, respectively, were co-chromatographed.

postmitochondrial fraction of soybean cotyledons. This supports our conclusion that the partially purified enzyme contains a complete set of the hypothesized leucyl-tRNA synthetase species.

The results of fractionation of the partially purified enzymes from the two tissues on HA columns are shown in Fig. 7.6. In the case of the cotyledon enzyme (solid line), the profile is typical in that three peaks of activity are always present. The relative magnitude of the peaks depends to a certain degree on the time elapsed between extraction and fractionation of the enzyme and on the presence of phenylmethyl sulfonylfluoride.[2] In contrast, the hypocotyl preparation (broken line) gives only 2 major peaks of synthetase activity which correspond to peaks 2 and 3 of the cotyledon synthetase.

The range of specificity of the individual enzyme peaks are examined using the technique of Freon column chromatography. Figure 7.7 shows the result of co-chromatography of the leucyl-tRNA's obtained using total enzyme from the DEAE-cellulose (broken line) and the first cotyledon peaks from the HA column (solid line). The enzyme eluting first from the HA column acylates tRNA's 5 and 6 exclusively (the presence of small amounts of peaks 1 and 4 can be ascribed to

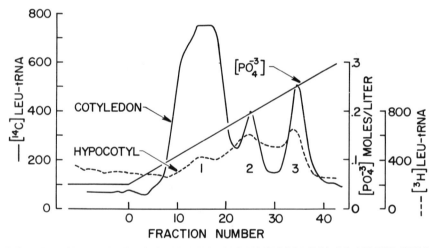

FIG. 7.6. HYDROXYLAPATITE COLUMN CHROMATOGRAPHY OF LEUCYL-tRNA SYNTHETASES FROM SOYBEAN TISSUE

Solid line: data obtained with a preparation from 50 gm cotyledons assayed using L-U-[^{14}C]-leucine. Broken line: data obtained with a preparation from 150 gm hypocotyls assayed using L-4,5-[^{3}H]-leucine. Reaction mixture (0.25 ml final volume) contained 0.05 ml enzyme (fractions of the eluate). After 20 min acylation, the entire reaction mixtures were mixed with trichloroacetic acid, filtered, and counted. The two experiments were run separately, assayed, and then the results superimposed.

---

[2]The extraction, purification, and fractionation of the cotyledon enzymes in the absence of leucine and phenylmethyl sulfonylfluoride gave similar results; however, the omission of the latter resulted in lower recovery of peak 2.

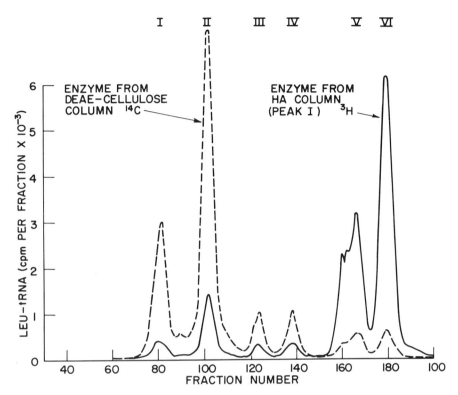

FIG. 7.7. FREON COLUMN CHROMATOGRAPHY OF LEU-tRNA'S PRODUCED BY
COTYLEDON ENZYME FROM THE HYDROXYLAPATITE COLUMN

Cotyledon tRNA (0.5 mg) was acylated using enzyme peak 1 from the HA column
(0.042 A_{260} units) and 50 μCi of L-4,5-[^{3}H]-leucine per milliliter. A sample corresponding to 0.95 ml of the above reaction mixture (solid line) was co-
chromatographed with a sample of [^{14}C]-Leu-tRNA (broken line) produced by the
unfractionated enzyme and corresponding to 0.78 ml of the reaction mixture 2
described under Fig. 7.5.

contamination of enzyme 1 with the neighboring enzyme fraction). In similar
experiments, the cotyledon synthetase fractions 2 and 3 and the 2 hypocotyl
enzymes all revealed specificity for the remaining 4 tRNA^{Leu} species producing
leucyl-tRNA^{Leu} peaks 1 through 4.

The same pattern of specificity was found (data not shown) when uncharged
tRNA was chromatographed on a Freon column and the eluate assayed for
leucine acceptor activity using isolated fractions of the cotyledon synthetase:
enzyme fraction 1 charged only 2 tRNA peaks located in the area of leucyl-
tRNA peaks 5 and 6 while fraction 3 acylated preferentially tRNA peaks corresponding to leucyl-tRNA peaks 1 through 4. Fraction 2 was not tested.

The Michaelis constant (Km) for the total cotyledon tRNA in the amino
acylation reaction with the enzyme fraction 1 (Km = 2.5 absorbance$_{260}$ units

per milliliter) was about 20 times higher than that with the other 2 enzymes. However, in terms of relative molar concentrations, the affinity of the 3 enzymes for their cognate tRNA's appear to be of the same order of magnitude if one considers that species 5 and 6 account for only 8–11% of the leucine acceptor activity of unfractionated tRNA while species 1 to 4 together make up the remaining 90% (calculated from the data in Fig. 7.5). This would argue against the possibility that one of the enzymes erroneously attaches leucine to tRNA's specific for an amino acid other than leucine.[3]. This possibility, although unlikely, cannot be excluded until the tRNA's involved are available in purity sufficient to directly test their specificity.

To our knowledge, the above results represent the first instance of isolation from a higher plant of an organ-specific aminoacyl-tRNA synthetase (peak 1, solid line in Fig. 7.6) with a distinct and strictly defined specificity (Fig. 7.7). It is interesting that a similar enzyme can hardly be detected in the hypocotyls, a tissue which has been shown to contain small amounts of both its cognate tRNA's.

DISCUSSION

As shown by Marcus and Feeley (1964, 1965) and Jachymczyk and Cherry (1968), a significant result of seed imbibition is the formation of polyribosomes *in vivo* (Fig. 7.3) and their associated capacity to support amino acid incorporation into protein. The fact that ribosomes isolated from dry seed are not capable of supporting the incorporation of amino acids into protein without the addition of mRNA suggests that the synthesis or release of mRNA is an essential process in the initiation of seed germination (Jachymczyk and Cherry 1968; Marcus and Feeley, 1964, 1965, 1966). To study this problem, we isolated a fraction of mRNA from imbibed cotyledons by use of a MAK column. Studies of amino acid incorporation show that the so-called mRNA has the greatest capacity to serve as template (Fig. 7.1).

As previously noted the mRNA fraction directed the greatest [^{14}C]-leucine incorporation into protein while the DNA-RNA fraction stimulated amino acid incorporation about 1/3 that of the messenger fraction. Thus, it seems that only these two fractions contain RNA with template activity. Herriman and Hunter (1965) in a study on the stimulatory effect of different fractions of nucleic acid isolated from mouse brain also found that only these two fractions stimulate protein synthesis in an *in vitro* system. It should be noted that an RNA fraction from dry seed, analogous to mRNA isolated from imbibed seed, did not stimulate the incorporation of [^{14}C]-leucine into protein (results not shown). Also, an analogous RNA fraction isolated from broad bean mosiac virus was inactive in stimulating amino acid incorporation. It had previously been shown that tobacco

---

[3] An elevated Km would be expected for the tRNA mischarged in such an erroneous reaction.

mosaic virus RNA failed to direct amino acid incorporation using ribosomes from hepatoma (Wilkinson and Kirby 1966) and ribosomes from rabbit reticulocytes (Nisman and Pelmont 1966). However, Marcus *et al.* (1968) have obtained polyribosome formation with tobacco mosaic virus with monoribosomes from seed. In fact this system is now being used to test various factors for protein synthesis (Marcus 1970).

The possibility of regulation of protein synthesis at the level of translation has been suggested by several workers (see Anderson and Cherry 1969; Kanabus and Cherry 1971). It is apparent that the rate of readout of a particular mRNA, that is, the rate of synthesis of a protein, can be controlled by rate-limiting quantities of tRNA's and/or synthetases.

The results presented in this paper clearly show that there are six leucyl-specific tRNA's in the cotyledon. On the other hand, only four major leucyl-tRNA's are found with the hypocotyl system. The inability of the hypocotyl enzyme preparation to acylate tRNA 5 and 6 seemed to us to be a result of the absence of a specific synthetase. In view of the possibility that these results may be a reflection of a part of the biochemical mechanism underlying plant differentiation, we investigated the differences between the enzymes from these two organs in more detail. The soybean cotyledon leucyl-tRNA synthetase was fractionated into 3 fractions which fall into 2 classes of exclusive, nonoverlapping ranges of specificity toward the 6 species of $tRNA^{Leu}$. Our results also indicate that soybean hypocotyls are deficient in 1 of the synthetase fractions, viz., the 1 responsible for specific aminoacylation of peaks 5 and 6 of $tRNA^{Leu}$.

SUMMARY

The following conclusions are drawn from the results presented in this paper:

(1) Preparations from cotyledons of unimbibed peanut seed contain few polyribosomes and show a low level of amino acid incorporation into protein. Similar preparations from imbibed seed contain high relative amounts of polyribosomes and exhibit high activity of amino acid incorporation.

(2) RNA extracted from germinating peanut cotyledons and then eluted after ribosomal RNA on the methylated albumin-Kieselguhr (MAK) column has the highest template activity. This fraction stimulates amino acid incorporation into protein in an *in vitro* system.

3) Charged leucyl-transfer ribonucleic acids of soybean cotyledons are fractionated into six radioactive peaks on a Freon column. Only four leucyl-tRNA peaks are observed with the homologous hypocotyl system. Hypocotyl synthetase preparations only slightly acylate the other two leucyl-tRNA's (species 5 and 6).

(4) The Leucyl-tRNA synthetase activity from cotyledons of 4-day-old soybean seedlings are fractionated into 3 components. One of these exclusively acylates 2 of the 6 $tRNA^{Leu}$ species present in this tissue. The remaining 2

enzyme fractions charge the other 4 tRNA[Leu] species equally well. Soybean hypocotyls appear to contain only the last two enzyme fractions.[4]

BIBLIOGRAPHY

ANDERSON, M. B., and CHERRY, J. H. 1969. Differences in leucyl-transfer RNA's and synthetase in soybean seedlings. Proc. Natl. Acad. Sci. U.S. *62*, 202-209.

CHERRY, J. H. 1964. Association of rapidly metabolized DNA and RNA. Science *146*, 1066-1069.

CHERRY, J. H., CHROBOCZEK, H., CARPENTER, J. G., and RICHMOND, A. 1965. Nucleic acid metabolism in peanut cotyledons, Plant Physiol. *40*, 582-587.

CHERRY, J. H., and OSBORNE, D. J. 1970. Specificity of leucyl-tRNA and synthetase in plants. Biochem. Biophys. Res. Commun. *40*, 763-769.

CHROBOCZEK, H., and CHERRY, J. H. 1966. Characterization of nucleic acids in peanut cotyledons. J. Mol. Biol. *19*, 28-37.

GIENTKA-RYCHTER, A., and CHERRY, J. H. 1968. De novo synthesis of isocitritase in peanut (Arachis hypogaea L.) cotyledons. Plant Physiol. *43*, 653-659.

GILBERT, W. 1963. Polypeptide synthesis in *Escherichia coli*. I. Ribosomes and the active complex. J. Mol. Biol. *6*, 374-388.

HARDESTY, B., ARLINGHAUS, R., SHAEFFER, J., and SCHWEET, R. 1963. Hemoglobin and polyphenylalanine synthesis with reticulocyte ribosomes. Cold Spring Harbor Symp. quant. Biol. *28*, 215-222.

HENSHAW, E. C., BOJARSKI, T. B., and HIATT, H. H. 1963. Protein synthesis by free and bound rat liver ribosomes *in vivo* and *in vitro*. J. Mol. Biol. *7*, 122-129.

HERRIMAN, J. D., and HUNTER, G. D. 1965. Cytoplasmic protein synthesis in mouse brain. J. Neurochem. *12*, 937-947.

HUYSTEE, van R. B., and CHERRY, J. H. 1966. Hybridization of messenger RNA with DNA from plants. Biochem. Biophys. Res. Commun. *23*, 835-841.

INGLE, J., KEY, J. L., and HOLM, R. E. 1965. Demonstration and characterization of a DNA-like RNA in excised plant tissue. J. Mol. Biol. *11*, 730-746.

JACHYMCZYK, W. J., and CHERRY, J. H. 1968. Studies on messenger RNA from peanut plants: *In vitro* polyribosome formation and protein synthesis. Biochim. Biophys. Acta *157*, 368-377.

KANABUS, J., and CHERRY, J. H. 1971. Isolation of an organ-specific leucyl-tRNA synthetase from soybean seedling. Proc. Natl. Acad. Sci. U.S. *68*, 873-876.

KIRBY, K. S. 1965. Isolation and characterization of ribosomal ribonucleic acid. Biochem. J. *76*, 266-269.

MANS, R. J., and NOVELLI, G. D. 1961. Measurement of the incorporation of radioactive amino acids into protein by a filter paper disk method. Arch. Biochem. Biophys. *94*, 48-53.

MARCUS, A. 1970. Tobacco mosaic virus ribonucleic acid-dependent amino acid incorporation in a wheat embryo system *in vitro*. J. Biol. Chem. *245*, 962-966.

MARCUS, A., and FEELEY, J. 1964. Activation of protein synthesis in the imbibition phase of seed germination. Proc. Natl. Acad. Sci. U.S. *51*, 1075-1079.

MARCUS, A., and FEELEY, J. 1965. Protein synthesis in imbibed seeds. II. Polysome formation during imbibition. J. Biol. Chem. *240*, 1675-1680.

MARCUS, A., and FEELEY, J. 1966. Ribosome activation and polysome formation *in vitro*: Requirements for ATP. Proc. Natl. Acad. Sci. U.S. *51*, 1770-1777.

MARCUS, A., LUGINBILL, B., and FEELEY, J. 1968. Polysome formation with mosaic virus RNA. Proc. Natl. Acad. Sci. U.S. *59*, 1243-1250.

MATTHAEI, J. H., and NIRENBERG, M. W. 1961. Characteristics and stabilization of DNAse-sensitive protein synthesis in *E. coli* extracts. Proc. Natl. Acad. Sci. U.S. *47*, 1580-1602.

---

[4] This research was supported by a grant from the National Science Foundation (GB-20483), a contract (COO-1313-28) from the U.S. Atomic Energy Commission, and a David Ross Fellowship from the Purdue Research Foundation.

NISMAN, B., and PELMONT, J. 1966. De novo protein synthesis *in vitro*. Progr. Nucleic Acid Res. *3*, 235-297.

WEISS, J. F., and KELMERS, A. D. 1967. A new chromatographic system for increased resolution of transfer ribonucleic acids. Biochemistry *6*, 2507-2513.

WILKINSON, B. R., and KIRBY, K. S. 1966. Rapidly labelled ribonucleic acid from rat liver studies in the attachment to ribosomes and stimulation of the incorporation of amino acids into polypeptides in cell-free systems. Biochem. J. *19*, 786-791.

WILLIAMS, G., and NOVELLI, G. D. 1964. Stimulation of an *in vitro* amino acid incorporating system by illumination of dark-grown plants. Biochem. Biophys. Res. Commun. *17*, 23-27.

ZUBAY, G. 1962. The isolation and fractionation of soluble ribonucleic acid. J. Mol. Biol. *4*, 347-356.

Bienvenido O. Juliano

Studies on Protein Quality and Quantity of Rice

Rice (*Oryza sativa* L.) is a principal source of calories and of protein among Asians. Other than starch, protein is the most abundant constituent of the rice grain. Brown (dehulled) rice usually has 8% protein (measured at 12% moisture); milled rice has 7% protein (Juliano 1966). The bran (aleurone layer plus germ) contains more protein than the starchy endosperm. So during milling the proportion of protein lost is higher than the proportion of weight lost.

PROPERTIES OF RICE PROTEIN AND ITS FRACTIONS

The protein of milled rice is unique among cereal proteins in that it contains at least 80% glutelin (alkali-soluble protein). The other Osborne protein fractions are 5% albumin (water-soluble protein), 10% globulin (salt-soluble protein), and less than 5% prolamin (alcohol-soluble protein) (Cagampang *et al.* 1966; Juliano 1965, 1967). Albumin and globulin are concentrated in the aleurone layers and in the germ. Glutelin is insoluble in water because of its high molecular weight which results from crosslinking of the smaller subunits through disulfide bonds. It is heterogeneous and consists of subfractions differing in solubility, electrophoretic mobility, and amino acid composition (Tecson *et al.* 1971). Kjeldahl nitrogen is multiplied by 5.95 to convert it to crude protein based on the 16.8% nitrogen content of glutelin.

Rice protein occurs in the endosperm mainly as discrete particles, 1–4 μ in size, called protein bodies (del Rosario *et al.* 1968; Mitsuda *et al.* 1967). Smaller but more numerous protein bodies are present in the peripheral cells of the endosperm. More protein bodies occur next to the endosperm cell walls. The electron micrograph of the protein bodies shows a limiting membrane and a layered structure inside the protein body (Mitsuda *et al.* 1967, 1969). The protein bodies have the same composition as whole milled-rice protein.

Among the milling fractions, bran, polish, and germ have higher levels of lysine and lower levels of glutamic acid than milled rice (Houston *et al.* 1969; Houston and Kohler 1970; Juliano 1966). This uneven distribution in some of the amino acids may be partly explained by the uneven distribution of the protein fractions that differ in amino acid composition. Albumin, which is concentrated in the bran layers, has the best amino acid balance, followed by glutelin, globulin, and prolamin (Table 8.1) (Tecson *et al.* 1971). Thus, the high lysine content of bran and brown rice compared with milled rice may be explained by the high level of albumin in the aleurone layer and in the germ.

TABLE 8.1

LEVELS OF ESSENTIAL AMINO ACIDS AND CYSTINE OF PROTEIN
FRACTIONS AND PROTEIN OF IR8 MILLED RICE
(GRAMS PER 16.8 GM N)

Amino Acid	Protein Fraction				Milled Rice Protein (%)
	Albumin	Globulin	Prolamin	Glutelin	
Isoleucine	4.05	3.03	4.68	5.27	4.13
Leucine	7.89	6.56	11.3	8.19	8.24
Lysine	4.92	2.56	0.51	3.47	3.80
Methionine	2.54	2.27	0.50	2.61	3.37
Methionine + cystine	5.40	2.27	0.80	4.09	4.97
Phenylalanine	2.97	3.32	6.26	5.42	6.02
Threonine	4.65	4.55	2.86	3.92	4.34
Tryptophan	1.88	1.34	0.94	1.16	1.21
Valine	8.72	6.18	6.97	7.31	7.21

Source: Tecson *et al.* (1971) except unpublished tryptophan data from International Rice Research Institute.

Among the protein fractions glutelin has the closest amino acid composition to milled rice protein probably because it is the major protein fraction. The high lysine content (3.5–4.0%) of rice protein may be ascribed to its low level of prolamin, the poor-quality fraction. However, lysine is the first limiting amino acid in rice protein as it is in other cereal proteins (Houston and Kohler 1970; Mossé 1966).

Few comparative nutritional studies on brown rice and its milling fractions and on the four protein fractions have been made. The data are difficult to compare because the protein levels, age, sex, and number of rats and feeding periods used differ. Losza and Koller (1954), however, reported the following biological values of milled rice protein and its fractions: whole protein, 81.4%, albumin-globulin, 85.4%; prolamin, 35.0%; and glutelin, 85.4%.

Because it is insoluble in water, rice protein affects the rate and extent of water absorption and the volume expansion of the starch granules of milled rice during cooking (Juliano *et al.* 1965). Milled rice treated with trypsin before cooking becomes very sticky when it is cooked (Barber 1969). The amylose-to-amylopectin ratio is the principal influence on the texture of cooked rice. But among the samples where the range of amylose content of milled rice is rather narrow (about 6%), protein content is more important (Juliano *et al.* 1965).

Many investigators have shown that grain protein content differs among varieties. The protein content of samples of any variety, however, may range up to 6 percentage points even when the samples are from the same location (Cagampang *et al.* 1966; Juliano *et al.* 1964). For example, the protein content of the high-protein variety BPI-76-1 may range from 8 to 14% (at 12% moisture) and the low-protein variety Intan may have from 5 to 11% protein. Direct comparison of varieties is complicated by differences in growth duration, temperature sensitivity, plant type, and tillering ability among varieties and by differences in soil fertility levels and cultural practices, such as rate and time of application of nitrogenous fertilizers, spacing, etc. (IRRI 1971). Solar radiation available during grain development also affects protein content. Herbicides like Simazine may increase protein content although they may at the same time decrease yield (Vergara *et al.* 1970).

The protein of the other *Oryza* species has the same composition as cultivated rice (Ignacio and Juliano 1968). The protein content (7.4–12.0%) of 17 samples of brown rice representing 11 wild species and 2 interspecific hybrids of genus *Oryza* was negatively correlated with lysine content (3.12–4.48%) and was positively correlated with tyrosine and phenylalanine.

EFFECTS OF HIGH PROTEIN CONTENT ON GRAIN PROPERTIES

A study of samples of the same varieties differing in protein content showed that changes in protein content involved principally the quantity per grain of glutelin and prolamin (Table 8.2) (Cagampang *et al.* 1966). A corresponding increase in the number of protein bodies in the endosperm accompanied the increase in protein content (del Rosario *et al.* 1968), but there was no change in ultrastructure of the protein bodies (IRRI 1970).

TABLE 8.2

MEAN WEIGHT RATIOS AND CONTENTS OF PROTEIN AND PROTEIN FRACTIONS OF MILLING FRACTIONS OF BROWN RICE[1] OF LOW- AND HIGH-PROTEIN SAMPLES OF THREE VARIETIES

Property	Moisture (%)	Low Protein Samples			High Protein Samples		
		Milled Rice (%)	Polish (%)	Bran (%)	Milled Rice (%)	Polish (%)	Bran (%)
Weight of brown rice		86.9	2.0	11.1	89.5	1.5	9.0
Protein	12	6.6	11.8	12.4	13.0	16.2	14.7
Albumin	12	0.40	2.81	3.41	0.44	3.65	4.01
Globulin	12	0.69	0.95	3.29	0.84	2.17	3.92
Prolamin	12	0.19	0.40	0.51	0.35	0.69	0.42
Glutelin	12	4.64	4.85	2.10	9.68	6.46	2.39

Source: Cagampang *et al.* (1966).
[1] Mean brown rice protein contents of 7.2 and 13.4%.

The distribution of protein in the endosperm of high protein rices is more even than in low protein rices. An 11% protein milled rice has outer layers with 19% protein, whereas a 5% protein milled rice has outer layers with about 12%

protein (Houston 1967). In addition, high protein rice seems more resistant to milling. High protein rice gives lower yields of bran and polish than normal protein rice (Table 8.2) (Cagampang *et al.* 1966). The polish fraction of high protein rice tends also to contain more protein than the bran from the same sample. Milled samples of high protein rices may be more translucent but darker tan than rice of the same variety with normal protein content.

High protein rices tend to have lower levels of some of the essential amino acids, particularly lysine, than low protein rices of the same variety (Cagampang *et al.* 1966). However, the drop in lysine content is less than proportional to the increase in protein content. These trends are not as consistent among high protein samples (>10%) as they are between low and high protein samples with more than 3 percentage points difference in protein content. For instance, 3 samples of BPI-76 brown rice with 8.0, 10.7, and 14.3% protein had lysine contents of 3.1, 2.9, and 2.9% in their protein.

In a cooperative study with Dr. Ricardo Bressani of the Instituto de Nutricion de Centro America y Panama at Guatemala on the effect of differences in protein content of milled rice on its nutritive quality (Bressani *et al.* 1971) we found that although protein quality tended to decrease as protein content increased, the decrease in quality was less than proportional to the increase in protein content (Table 8.3). Higher growth responses (nitrogen growth indices) were obtained with milled rice diets with 0, 1, 2, 3, 4, and 5% protein than when only the data showing weight gain (diets with at least 3% protein) were considered (Table 8.3). Based on a relative quality value of 75 for casein, the quality

TABLE 8.3

SUMMARY OF PROTEIN QUALITY INDICES FOR FOUR MILLED RICE SAMPLES AND CASEIN BASED ON WEIGHT GAIN IN WHITE RATS

| Protein Source | Protein Content[1] at 12% Moisture | 0–5% Dietary Protein | | | | Data Showing Net Growth | | |
		PER[2]	NPR[3]	N Growth Index	Protein Quality[4]	PER[5]	N Growth Index	Relative Quality[4]
Intan	5.68	2.56	3.71	3.49	80	2.04	2.37	47
IR8	7.32	2.20	3.36	3.25	75	2.02	2.30	46
IR8	9.73	1.94	3.07	3.04	71	2.02	2.17	43
BPI-76-1[6]	14.3	1.50	2.57	2.47	57	1.84	2.12	42
Casein	86.2	2.20	3.36	3.23	75	–	3.78	75

Source: Bressani *et al.* (1971).
[1] N X 5.92 for rice protein and N X 6.25 for casein.
[2] Protein efficiency ratio at 5% protein.
[3] Net protein ratio at 5% protein.
[4] Based on a value of 75 for casein.
[5] Protein efficiency ratio of 90% rice diet.
[6] Corrected for differences in casein values between the two feeding experiments.
BPI-76-1 was tested later than the other rice samples.

values for milled rice in the diets with 5% protein or less were about the same as those for milled rice protein which has biological values of 70–75% (Juliano 1966).

In diets containing 90% rice, the protein efficiency ratio (PER) of the 4 milled rice samples showed less difference (1.84–2.04) than at 0–5% in the diets (1.50–2.56) (Table 8.3). Based on a relative quality value of 75 for casein, the quality values for the high rice diets were similar to the relative nutritional value of rice protein of 50% found by Hegsted (1969). In fact, Hegsted (1971) found the relative nutritional value of the BPI-76-1 rice to be 40%, which is very close to our value of 42%.

The apparent utilizable protein (protein content x relative nutritional value) in Intan was 2.9%; in low protein IR8, 3.2%; in high protein IR8, 4.2%; and in BPI-76-1, 6.0%. The corresponding net protein values (net protein ratio x 16.8 x protein content) were Intan, 2.3%; low protein IR8, 4.1%; high protein IR8, 5.0%; and BPI-76-1, 6.2%. These two indices again show that the nutritional value of milled rice increases with protein content.

Carcasses of rats fed rice diets with 5% protein or less generally had lower nitrogen content but higher crude fat content than carcasses of rats fed casein diets (Bressani *et al*. 1971). Hence, protein quality indices of milled rice based on the weight gain of rats are overestimated, since they assume that the nitrogen content of carcasses is the same regardless of protein source. The nitrogen balance index for Intan was 0.49; for low protein IR8, 0.47; for high protein IR8, 0.36; and for casein, 0.53 at protein levels in the diets of 5% and less.

The varietal differences in protein quality may be explained by the decreasing levels of lysine, threonine, sulfur amino acids, and tryptophan in rice protein as protein content increases (Table 8.4) (Bressani *et al*. 1971). The prolamin fraction tended to increase in the protein of IR8 and BPI-76-1 as protein content increased. However, Intan protein had a higher prolamin content than IR8 protein. In addition to the ratio of protein fractions, varietal differences in amino acid composition of prolamin and glutelin from IR8 and BPI-76-1 have been reported (Palmiano *et al*. 1968; Tecson *et al*. 1971). Thus, the amino acid composition of a protein fraction may differ among rice varieties.

The observations on white rats were confirmed by experiments on seven adult human subjects by Dr. Helen Clark at Purdue University. (Clark *et al*. 1971). Clark compared a sample of the BPI-76-1 (14.5% protein) milled rice with Bluebonnet rice (7.9% protein). When equal weights of milled rice of each variety were fed to the human subjects for a week, BPI-76-1 rice caused a highly significant improvement in nitrogen retention (Table 8.5). The mean nitrogen balance was +1.41 gm per day N for BPI-76-1 rice, +0.24 gm per day for Bluebonnet rice, and +0.48 gm per day for Bluebonnet rice plus sufficient nonspecific nitrogen to bring the nitrogen level to that of BPI-76-1. Adding nitrogen to Bluebonnet rice did not influence nitrogen retention significantly. Clark attributed the high nitrogen retention from the BPI-76-1 rice to its high

TABLE 8.4

LEVELS OF ESSENTIAL AMINO ACIDS AND CYSTINE AND RATIO
OF FRACTIONS OF FOUR MILLED RICE SAMPLES USED FOR
PROTEIN QUALITY STUDIES AT THE INSTITUTO
DE NUTRICION DE CENTRO AMERICA Y PANAMA

Property	Intan	Low Protein IR8	Low Protein IR8	BPI-76-1	LSD (5%)
Essential Amino Acids (Gm/16.8 Gm N)					
Arginine	7.90	8.78	8.82	8.58	NS
Histidine	2.45	2.45	2.48	2.48	NS
Isoleucine	4.89	5.10	4.78	5.34	NS
Leucine	7.84	8.32	8.20	9.63	0.92
Lysine	4.27	3.77	3.69	3.35	0.32
Methionine	3.45	3.02	2.44	1.80	0.36
Methionine + cystine	5.26	4.83	3.89	3.32	0.81
Phenylalanine	5.55	5.74	5.66	6.25	0.41
Threonine	4.10	4.10	3.78	3.75	0.20
Tryptophan	1.35	1.26	1.05	0.97	0.08
Valine	6.24	6.55	6.51	7.30	NS
Ratio of Protein Fractions (Parts per Hundred)					
Albumin	3.6	4.5	3.7	2.6	NS
Globulin	9.3	8.6	10.8	8.9	NS
Prolamin	2.4	1.2	1.9	6.2	0.7
Glutelin	84.7	85.7	83.6	82.3	NS
Protein at 12% moisture (%)	5.68	7.32	9.73	14.3	

Source: Bressani *et al.* (1971).

level of all essential amino acids per unit weight as found by amino acid analysis.

The introduction of high yielding semidwarf varieties like IR8 caused some apprehension among nutritionists who thought these varieties might contain less protein than the traditional varieties in farmers' fields. However, IR8 from farmers' fields was found to have the same protein content (8%) as lower yielding traditional varieties (IRRI 1968, 1970, 1971). We found that the protein contents and aminograms of the other IRRI varieties, IR5, IR20, and IR22, are similar to those of IR8.

A study of sources of variation in Kjeldahl protein showed that in unfertilized plots, the variation among panicles from the same hill was as high (6.0% of the mean) as that between hills (IRRI 1971). The variation in protein content of grain from different branches within each panicle was 3%. In the plots fertilized with 120 kg per ha N, the variation in protein content among panicles within a hill was considerably larger than that among hills. Unlike traditional varieties, the percentage of unfilled grains was not always higher in the lower branches of an IR8 panicle so that the grains at the lower branches did not always contain more protein.

TABLE 8.5

NITROGEN AND ESSENTIAL AMINO ACIDS IN EXPERIMENTAL
MILLED RICE DIETS AND THE CORRESPONDING MEAN DAILY
NITROGEN BALANCE OF SEVEN HUMAN SUBJECTS
(GRAMS PER DAY)

| Property | Experimental Diets[1] | | | |
	480BPI	480BB	480BB-N	320BB-N
Nitrogen in Diet				
Rice[2]	11.71	6.37	6.37	4.25
Total	12.06	6.72	12.06	6.72
Amino Acids				
Isoleucine	3.60	1.79	1.79	1.19
Leucine	7.77	3.92	3.92	2.61
Lysine	2.26	1.50	1.50	1.00
Methionine	1.32	1.44	1.44	0.96
Methionine + cystine	2.55	2.31	2.31	1.44
Phenylalanine	4.88	2.30	2.30	1.53
Tyrosine	4.92	2.31	2.31	1.54
Threonine	3.07	1.77	1.77	1.18
Tryptophan	0.91	0.55	0.55	0.37
Valine	5.68	2.84	2.84	1.90
Mean Nitrogen Balance	1.41 ± 0.34[3]	0.24 ± 0.11	0.48 ± 0.12	−0.58 ± 0.12

Source: Clark *et al*. (1971).
[1] Diet description states daily intake in grams of each variety of rice, BPI-76-1
or Bluebonnet, and the presence or absence of nonspecific nitrogen source
(N: glycine, glutamic acid, and diammonium citrate).
[2] BPI-76-1 contained 2.44% nitrogen and Bluebonnet rice, 1.33%.
[3] Standard error.

A 10-grain sample was found to be optimal for measuring Kjeldahl protein; it showed less variance than a 5-grain sample and was comparable to samples with up to 100 grains. We digested samples manually for 20 min and determined the ammonia in the digested samples with sodium phenate and hypochlorite reagents in an AutoAnalyzer (Juliano *et al.* 1968).

BREEDING FOR HIGH PROTEIN CONTENT

Since the chief nutritional limitation of rice is its low protein content, breeding efforts to improve its nutritive value have concentrated on increasing its protein content without decreasing protein quality (IRRI 1968, 1970, 1971). Screening 7760 entries for high protein content in IRRI's rice collection revealed 126 entries that consistently had a high protein content of at least 13.5% in 2 samples or a mean level of 14% (Juliano *et al.* 1968). The overall mean protein content was 10.6 ± 1.6% at 12% moisture. Many of these entries were *japonica* varieties, which may have high protein content at Los Baños because of their short growth duration and sensitivity to high ambient temperatures.

Six of the high protein entries were crossed by IRRI breeders to IR8. The lines with good plant type and high protein content were selected through several generations. Some of the F_7 IR1103 lines (IR8 x Chow-sung) may have good yield potential. Some of them gave 2% more protein than IR8 (IRRI 1970). By screening lines in yield trials, several other semidwarf sources of high protein content were identified: IR480-5-9-3-3 (Nahng Mon S-4/2 x Taichung Native 1) and some IR1163 lines (IR8/2 x BPI-76-1). IRRI agronomists first identified IR480-5-9-3-3 as a high protein line. Our breeders will make crosses to combine high protein content from these sources. Our goal is to produce lines with 2% higher protein content than IR8 while maintaining its high yield.

The most difficult aspect of the breeding program is the variable effect of environment on protein content. The better lines must be tested at several locations, and only those lines that consistently show high protein content should be considered genetically high protein.

Low protein, semidwarf lines of the IR1103 cross and of crosses involving a low protein Philippine variety, Intan, have been identified for genetic and biochemical studies. In a cooperative study with IRRI agronomists and statisticians, we are trying to determine the effect of management practices, such as spacing and nitrogen level, on protein content of these high protein and low protein rices.

EFFECT OF PROTEIN CONTENT ON GRAIN PROPERTIES IN BREEDING LINES

Amino acid analysis of F_4 brown rice of the lines with the lowest and the highest protein contents among the six crosses between IR8 and a high protein variety (Table 8.6) showed a decrease in lysine and tryptophan values in the protein as protein content increased (IRRI 1970). These results indicate that genetics and environment have similar effects on the amino acid composition of rice protein. If both parents have normal lysine content in their protein, however, an increase of 2% may not change the lysine content (as reflected by the two IR8 samples in Table 8.4).

A taste panel evaluation of cooked F_7 milled rice of some of these breeding lines conducted by Dr. Luz U. Onate of the University of the Philippines College of Agriculture showed no significant differences in color scores among lines from the same cross that differed by as much as 4.1% protein and had similar amylose content (Table 8.7) (IRRI 1971). Presumably an increase in protein content of 2% has a negligible effect on the color of milled rice. However, the higher color score for BPI-76-1, as compared to IR8, at similar protein and amylose contents, verify the observed darker color of BPI-76-1 (Juliano *et al.* 1965). Crosses involving BPI-76-1 should be screened for this defect.

Milled rice of 5 lines representing 3 of the 6 crosses between IR8 and high protein varieties was analyzed for amino acid composition. The samples, which had 11.4–11.8% protein, contained 3.3–4.0% lysine, 3.4–4.1% threonine, 0.9–1.2% tryptophan, and 3.0–4.6% sulfur amino acids in the protein. The three

TABLE 8.6

CONTENTS OF ESSENTIAL AMINO ACIDS AND CYSTINE OF F_4
BROWN RICE FROM LOW- AND HIGH-PROTEIN PAIRS
FROM CROSSES BETWEEN IR8 AND SIX HIGH-PROTEIN VARIETIES
(GRAMS PER 16.8 GM N)[1]

Amino Acid	Low Protein		High Protein		r^2 (n = 12)
	Range	Mean	Range	Mean	
Isoleucine	4.29–4.81	4.58	4.44–5.06	4.72	0.360
Leucine	7.79–8.02	7.90	7.95–8.82	8.38	0.791
Lysine	4.04–4.57	4.38	3.47–3.93	3.68	−0.838
Methionine	2.16–3.01	2.58	1.89–2.45	2.26	−0.566
Methionine + cystine	3.46–5.05	4.29	3.48–4.34	4.00	−0.351
Phenylalanine	5.17–5.79	5.48	5.34–5.88	5.62	0.310
Threonine	3.91–4.23	4.04	3.55–4.30	3.83	−0.532
Tryptophan	1.14–1.57	1.37	0.92–1.32	1.06	−0.752
Valine	6.44–7.47	6.73	6.22–7.27	6.80	0.151
Protein at 12% moisture (%)	5.4 –7.6	6.7	12.6–15.3	14.4	

Source: IRRI (1970).
[1] Recalculated to 95% nitrogen recovery; mean nitrogen recovery was 93%.
[2] Correlation coefficient with protein content of brown rice.

TABLE 8.7

PANEL SCORES FOR COLOR
FOR MILLED RICE OF HIGH-PROTEIN
LINES DIFFERING IN PROTEIN CONTENT

Line of Variety Name	Protein at 12% Moisture (%)	Amylose, Dry Basis (%)	Mean Color Scores[1]
IR1100-185-7-4	9.1	26.2	2.2
IR1100-89-2-6	11.1	23.1	2.2
IR1104-64-4-1	8.8	27.1	1.5
IR1103-15-8-5	11.6	25.2	2.2
IR1103-15-9-8	12.9	25.0	1.8
IR8	9.9	26.0	1.5
BPI-76-1	10.0	26.2	2.8
LSD (5%)			1.3

Source: IRRI (1971).
[1] Mean scores of a panel of six. Numerical scores of 1
to 9 were assigned, a score of 1 representing the least
expression of color (white); 3 (cream); and 9, the
highest expression (brown). Milled rice (100 gm) was
cooked in 220 gm of water in electric rice cookers.

IR1103 lines (IR8 x Chow-sung) showed a high level of these essential amino acids. The protein of BPI-76-1 has 0.5% less lysine than the protein of most other rice varieties (including IR8) at the same protein level (Palmiano *et al.* 1968).

Because of such genetic differences in amino acid composition at similar protein levels, brown rice of varieties in IRRI's world collection is being screened for lysine based on dye-binding capacity (DBC). An AutoAnalyzer dilutes the supernatant dye solution and records its color intensity. DBC values are correlated with the lysine content of rice protein (IRRI 1970). Lysine is the only basic amino acid of rice protein that changes in concentration with changes in protein content (Cagampang *et al.* 1966; Juliano 1965, 1967). From the linear regression equation of DBC (absorbance units) as a function of protein content, entries are being selected that have higher DBC than the value derived from the regression equation. The lysine content of the selected entries will be rechecked by direct chemical methods. The slope of regression line (DBC as a function of protein) was lower for entries with 10% and higher protein contents than for the low protein entries. This change in slope reflects the greater dependence of lysine content on protein content at low-protein than at high-protein levels.

BIOCHEMICAL STUDIES ON RICE PROTEIN

Studies on protein accumulation in the developing IR8 rice grain showed that protein nitrogen does not change in concentration but increases in amount per grain (Palmiano *et al.* 1968). Albumin-globulin increases in amount per grain during the first 2 weeks of ripening, reaches an optimum level between 14 and 21 days, and decreases progressively toward maturity. The major protein change is the large increase in glutelin between 4 and 21 days. Prolamin increases sixfold per grain during this period. Aminograms of the developing caryopsis showed that the 4-day grain had a higher level of lysine (6.7%) and a lower level of glutamic acid (17.2%) than the 14-day and mature grains. These changes are consistent with the increase in prolamin after the fourth day from flowering.

Developing grains of lines with high protein content tended to have a higher level of free amino nitrogen and RNA, and a faster rate of amino acid incorporation than those with normal protein content (Cruz *et al.* 1970). The bleeding sap obtained from the culm ten days after flowering also tended to have a higher level of free amino nitrogen in high protein lines (IRRI 1971). Since physiologists find that little net nitrogen absorption occurs during grain development, protein nitrogen in the grain must originate from translocation of nitrogen already in the rice plant (principally the leaf blades) at flowering. Preliminary results indicate that leaf blades of high protein lines have high protease activity (IRRI 1971). We are checking the nature of leaf proteins and their changes during leaf growth and senescence, and, in samples differing in protein content, the levels of the protease and ribulose 1,5-diphosphate carboxylase activities and the rate of amino acid incorporation in the leaf blades during grain development.

SUMMARY

Rice protein is unique among cereal proteins in that it contains at least 80% glutelin. Albumin and globulin are concentrated in the germ and aleurone layer. Endosperm protein occurs in protein bodies 1–4 μ in size. An increase in protein content in a variety results mainly from an increase in the glutelin fraction with a corresponding increase in the number of protein bodies.

The improved high-yielding semidwarf varieties and the traditional varieties have similar protein content and aminograms, even in farmers' fields. The grain protein content of a variety may vary from field to field, between hills, between panicles from the same hill, and between grains from the same panicles.

Rat feeding experiments with 4 milled rice samples (5.7–14.3% protein) showed that the protein quality of milled rice (at ≤5% dietary protein levels) decreased with increasing protein content of rice. The decrease in quality was less than proportional to the increase in protein content, however. The contents of lysine, threonine, sulfur amino acids, and tryptophan of the protein also decreased slightly with increasing protein content of milled rice. Human subjects fed milled rice containing 14.5% protein retained more nitrogen than subjects fed rice containing 7.9% protein (even when sufficient nonspecific nitrogen was added to make the diets isonitrogenous) because of the higher levels of essential amino acids in the high-protein sample.

IRRI plant breeders are attempting to improve the protein content of an IR8-type rice by at least 2% by crossing IR8 (8% protein) with high-protein varieties. Some promising lines of good plant type and that have consistently higher protein than IR8 in four generations are being tested in yield trials. The changes in aminograms resulting from differences in protein content among these lines were similar to changes observed previously with samples of the same variety differing in protein content.

BIBLIOGRAPHY

BARBER, S. 1969. Basic studies on aging of milled rice and application to discriminating quality factors. Final Rept., Inst. Agroquim. Tecnol. Alimentos, Valencia, Spain.

BRESSANI, R., ELIAS, L. G., and JULIANO, B. O. 1971. Evaluation of the protein quality of milled rices differing in protein content. J. Agr. Food Chem. (in press).

CAGAMPANG, G. B. et al. 1966. Studies on the extraction and composition of rice proteins. Cereal Chem. *43*, 145-155.

CLARK, H. E., HOWE, J. M., and LEE, C. J. 1971. Nitrogen retention of adult human subjects who consumed a high-protein rice. Am. J. Clin. Nutr., *24*, 324-328.

CRUZ, L. J., CAGAMPANG, G. B., and JULIANO, B. O. 1970. Biochemical factors affecting protein accumulation in the rice grain. Plant Physiol. *46*, 743-747.

del ROSARIO, A. R., BRIONES, V. P., VIDAL, A. J., and JULIANO, B. O. 1968. Composition and endosperm structure of developing and mature rice kernel. Cereal Chem. *45*, 225-235.

HEGSTED, D. M. 1969. Nutritional value of cereal proteins in relation to human needs. *In* Protein-enriched Cereal Foods for World Needs, M. Milner (Editor). Am. Assoc. Cereal Chemists, St. Paul.

HEGSTED, D. M. 1971. Personal communication. Nutrition Dept., Harvard University, Boston.

HOUSTON, D. F. 1967. High-protein flours can be made from all types of milled rice. Rice J. *70*, No. 9, 12-15.

HOUSTON, D. F., ALLIS, M. E., and KOHLER, G. O. 1969. Amino acid composition of rice and rice by-products. Cereal Chem. *46*, 527-537.

HOUSTON, D. F., and KOHLER, G. O. 1970. Nutritional properties of rice. Natl. Acad. Sci. U.S., Washington, D.C.

IGNACIO, C. C., and JULIANO, B. O. 1968. Physicochemical properties of brown rice from *Oryza* species and hybrids. J. Agr. Food Chem. *16*, 125-127.

IRRI. 1968. 1968 annual report. International Rice Research Institute, Los Baños, Laguna, Philippines.

IRRI. 1970. 1969 annual report. International Rice Research Institute, Los Baños, Laguna, Philippines.

IRRI. 1971. 1970 annual report. International Rice Research Institute, Los Baños, Laguna, Philippines.

JULIANO, B. O. 1965. The proteins of rice grain. Trans. 2nd Far East Symp. Nutr. Taipei, Taiwan, May 18-24, 1964.

JULIANO, B. O. 1966. Physicochemical data on the rice grain. Intern. Rice Res. Inst. Tech. Bull. *6*.

JULIANO, B. O. 1967. Physicochemical studies of rice starch and protein. Intern. Rice Res. Inst. Commun. Newsletter (Special Issue) 93-105.

JULIANO, B. O., ALBANO, E. L., and CAGAMPANG, G. B. 1964. Variability in protein content, amylose content, and alkali digestibility of rice varieties in Asia. Philippine Agriculturist *48*, 234-241.

JULIANO, B. O., IGNACIO, C. C., PANGANIBAN, V. M., and PEREZ, C. M. 1968. Screening for high protein rice varieties. Cereal Sci. Today *13*, 299-301, 313.

JULIANO, B. O., ONATE, L. U., and del MUNDO, A. M. 1965. Relation of starch composition, protein content, and gelatinization temperature to cooking and eating qualities of milled rice. Food Technol. *19*, 1006-1011.

LOSZA, A., and KOLLER, K. 1954. Estimation of the biological value of rice proteins. Acta Physiol. Acad. Sci. Hung. *5*, 477-487.

MITSUDA, H. *et al.* 1967. Studies on the proteinaceous subcellular particles in rice endosperm: electron-microscopy and isolation. Agr. Biol. Chem. (Tokyo) *31*, 293-300.

MITSUDA, H., MURAKAMI, K., KUSANO, T., and YASUMOTO, K. 1969. Fine structure of protein bodies isolated from rice endosperm. Arch. Biochem. Biophys. *130*, 678-680.

MOSSE, J. 1966. Alcohol-soluble proteins of cereal grains. Federation Proc. *25*, 1663-1669.

PALMIANO, E. P., ALMAZAN, A. M., and JULIANO, B. O. 1968. Physicochemical properties of protein of developing and mature rice grain. Cereal Chem. *45*, 1-12.

TECSON, E. M. S., ESMAMA, B. V., LONTOK, L. P., and JULIANO, B. O. 1971. Studies on the extraction and composition of rice endosperm glutelin and prolamin. Cereal Chem. *48*, 168-181.

VERGARA, B. S., MILLER, M., and AVELINO, E. 1970. Effect of Simazine on protein content of rice grain (*Oryza sativa* L.). Agron. J. *62*, 269-272.

CHAPTER 9

Virgil A. Johnson
Paul J. Mattern
and
John W. Schmidt

Genetic Studies of Wheat Proteins

INTRODUCTION

The average protein content of wheat usually exceeds 12%. This is higher than the protein content of rice, maize, or sorghum, the other widely utilized food cereals. Production environment exerts a strong influence on protein level of wheat. Protein values as high as 18% and as low as 7% are relatively common in the United States.

The cereal grains lack the balance of essential amino acids required for maximum utilization of their protein when they are the main or sole source of daily nutrition. Lysine is in shortest supply in the cereal proteins; thus, it largely fixes the degree of protein utilization.

Increases in the quantity of protein in the grain of wheat and/or improvement in the ratio of essential amino acids in the protein would enhance significantly its nutritional value. The discovery in 1963 (Mertz *et al.* 1964) of the effect of the *opaque-2* gene in maize on the lysine and tryptophan content of its seed protein pointed to the opportunity for significant genetic improvement of protein quality in the cereal species. A gene in barley with effect on lysine similar to that of *opaque-2* was subsequently reported (Hagberg and Karlsson 1969).

This presentation will be concerned with wheat protein research at the University of Nebraska and the outlook for enhancement of nutritional value in wheat by genetic means.

THE NEBRASKA-ARS RESEARCH PROGRAM

Our efforts to improve the nutritional value of wheat by breeding date back to 1954 when Atlas 66, a soft winter wheat from North Carolina was introduced into our wheat breeding program. Atlas 66, selected from a cross Redhart/No11/2/Frondoso, already had been determined to be higher in grain protein than other soft wheats. Our early research was supported in part by funds from the Nebraska Wheat Commission. It encompassed investigation of the heritability of grain protein, magnitude and stability of the genetic effect, the relationship of grain protein to grain yield, and physiological aspects of the high protein phenomenon.

The Nebraska-ARS protein research was broadened in 1966 to include investigation of the amino acid composition of wheat protein. Funds from the Agency for International Development, U.S. Dept. of State, were utilized to screen the

World Collection of Wheats for protein and lysine differences and to enlarge our breeding program for wheat with improved nutritional value. An International Winter Wheat Performance Nursery has been established in 26 countries to identify superior winter wheat genotypes and measure the impact of environment on nutritional quality.

RESEARCH RESULTS

Protein Content

Grain protein content of wheat is an heritable trait. We have computed estimates of heritability as high as 0.83 (Stuber *et al*. 1962). The high protein trait is readily transferable among common wheats. Its inheritance is relatively simple. Our data suggest the operation of two major genes from Atlas 66, CI 12561, our main genetic source of high protein to date. One of the genes is linked genetically with a gene that conditions adult plant leaf rust resistance (Johnson *et al*. 1963).

We have achieved protein increases from the Atlas 66 genetic source that range from 20 to 25% (Johnson *et al*. 1968A, 1969). Thus, varieties with 15% protein are possible in production situations in which ordinary varieties produce only 12%.

Our studies of nitrogen relations in high and low protein winter wheats show that high grain protein results from more efficient and complete translocation of nitrogen from the wheat plant to its grain (Johnson, *et al*. 1967, 1968A). We have been unable to demonstrate differential nitrogen uptake as a basis for high grain protein in Atlas-derived lines. The nitrate reductase enzyme transport system has been implicated by Illinois researchers (Croy and Hageman 1970).

Protein Stability

Grain protein content cannot be fixed at a specified high level by breeding. Production environment, particularly soil fertility, has strong influence on protein content as well as on grain yield. However, our data do show that high protein varieties can be expected to be superior in protein level to ordinary varieties in an array of environments and soil fertility situations.

Atlas 66 and a hard winter wheat experimental variety, NB 67730, with genes for high protein from Atlas 66 were consistently higher in protein content than other varieties tested in an International Winter Wheat Performance Nursery (IWWPN) in 1969. Atlas 66 averaged 17.4% and NB 67730 16.4% protein over 16 international test sites (Table 9.1). No ther variety shown in Table 9.1 averaged higher than 14.6%.

The high protein experimental variety NB 65307 was compared with the popular Lancer variety at sites in western Nebraska in 1968. Nitrogen fertilizer ranging from 20 to 80 lb per A was applied. NB 65307 maintained approximately 2% protein advantage over Lancer in all fertility levels (Johnson *et al*. 1970).

TABLE 9.1

AVERAGE GRAIN YIELD AND PROTEIN CONTENT
OF 12 VARIETIES OF WINTER WHEAT GROWN IN AN
INTERNATIONAL WINTER WHEAT PERFORMANCE
NURSERY AT 16 SITES IN 1969

Variety	Grain Yield (Bu/A)	Protein Content (%)
Atlas 66	50	17.4
Atl 66/Cmn, NB 67730	53	16.4
Triumph 64	54	14.6
Gage	56	14.4
Parker	60	14.3
Sturdy	62	14.2
Scout 66	57	14.0
Lancer	56	13.9
Shawnee	56	13.9
Winalta	48	13.7
Bezostaia	68	13.3
Gaines	47	12.8

High Protein and High Grain Yield

The high protein trait was transferred to many experimental lines of hard winter wheat. Lines selected from the first cycle of breeding were not sufficiently productive for consideration as commercial varieties. Their yields frequently equalled the parent varieties but were generally lower than the commercially-grown Nebraska varieties. Additionally, there were other agronomic as well as milling and baking deficiencies that were unacceptable in U.S. hard winter wheats.

The performance of selections from the second cycle of breeding are highly encouraging. Our data indicate that there can be simultaneous advances in both grain yield and grain protein content (Table 9.2). Atlas 66/Comanche//Lancer, NB 701132, produced an average yield of 59.4 bu per A at 3 test sites in 1970 compared to 51.8 bu for the widely grown Scout variety. The grain protein content of NB 701132 was 14.4% compared to only 11.8% for Scout. This is a 23% advance in protein content combined with 13% yield increase.

Seed of 12 experimental varieties from the second breeding cycle is being increased in 1971 in Nebraska and Arizona. Some may have good potential as commercial varieties.

Effect of High Protein on Lysine

Does the high protein trait in wheat alter the ratio of essential amino acids? The Nebraska-ARS research provides useful information. Amino acid profiles for several high protein Atlas 66-derived lines from the first breeding cycle were

TABLE 9.2

GRAIN YIELD AND PROTEIN CONTENT OF SECOND CYCLE ATLAS 66-DERIVED
LINES GROWN AT THREE SITES IN NEBRASKA IN 1970

Variety	CI or Sel. No.	Grain Yield (Bu/A)	Grain Protein Content		Increase in Protein over Scout 66 (%)
			(%)	Advantage over Scout 66	
Atl 66/Cmn//Lancer	NB 701132	59.4	14.4	2.3	23
Atl 66/Cmn//Lancer	NB 701134	55.2	14.3	2.2	22
Atl 66/Cmn//Lancer	NB 701137	53.9	14.1	2.0	20
(Wrr/2/Atl 66/Cmn)					
F₁/3/Lancer	NB 701154	53.4	14.5	2.4	24
Lancer//Atl 66/Cmn	NB 70654	50.7	14.4	2.3	23
Scout 66	13996	51.8	11.8	–	–

determined in our laboratory (Johnson *et al*. 1968B; Mattern *et al*. 1968). Several were comparable to their low protein parent in lysine, methionine, and threonine expressed as percentage of protein. Others showed some depression in these amino acids. Most important, in every line there was an increase in the amount of the three amino acids per unit weight of grain. Other things equal, such lines would be more nutritional than lower protein varieties.

A limited number of high protein lines from the second breeding cycle were analyzed for amino acid content. These lines had been selected only for productivity, desirable agronomic traits, and high grain protein. There were no prior analyses for amino acid profiles of their protein. The amino content of 3 of the lines grown at 2 Nebraska sites is compared with that of a check variety, NB 66403, in Table 9.3. At Mead, Nebraska, all of the high protein lines had lower lysine content of their protein than the check variety. Lysine per grain weight was higher than the check in two of the lines but lower in the third, NB 68570. The lysine content of protein in NB 68513 and NB 701137 grown at Sidney, Nebraska was no different than the check variety. As at Lincoln, NB 68570 was lower. All of the lines produced substantially more lysine per grain weight than did the check. There was some variation in methionine and threonine at both sites.

Lysine Variability

Fifteen thousand hexaploid and tetraploid wheats in the World Collection maintained by the Agricultural Research Service, USDA, have been screened for protein and lysine in the Nebraska laboratory. Complete amino acid profiles have been done on a limited number. Mean protein was approximately 13.0% with a range from 7.0 to 22.0%. The range in lysine was from 2.2 to 4.2% with a mean of approximately 3.0%.

TABLE 9.3

EFFECT OF HIGH PROTEIN ON THE AMINO ACID CONTENT OF THREE
ATLAS 66-DERIVED WHEATS GROWN AT TWO SITES IN NEBRASKA IN 1970

Variety	Selection Number	Protein Content (%)	Lysine Protein (%)	Lysine Dry Wt. (%)	Methionine Protein (%)	Methionine Dry Wt. (%)	Threonine Protein (%)	Threonine Dry Wt. (%)
			Mead, Nebraska					
Standard	NB 66403	13.6	2.9	.40	1.5	.20	3.2	.43
High Protein Sel.	NB 68513	17.5	2.5	.43	1.4	.24	3.0	.53
High Protein Sel.	NB 701137	17.0	2.6	.44	1.2	.20	3.0	.51
High Protein Sel.	NB 68570	17.1	2.3	.39	1.4	.24	3.1	.53
			Sidney, Nebraska					
Standard	NB 66403	14.0	2.7	.38	1.3	.18	2.9	.40
High Protein Sel.	NB 68513	16.2	2.7	.43	1.4	.23	2.9	.47
High Protein Sel.	NB 701137	15.5	2.7	.42	1.5	.23	3.0	.47
High Protein Sel.	NB 68570	17.3	2.5	.43	1.4	.23	2.9	.49

Lysine expressed as percentage of protein is negatively correlated with protein. The correlation coefficient for 7000 wheats was −0.63. Lysine per protein decreased from 3.2 to 2.7% as protein increased from 10 to 20%. Lysine expressed as percentage of grain weight is positively correlated with protein. The coefficient for 7000 wheats was +0.83. Lysine per dry grain weight increased from 0.33 to 0.55% as protein increased from 10 to 20%. The +0.83 value is particularly significant in assessment of nutritional impact, because it indicates that high protein wheats can be expected to provide more lysine per unit weight of grain than lower protein wheats; this despite the tendency for high protein to be associated with depressed lysine per protein.

Selected high protein wheats are compared with a group of high lysine wheats from the World Collection in Table 9.4. The high protein group had only 2.75% lysine per protein but provided substantially more lysine, methionine and threonine per unit weight of grain than did the high lysine group or ordinary winter wheat. In contrast, the low protein content of the high lysine group resulted in no increase of the three amino acids per weight of grain.

The high protein experimental variety NB 67730 and three other varieties grown in the 1969 International Winter Wheat Performance Nursery are compared in Table 9.5. Although its lysine per protein content was lower, NB 67730 provided more lysine per grain weight than the other three varieties.

Reevaluation of high lysine wheats from the World Collection following their propagation in an array of environments has shown that most of the variation initially measured was environmentally induced. The genetic component of variation appears to be only approximately 0.5%. This is shown in Table 9.6 in

TABLE 9.4

AVERAGE LYSINE, METHIONINE, AND THREONINE CONTENT OF
SELECTED HIGH-PROTEIN AND HIGH-LYSINE WHEATS FROM THE
WORLD COLLECTION COMPARED WITH AVERAGE VALUES
FOR WINTER WHEAT

Measurement	High Protein Samples (N = 46)	High Lysine Samples (N = 72)	Avg. for Winter Wheat[1]
Protein (%)	18.9	9.6	12.3
Lysine (% of protein)	2.75	3.40	2.76
Lysine (gm/100 gm wheat)	0.52	0.33	0.34
Methionine (% of protein)	1.32	1.52	1.54
Methionine (gm/100 gm wheat)	0.25	0.15	0.19
Threonine (% of protein)	2.94	3.32	2.85
Threonine (gm/100 gm wheat)	0.56	0.32	0.35

[1] Amino acid content of foods, USDA Publication.

TABLE 9.5

AVERAGE PROTEIN AND LYSINE PRODUCTION
OF FOUR VARIETIES GROWN
IN THE INTERNATIONAL WINTER WHEAT
PERFORMANCE NURSERY IN 1969

Variety	Protein Content (%)	Lysine Protein (%)	Lysine Gm/100Gm Grain
Atl 66/Cmn, NB 67730	16.4	2.75	0.45
Triumph	14.7	2.88	0.42
Scout 66	14.1	2.88	0.41
Bezostaia	13.4	2.90	0.39

which Aniversario, a high protein wheat, and Pearl, a high lysine wheat, are compared with Nap Hal which we believe to be high in both protein and lysine. The difference in lysine per protein between Nap Hal and Aniversario at a comparable level of protein is about 0.5%. The impact of combined high protein and high lysine can be noted for Nap Hal. It produced 0.60 gm of lysine, 0.33 gm of methionine, and 0.67 gm of threonine per 100 gm of wheat. The amounts of these amino acids were much lower for Aniversario and Pearl.

The large effect of environment on lysine content suggests the possibility that changes in kernel morphology induced by environment could be involved. Eight

TABLE 9.6

LYSINE, METHIONINE, AND THREONINE CONTENTS
OF SPRING WHEAT VARIETIES GROWN IN THE GREENHOUSE
AT LINCOLN, NEBRASKA IN 1969

Measurement	Aniversario (High Protein)	Pearl (High Lysine)	Nap Hal (High Protein and High Lysine)
Protein (%)	20.7	19.4	21.3
Lysine (% of protein)	2.36	2.83	2.81
Lysine (gm/100 gm wheat)	0.48	0.54	0.60
Methionine (% of protein)	1.32	1.91	1.57
Methionine (gm/100 gm wheat)	0.27	0.37	0.33
Threonine (% of protein)	2.88	3.11	3.18
Threonine (gm/100 gm wheat)	0.59	0.60	0.67

seed morphology traits were correlated with protein and lysine in 280 wheats from the World Collection that exhibited wide differences in protein and lysine content. Protein level and lysine per grain weight were most strongly affected by degree of kernel shriveling. R-values were 0.54 and 0.55, respectively. Degree of kernel shriveling did not affect lysine per protein. Other seed traits showed little or no correlation with protein and lysine content of the seed.

Genetic Sources of High Protein and High Lysine

We are utilizing various wheat varieties as genetic sources of high protein and high lysine. Our classification should be considered tentative—particularly for high lysine—until there is additional confirmative data. The most promising are listed:

High Protein		High Lysine	
Atlas 66	CI 12561	Nap Hal	PI 176217
Nap Hal	PI 176217	Pearl	CI 3285
Hybrid English	CI 6225	Hybrid English	CI 6225
Fertility Restorer	NB 542437	April Bearded	CI 7337
Aniversario	CI 12578	Fultz/Hungarian	CI 11849
Atlas-derived lines		Fultz Sel/Hungarian//	
Aniversario-derived lines		Minturki/Fultz Sel	CI 12756

Detailed protein and lysine analyses of hybrid populations involving the above wheats are in progress. Nap Hal is of particular interest because it has consistently produced grain in greenhouse studies with both above-normal protein and lysine. Hybrid English has produced grain with higher-than-normal protein and lysine but of less magnitude than Nap Hal.

DISCUSSION AND OUTLOOK

Estimated amino acid requirments for man were published by FAO in 1957.

These were compared with availability of the amino acids in corn, rice, soybeans and wheat (VanEtten *et al*. 1967). We have computed for wheat the deviation of essential amino acids from requirements of man from these data (Table 9.7).

TABLE 9.7

DEVIATION OF ESSENTIAL AMINO ACIDS
IN WHEAT PROTEINS FROM
REQUIREMENTS OF MAN, FAO-1957

Amino Acid	Deviation from Requirement (%)
Lysine	−55.0
Methionine[1]	−12.5
Isoleucine	−16.7
Leucine	+30.4
Tyrosine	+ 9.7
Phenylalanine	+40.4
Threonine	− 3.4
Valine	+ 4.5

[1] Adjusted provisional amino acid pattern (WHO 1965).

Clearly, lysine is the critical essential amino acid in wheat. It would appear that a substantial increase in lysine could take place before any other amino acid would emerge as the limiting amino acid in wheat protein utilization.

The literature usually lists lysine, methionine, and threonine as the amino acids in shortest supply in wheat protein. The calculations would indicate that isoleucine may be more critical in wheat than either methionine or threonine.

Data have been reported (Benton *et al*. 1956) indicating that leucine in excess of requirements can interfere with utilization of isoleucine and valine. If so, reduction of leucine in wheat protein could possibly achieve a beneficial effect on its utilization.

A gene for lysine in wheat with the impact of *opaque-2* in maize has not been identified. Lysine differences that appear to be genetic were detected in our search of the World Wheat Collection. Their magnitude is relatively small compared to the total lysine variation encountered. However, the 0.5% genetic component suggested by our data represents a 17% potential advance in lysine content by breeding.

Segregating populations from crosses in which the potentially best genetic sources of higher lysine were combined with high protein varieties are being propagated. Protein and lysine analyses of selections from such populations should provide valuable information on high lysine-high protein breeding expectations.

Common wheat is a hexaploid plant whereas durum wheat has a tetraploid genetic constitution. The presence of more than one genome in common and durum wheat may have contributed to our failure to turn up large genetic differences in lysine content. A gene in one genome with large effect on lysine could be masked by genes in the other genomes. It may be significant that maize and barley in which genes with large effect on lysine have been identified, are both diploid species.

The negative relationship between level of protein and lysine per protein adds to the problem of making simultaneous advances in both. Fortunately, depression in lysine per protein associated with increased protein content is of insufficient magnitude to offset the increase in lysine per unit weight of grain due to higher protein content. Thus, the net effect of higher protein on amount of lysine per weight of grain is positive.

Our data point to increased protein content as the key to nutritional improvement of wheat. Maximum enhancement of nutritional value should come from combining genes for high protein with genes for higher lysine. Variations of lysine content among our high protein selections indicate that selection for high protein must be accompanied by lysine analyses to achieve the highest lysine per protein. Wheats with 25% more protein have been achieved. Lysine increases of 15 − 17% also may be possible. Combination of the two would represent a very significant nutritional step in wheat.

BIBLIOGRAPHY

BENTON, D. A., HARPER, A. E., SPIVEY, H. E., and ELVEHJEM, C. A. 1956. Leucine, isoleucine and valine relationships in the rat. Arch. Biochem. 60, 147-155.

CROY, L. I., and HAGEMAN, R. H. 1970. Relationship of nitrate reductase activity to grain protein production in wheat. Crop Sci. 10, 280-285.

HAGBERG, A., and KARLSSON, K. E. 1969. Breeding for high protein content and quality in barley. Symp. New Approaches to Breeding for Improved Plant Protein. Röstanga, Sweden, June 17-21. 1968. Intern. Atomic Energy Agency, Vienna. STI/PUB/212.

JOHNSON, V. A., SCHMIDT, J. W., MATTERN, P. J., and HAUNOLD, A. 1963. Agronomic and quality characteristics of high protein F_2-derived families from a soft red winter-hard red winter wheat cross. Crop Sci. 3, 7-10.

JOHNSON, V. A., MATTERN, P. J., and SCHMIDT, J. W. 1967. Nitrogen relations during spring growth in varieties of Triticum aestivum L. differing in grain protein content. Crop Sci. 7, 664-667.

JOHNSON, V. A., SCHMIDT, J. W., and MATTERN, P. J. 1968A. Cereal breeding for better protein impact. Econ. Botany 22, 16-25.

JOHNSON, V. A., WHITED, D. A., MATTERN, P. J., and SCHMIDT, J. W. 1968B. Nutritional improvement of wheat by breeding. Proc. 3rd Intern. Wheat Genetics Symp., Canberra, Australian Acad. Sci. 457-461.

JOHNSON, V. A., MATTERN, P. J., WHITED, D. A., and SCHMIDT, J. W. 1969. Breeding for high protein content and quality in wheat. Symp. New Approaches to Breeding for Improved Plant Protein. Röstånga, Sweden, June 17-21, 1968. Intern. Atomic Energy Agency, Vienna. STI/PUB/212.

JOHNSON, V. A., MATTERN, P. J., and SCHMIDT, J. W. 1970. The breeding of wheat and maize with improved nutritional value. Proc. Nutr. Soc. (Engl. Scot.) 29, 20-31.

MATTERN, P. J., SALEM, A., JOHNSON, V. A., and SCHMIDT, J. W. 1968. Amino acid composition of selected high-protein wheats. Cereal Chem. *45*, 437-444.

MERTZ, E. T., BATES, L. S., and NELSON, O. E. 1964. Mutant gene that changes protein composition and increases lysine content of maize endosperm. Science *145*, 279-280.

STUBER, C. W., JOHNSON, V. A., and SCHMIDT, J. W. 1962. Grain protein content and its relationship to other plant and seed characters in the parents and progeny of a cross of *Triticum aestivum* L. Crop Sci. *2*, 506-508.

VANETTEN, C. H., KWOLEK, W. F., PETERS, J. E., and BARCLAY, A. S. 1967. Plant seeds as protein sources for food or feed. Evaluation based on amino acid composition of 379 species. Agr. Food Chem. *15*, 1077-1089.

WHO. 1965. Protein requirements: Report of a joint FAO/WHO expert group. World Health Organ. Tech. Rept. Ser. *301*. Geneva.

CHAPTER 10

Edwin T. Mertz

Recent Improvements
In Corn Proteins

Studies on the nature of corn proteins started by the author and his graduate students in 1946 led to the discovery in 1963 (Mertz 1963; Mertz *et al.* 1964; Nelson *et al.* 1965) of two types of high lysine corn: *opaque-2* and *floury-2*. Before discussing these two maize mutants, I would like to briefly explain the gross anatomy of the corn kernel. The kernel is divided into three parts. The endosperm, which is the major part and contains 80% of the protein and all the starch; the embryo, which is much smaller containing about 15% of the protein and nearly all of the oil; and the pericarp, which contains about 3% of the protein and most of the fiber.

Under a low power microscope, a cross section of ordinary yellow dent corn reveals a cortex of vitreous yellow endosperm which surrounds the center of floury endosperm, whereas a cross section of *opaque-2* and *floury-2* corn reveals no vitreous or crystalline cortex and the entire endosperm has a floury texture. Under the electron microscope (Wolf *et al.* 1967) one observes a major decrease in the size of the protein bodies in the *opaque-2* and *floury-2* mutant types. The protein bodies in a section of normal corn have 20 times the diameter of the protein bodies in a section of *opaque-2* endosperm.

Along with the decrease in the size of the protein bodies there is a marked change in the protein composition of these two mutants. In the endosperm there is a marked increase in the level of albumins and globulins and in the alkali-soluble protein or glutelin fraction. At the same time, there is a marked decrease in the amount of alcohol-soluble protein or zein (Mertz *et al.* 1964; Jimenez 1966). This is also found in the developing endosperm. In the normal there is a marked rise in zein postpollination, but in the *opaque-2* endosperm there is a much slower rise (Dalby 1966).

Zein is practically devoid of the essential amino acids, lysine and tryptophan. On the other hand, glutelin and the albumin and globulin fractions contain a satisfactory level of both. Thus, the marked decrease in zein with a concomitant increase in glutelin, albumin, and globulin proteins tends to increase the lysine and tryptophan content in the endosperm. In fact, in the endosperm of isogenic lines the lysine and tryptophan are actually doubled by the introduction of the *opaque-2* gene, as shown in Table 10.1. The net result of the mutation on the amino acid pattern of the whole kernel is indicated in Table 10.2. With the introduction of the *opaque-2* mutant gene, lysine increases from 3 to 4.8%, tryptophan from 0.7 to 1.3%, and leucine decreases from 14.6% to 9.1% of the protein.

The *opaque-2* gene does not seem to have a marked effect on the amino acid composition of the embryo (Mertz *et al.* 1966) as shown in Table 10.3.

Changes also occur in the level of free amino acids and of certain trace

TABLE 10.1

AMINO ACIDS IN ENDOSPERMS OF ISOGENIC
LINES OF MAIZE
(GRAMS PER 100 GM PROTEIN)

Amino Acid	W64A+	W64A0$_2$
Lysine	1.6	3.7
Tryptophan	0.6	1.2
Histidine	2.9	3.2
Arginine	3.4	5.2
Aspartic Acid	7.0	10.8
Glutamic Acid	26.0	19.8
Threonine	3.5	3.7
Serine	5.6	4.8
Proline	8.6	8.6
Glycine	3.0	4.7
Alanine	10.1	7.2
Valine	5.4	5.3
Cystine	1.8	1.8
Methionine	2.0	1.8
Isoleucine	4.5	3.9
Leucine	18.8	11.6
Tyrosine	5.3	3.9
Phenylalanine	6.5	4.9
Protein (%)	12.7	11.1

TABLE 10.2

AMINO ACIDS IN WHOLE KERNELS OF NORMAL
AND Opaque-2 MAIZE
(GRAMS PER 100 GM PROTEIN)

Amino Acid	Normal	Opaque-2
Lysine	3.0	4.8
Tryptophan	0.7	1.3
Histidine	2.6	3.3
Arginine	4.9	8.5
Aspartic Acid	9.2	10.8
Glutamic Acid	22.6	17.5
Threonine	4.1	4.0
Serine	5.6	4.8
Proline	9.6	7.6
Glycine	4.7	4.8
Alanine	9.2	6.6
Valine	5.7	5.1
Cystine	1.7	1.7
Methionine	1.3	2.1
Isoleucine	4.2	3.4
Leucine	14.6	9.1
Tyrosine	5.2	4.0
Phenylalanine	5.8	4.5
Protein (%)	9.0	11.6

TABLE 10.3

AMINO ACIDS IN MAIZE EMBRYOS (GRAMS PER 100 GM PROTEIN)

Amino Acid	Normal	Opaque-2
Lysine	6.1	5.9
Tryptophan	1.3	1.3
Histidine	2.9	2.9
Arginine	9.1	9.2
Aspartic Acid	8.2	9.2
Glutamic Acid	13.1	13.9
Threonine	3.9	3.7
Serine	5.5	5.0
Proline	4.8	5.3
Glycine	5.4	5.5
Alanine	6.0	5.8
Valine	5.3	4.4
Cystine	1.0	0.9
Methionine	1.7	1.5
Isoleucine	3.1	2.5
Leucine	6.5	5.6
Tyrosine	2.9	2.2
Phenylalanine	4.1	3.6
Protein (%)	20.0	20.0

proteins in the *opaque-2* endosperm. These changes are shown in Table 10.4. It can be seen that the level of free amino acids increases about ten-fold in the mature *opaque-2* endosperm when compared with the normal endosperm. Complete amino acid analysis of the free amino acids shows that the two patterns (*opaque-2* and normal) are very similar with about 1/2 of the total amino acids accounted for as aspartic acid and its amide, asparagine. This large increase in the level of free amino acids in the *opaque-2* endosperm suggests that the *opaque-2* mutant gene suppresses zein synthesis and causes a damming up of the building blocks for protein synthesis.

The level of ribonuclease is also about four times higher in the mature *opaque-2* endosperm than in normal endosperm (Dalby 1966). The ratio is seven-fold in the immature endosperm. The significance of these changes were discussed in Chap. 4.

One of my graduate students, Mr. John Knox, found that the *opaque-2* endosperm contains in its water soluble fraction four times more trypsin inhibitor than normal endosperm. This difference appears to be present throughout seed development. He isolated the pure trypsin inhibitor from *opaque-2* endosperm by affinity chromatography.

An immunochemist, Dr. Arthur Lietze, of the Samuel Merritt Hospital in Oakland, California has suggested that *opaque-2* corn meal may be well tolerated by patients who have an allergy to normal corn meal. *Opaque-2* corn meal

TABLE 10.4

FREE AMINO ACIDS AND TRACE PROTEINS
IN MAIZE ENDOSPERMS

Components	Normal	Opaque-2
Free amino acids (% of protein)	0.5	5.7
Ribonuclease (units per endosperm)	1000	4000
Trypsin inhibitor (units per mg N)	47	210
Antigens (relative units)		
A 1	2	1
A 2	4	0
A 3	6	0
A 4	0	12
A 5	12	0
A 8	60	6

Source: Mertz *et al.* (1970).

appears to lack or be deficient in antigens of corn most likely to produce allergic reactions. Table 10.4 shows that the salt extract of *opaque-2* corn meal is devoid of antigens 2, 3, and 5 found in normal corn meal. Also, antigen 8, which is probably the strongest allergin in corn protein, is present in much lower amounts.

NUTRITIONAL VALUE OF OPAQUE-2 AND FLOURY-2 MAIZE

Young white rats will gain nearly four times as much weight on *opaque-2* corn when compared with normal corn (Mertz *et al.* 1965). Dr. Beeson and co-workers at Purdue showed that pigs which received *opaque-2* corn gained five times as much weight as pigs receiving normal corn (Cromwell *et al.* 1967). Furthermore, animals weighing 125 lb could be fed *opaque-2* corn alone without a protein supplement.

The reduced level of leucine in *opaque-2* corn may also contribute to its superior nutritional value. Workers in India (Gopalan *et al.* 1969) showed that the high levels of leucine found in normal corn and in sorghum produce pellagra in humans, when these cereals form the major part of the diet. Apparently the high level of the leucine interferes with the normal utilization of the niacin in the diet, thus producing a niacin deficiency. These workers produced a pellagra-like condition (black tongue) in puppies by feeding the puppies on a diet which had about 65% normal corn meal. On this diet all the puppies developed black tongue and the disease was cured by the addition of niacin. When puppies were raised on an identical diet in which *opaque-2* corn meal replaced normal corn meal, only 1 out of 7 dogs developed black tongue. However, when leucine was added to the *opaque-2* corn meal diet to bring its level in the diet to that found in the normal corn meal diet, all of the dogs on the diet developed black tongue. Leucine is, therefore, an important factor in the development of black tongue and pellagra. Human pellagra is still endemic in certains parts of India where sorghum is the major food source. Possibly pellagra could be eliminated by replacing the sorghum with *opaque-2* corn.

Opaque-2 corn also has a high nutritive value for humans. Work in Guatemala by Dr. Bressani (Bressani 1966) and by Dr. Pradilla in Colombia (Byrnes 1969) demonstrates that both the whole kernel and the endosperm of *opaque-2* corn are equivalent to skim milk when fed at an adequate level in the diet.

Feeding tests with *floury-2* maize are limited but suggest that the response to the *floury-2* maize will be in proportion to the level of lysine (based on protein) in this cereal. Cromwell *et al*. (1968) reported that chicks fed on *floury-2* corn gained faster than on either normal corn or *opaque-2* corn. The response over *opaque-2* corn was attributed to the greater methionine content of *floury-2* (Nelson *et al.* 1965). Gipp *et al.* (1968) reported that *floury-2* corn appeared to be intermediate to normal and *opaque-2* corn value when fed to rats.

Recently, Mr. Lester Schaible of PAG Hybrid Corn Company, Aurora, Illinois, placed at my disposal isogenic hybrid lines of *opaque-2* and *floury-2* corn derived from their normal hybrid SX-52. The isogenic lines were derived from the normal hybrid by 7–8 back crosses. These 3 corns were ground and made up in a diet containing 95% of the ground corn, 4% of a mineral mix (Hawk-Oser Salt Mixture No. 4) and 1% of vitamin fortification mixture (General Biochemicals). Each of the diets was fed to 10 rats for 21 days. In Table 10.5 we have listed the protein contents of the three corns, the lysine content based on protein, the daily average gain of the rats and the average feed efficiency of the rats. It can be seen from this Table that the lysine content based on protein of the *floury-2* was intermediate between that of the *opaque-2* and the normal and that the daily gain of the rats was also intermediate. In addition, the feed efficiency was also intermediate. These results, therefore, confirm the data of Gipp and suggest that the nutrient value of *floury-2* is related to its lysine content.

Dr. O. E. Nelson at the University of Wisconsin has provided me with *floury-2* hybrids containing 16–17% protein and 3.6–3.8 gm of lysine per 100 gm of protein. In feeding tests with rats, these hybrids gave about the same weight gains and feed efficiencies as the PAG *opaque-2* shown in Table 10.5 which contained 5% less protein. These results suggest that the lysine content of *floury-2* lines must be raised before a response will be obtained to the extra protein in these lines.

Several hybrid corn companies are offering *floury-2* hybrids as an attractive alternative to *opaque-2* hybrids because of somewhat better yield characteristics. In my opinion this is dangerous, because our experience so far suggests that

TABLE 10.5

GAINS OF WEANLING RATS ON ISOGENIC LINES
OF MUTANT AND NORMAL MAIZE

Maize Type	Protein (%)	Lysine (Grams/100 Gm Protein)	Avg. Daily Gain[1]	Feed/Gain[2]
PAG SX-52 (normal)	13.1	2.6	1.0	7.4
PAG floury-2	12.0	3.3	1.8	5.1
PAG opaque-2	11.4	3.8	3.5	3.7
Glover Oh 43 bt_2/o_2	17.0	4.4	4.8	2.2
Glover Oh 43 bt_2/bt_2	15.7	3.8	3.0	2.7
Glover Ch 43 su_2/o_2	13.8	4.1	3.5	2.8
Glover Oh 43 ae/o_2	13.6	4.6	3.6	3.1
Glover Oh 43 wx/o_2	11.9	3.6	3.2	3.0
Glover Oh 43 o_2/o_2	10.6	4.3	3.0	3.2
PAG SX-52 (normal)	13.1	2.6	0.4	11.1

[1] Grams per day. Experiment No. 1 (3 PAG lines): 10 rats on each type of maize for 21 days. Experiment No. 2 (Glover mutants + PAG normal as control): 2 rats on double mutants, and 4 rats on Glover o_2/o_2 and PAG normal, for 10 days.
[2] Grams of feed required for 1 gm of gain.

floury-2 hybrids, as shown in Table 10.5, are quite inferior nutritionally to *opaque-2* hybrids, even though superior to most normal hybrids.

In addition to our studies on *floury-2* mutants, we have been evaluating double mutants of *opaque-2* with a variety of starch-modifying genes supplied by Dr. David Glover of Purdue. In Table 10.5 we have compared results obtained with several double mutants in the same genetic background (Ohio 43). Because of the small amounts of seed available, the feeding test was reduced to 10 days with only 2 animals in each group (4 animals in the control groups). These preliminary data suggest that the double mutant of *brittle-2* and *opaque-2* has remarkably good biological value and could possibly serve as a complete infant weaning food when supplemented with certain minerals and vitamins. One interesting finding was the high value of the *brittle-2* mutant itself. Although only 2/3 as effective in promoting growth as its double mutant counterpart (bt_2/o_2), it was as effective as the other double mutants. We do not know whether this is due to an exceptionally large germ (whole kernels were fed) or to a higher than normal level of lysine in the endosperm. Another study will be made on this type of kernel.

SEED PRODUCTION

The plant breeders in the developing countries need simple inexpensive methods for determining lysine and tryptophan so that they can make rapid progress in the production and screening of locally adapted varieties of *opaque-2* corn. In cooperation with Dr. E. Villegas at CIMMYT in Mexico City, we have tested various simple colorimetric methods and are now recommending that initial screening of corn samples be made on the separated endosperm using a colorimetric tryptophan assay. Usually the lysine value will be four times the tryptophan value. To confirm the lysine value, a simple colorimetric method developed originally at Purdue and improved by Dr. Villegas is recommended. In her laboratory at CIMMYT, over 100 samples can be analyzed in duplicate for tryptophan and over 30 in duplicate for lysine per day using these two color-imetric methods. Details on these methods are available at either Purdue or CIMMYT.

DRY MILLING *opaque-2* CORN

Because of the floury nature of the *opaque-2* corn endosperm, dry millers are reluctant to use this nutritionally superior type of corn in place of normal crystalline types. W. R. Wichser milled 400 lb of a synthetic yellow dent *opaque-2* hybrid produced at Purdue (Wichser 1966) and reached the following conclusions based on this sample: (A) Lack of vitreous endosperm in *opaque-2* corn makes it nearly impossible to obtain flaking grits and coarse brewers' grits. (B) The yield of regular grits is 10% lower than from normal corn. (C) Due to difficulties in bran and germ separation along with increased percentages of these two fractions, the amount of *opaque-2* material going to hominy feed and to oil processing will be increased if low fat, low fiber products are required. The dry miller would therefore welcome the development of a more vitreous type of *opaque-2* maize to reduce these problems.

Plant breeders in the tropics also are interested in developing vitreous types of *opaque-2* corn since these would be more readily acceptable to farmers. Two research centers are actively carrying on work in this area CIAT in Cali, Colombia and CIMMYT in Mexico City.

At CIAT, Dr. C. A. Francis has separated by hand vitreous kernels of a commercial harvest of two Colombian *opaque-2* hybrids, yellow ICA H208 and white ICA H255. The protein and lysine contents of opaque H208, vitreous H208, and opaque H255 were very similar. Although the percentage of protein in vitreous H255 was similar to the opaque H255, the lysine content was lower. Rat feeding tests similar to those conducted at Purdue in which the different maize types supplied all of the protein indicated that the vitreous H-208 is as nutritious as the opaque H-208, but that the vitreous H255 is inferior to the opaque H255. Dr. Francis has concluded that the *opaque-2* gene may have one of its characteristic expressions (i.e., "opaqueness") changed by one or more modifier genes, while its zein-suppressing character remains unchanged in H-208. In H-255, however, the genetic modifiers for vitreous character suppress not only the "opaqueness" associated with the *opaque-2* gene but also the control of the latter on zein synthesis.

The discovery of gene modifiers that can separate the "opaqueness" charac-teristics from the "high lysine" characteristic in *opaque-2* maize is an important advance and may lead to new types of high lysine corn more acceptable to the dry miller.

Dr. Elmer Johnson and co-workers at CIMMYT have also isolated vitreous types of *opaque-2* maize that are high in lysine. They are now increasing their seed for animal feeding tests.

SUMMARY

Research on high lysine types of maize is now moving forward on a broad front both in the United States and around the world. Significant benefits from this research in both animal production and human nutrition should result well before the end of this decade.[1]

BIBLIOGRAPHY

BRESSANI, R. 1966. Protein quality of *opaque-2* maize in children. Proc. High Lysine Corn Conf., E. T. Mertz and O. E. Nelson (Editors). Corn Ind. Res. Found., Washington, D.C.

BYRNES, F. C. 1969. A matter of life and death. The Rockefeller Foundation Quart. *I* 4-20.

CROMWELL, G. L., CLINE, T. R., PICKETT, R. A., and BEESON, W. M. 1967. Growth and nitrogen balance studies with *opaque-2* corn. J. Animal Sci. *26*, 905.

CROMWELL, G. L., ROGLER, J. C., and FEATHERSTON, W. R. 1968. A comparison of the nutritive value of *opaque-2, floury-2* and normal corn for the chick. Poultry Sci. *57*, 840.

DALBY, A. 1966. Protein synthesis in maize endosperm. Proc. High Lysine Corn Conf., E. T. Mertz and O. E. Nelson (Editors). Corn Ind. Res. Found., Washington, D.C.

GIPP, W. F., CLINE, T. R., and ROGLER, J. C. 1968. Nutritional Studies with *opaque-2* and high protein *opaque-2* corn. J. Animal Sci. *27*, 1775 (Abstr).

GOPALAN, C., BELAVADY, B., and KRISHNAMURTHI, D. 1969. The role of leucine in the pathogenesis of canine black-tongue and pellagra. Lancet *II*, 956-957.

JIMENEZ, J. R. 1966. Protein fractionation studies of high lysine corn. Proc. High Lysine Corn Conf., E. T. Mertz and O. E. Nelson (Editors). Corn Ind. Found., Washington, D.C.

MERTZ, E. T. 1963. Corn proteins—A chemical and nutritional perspective. Proc. 18th Ann. Hybrid Corn Industry Research Conf. Dec. 12–13. Publ. *18*, American Seed Trade Assoc. Washington, D.C.

MERTZ, E. T., DOELLGAST, G., and KNOX, J. N. 1970. Unpublished data. Biochemistry Dept., Purdue Univ.

MERTZ, E. T., BATES, L. S., and NELSON, O. E. 1964. Mutant gene that changes protein composition and increases lysine content of maize endosperm. Science *145*, 279-280.

MERTZ, E. T., NELSON, O. E., BATES, L. S., and VERON, O. A. 1966. Better protein quality in maize. Advan. Chem. *57*, 228-242.

MERTZ, E. T., VERON, O. A., BATES, L. S., and NELSON, O. E. 1965. Growth of rats fed on *opaque-2* maize. Science *148*, 1741-1742.

NELSON, O. E., MERTZ, E. T., and BATES, L. S. 1965. Second mutant gene affecting the amino acid pattern of maize endosperm proteins. Science *150*, 1469-1470.

WICHSER, W. R. 1966. Comparison of the dry-milling properties of *opaque-2* and normal dent corn. Proc. High Lysine Corn Conf., E. T. Mertz and O. E. Nelson (Editors). Corn Ind. Res. Found., Washington, D.C.

WOLF, M. J., KHOO, U., and SECKINGER, H. L. 1967. Subcellular structure of endosperm protein in high lysine and normal corn. Science *157*, 556-557.

---

[1] The research on high-lysine corn at Purdue University is being supported by an Agency for International Development Grant AID/csd-2809 entitled The Inheritance and Improvement of Protein Quality and Content in Maize.

CHAPTER 11

Lars Munck | **Barley Seed Proteins**

From the scientific point of view, barley is one of the most thoroughly investigated cereals with regard to biochemistry, plant physiology, and genetics with applications in agronomy, and in the brewing, feed, and food industries. Smith (1951) and Nilan (1964) reviewed the substantial genetic knowledge, including gene maps of the seven pairs of chromosomes. Cook (1962) has likewise gathered abundant data on barley and malt, but with emphasis especially on biochemical and technological aspects. A compilation of animal nutritional data and knowledge of barley integrating information from other disciplines, however, is still lacking. An attempt will be made in the present review to give some hints of the inherent advantage of interdisciplinary scientific approaches to a given problem—in this case barley seed proteins: synthesis, properties, and processing. Selected references from the literature will be illustrated with results obtained by our research group and co-workers at the Swedish Seed Association (Svalof), Institute of Genetics (Lund), Institute of Biochemistry (Uppsala), Weibullsholm Plant Breeding Institute (Landskrona), and the National Institute of Animal Science (Copenhagen).

About 95% of the barley seed is made up by a storage organ—the endosperm—which supplies the embryo with protein and energy to produce a vigorous seedling under favorable growing conditions. The brewing industry is interested mainly in the carbohydrates and the carbohydrases produced during germination of the seed to yield a high extract of sugars suitable to nourish the yeast. Storage proteins reduce the starch content of the seed, and some of them also give rise to turbidity in the final product—the beer. In animal feeding—for instance, pigs and poultry—and for human growth and maintenance, barley proteins, although better balanced than wheat and common maize, lack sufficient available amounts of the essential amino acids, especially lysine and methionine. The total protein level is also too low to obtain sufficient feed efficiency and carcass quality in animal feeding from the seed proteins alone.

There are two important questions that are currently relevant: (1) How can man best benefit from and change the barley protein synthesis for his purposes? (2) To what extent can the different functional factors be combined in the same barley seed for a multipurpose use?

SYNTHESIS

Let us first consider the crude amino acid composition of proteins from barley—namely leaf and stem proteins at maturity and proteins from the nearly ripe seed with the endosperm and embryo plus the scutellum (Table 11.1) [Amino acid analyses in this paper were made by Eaker, according to Eaker (1970).] Lysine, arginine, asparagine, threonine, glycine, and alanine are all

TABLE 11.1

AMINO ACID COMPOSITION (MOL %)
OF NORMAL BARLEY PLANT PARTS

	Endosperm	Embryo + Scutellum	Leaf	Stem
Lysine	2.2	5.5	4.9	4.1
Histidine	1.4	2.2	1.5	1.2
NH_3	22.5	15.2	14.1	22.1
Arginine	2.5	6.2	3.5	3.7
Cysteine	1.8	1.6	1.6	1.3
Methionine	1.2	1.9	1.4	1.0
Asparagine	3.9	7.5	8.0	7.4
Threonine	2.8	4.5	4.8	4.2
Serine	4.5	5.5	5.2	5.0
Glutamine	19.2	10.1	10.1	13.1
Proline	12.7	4.2	5.2	5.1
Glycine	4.6	9.4	8.1	6.4
Alanine	4.0	8.4	8.6	7.2
Valine	4.2	5.7	5.9	4.9
Isoleucine	2.9	3.2	3.7	3.0
Leucine	5.3	5.8	7.3	5.8
Tyrosine	1.6	1.7	2.4	1.7
Phenylalanine	3.8	2.8	3.7	2.7
mMolar sum	1503	2790	867	349

Source: According to Eaker (1970).

strongly reduced in the endosperm protein, whereas glutamine and proline are increased at least twice. The embryo plus scultellum is especially rich in arginine and lysine, but otherwise quite comparable to the leaf and stem protein composition. Nelson (1969A) discussed the relative constancy of amino acid composition of active tissues as leaves due to the rather restricted possibilities for a change in the amino acid code for proteins with enzymatic functions with loss of viability of the organism. In the cells of the triploid endosperm, mutations affecting the amino acid message in structural proteins could have a higher degree of freedom for survival during the phylogenetic development of the seed. Changes in the protein composition of the endosperm favorable for amino acids such as glutamine more readily available for the seedling than, for instance, lysine—could thus tentatively constitute a positive trait for survival.

Environmental Effects

Various characteristics of the vegetative parts of the barley plant such as nitrogen uptake from the soil by the roots (Wray *et al.* 1970), assimilation, and translocation of the assimilated products (Stoy 1966) are of decisive importance for the development and nutrition of the seed, especially regarding quantitative traits: yield, seed size, and crude protein content. Harris (1962) in his review of the structural chemistry of barley and malt treated, inter alia, the classical works of Osborne and Bishop. The lysine-rich Osborne fractions, albumins and globulins, containing proteins of enzymatic character are synthesized at a faster rate during the early stages of seed formation, while lysine-poor storage proteins such as prolamins dominate later stages. Glutelins, intermediate in lysine, are formed in a linear rate with increasing maturation. The amino acid nitrogen of the grain passes through a maximum value 4–5 weeks after anthesis and thereafter drops sharply. It is obvious that factors like seed size (endosperm:embryo ratio) and the balance between protein and carbohydrate synthesis in the seed as a function of time will greatly affect seed nitrogen content, ratio of Osborne protein fractions, and consequently the overall amino acid composition. To demonstrate extreme fertilizer effects on protein content in barley and lysine composition from variety Sv 59132, a pot trial in perlite was performed at different N, P, and K levels (See Fig. 11.1). The standard dose S equal to $2N_2 2P_2 2K_2$ was varied with lower $(N_1 P_1 K_1)$ and higher $(N_2 P_2 K_2)$ doses. Crude protein varied from 6–20% and was negatively correlated with lysine per 16gm N, varying from 4.5 to 2.6 gm. A maximum in yield per plant was obtained with N_2 and in crude protein and lysine yield per plant with S. The overall yield minimum was with N_1. Seed size was high with N_2, $3K_1$, P_2, and $4P_2$, and especially low at 2S, S, and $4K_2$. Thus, lysine content per 16gm N is negatively correlated to crude protein and lysine yield and positively correlated with seed size under these artificial conditions.

In a phytotron experiment, the day length insensitive, highly adaptable mutant Mari (derived from the variety Bonus) was grown by Gustafsson (1969) at 25°, 20°, 15°, and 10°C and 24, 16, and 8 hr of day length (Table 11.2). Water and nutrients were equally supplied to the different treatments. Long days and high temperatures produced a short vegetative period. Seed size was thus larger owing to longer vegetative periods at low temperatures. Crude protein content varied from 19.1 to 11.6%, and was negatively correlated with seed size, which is in good agreement with the fertilizer experiment (Fig. 11.1). The crude protein content was also positively related to the increased vegetative periods under the short day lengths. Lysine per 16gm N varied from 3.14 to 4.16 gm with minima at 25°C (8, 16, and 24 hr) and maxima at 15°C (16 and 24 hr), the latter corresponding to yield optima. Lysine (milligrams per seed) was remarkably constant under the different conditions, with a slight indication of a maximum at 10°C independent of day length. It is very difficult to interpret these results from the phytotron trials—with the plants grown separately in pots—and to compare them with results obtained under field conditions. Differences in uptake of fertilizer owing to climate produce different soil conditions, although the supply of mineral nutrients is moderated in accordance with a fixed schedule. The effects of climate on protein synthesis under these circumstances, therefore, to some extent, can be considered to give similar effects as a pure fertilizer experiment.

TABLE 11.2

LYSINE CONTENT OF BARLEY PROTEIN FROM EARLY
VARIETY MARI GROWN IN DIFFERENT ARTIFICIAL CLIMATES

Temperature (°C)	Day Length (Hr)	Seed Weight (Mg)	Protein Content (%)	Lysine (Gm/16Gm N)	Lysine per Seed (Mg)
25	24	37	16.2	3.28	0.19
20	24	46	12.4	3.69	0.21
15	24	44	11.6	4.16	0.21
10	24	56	11.7	3.65	0.24
25	16	33	18.9	3.35	0.21
20	16	36	13.2	3.59	0.17
15	16	44	11.7	3.73	0.19
10	16	58	13.2	3.22	0.25
25	8	30	19.1	3.14	0.18
20	8	31	16.5	3.38	0.17
15	8	35	14.3	3.48	0.17
10	8	45	14.6	3.57	0.23

Source: According to Eaker (1970).

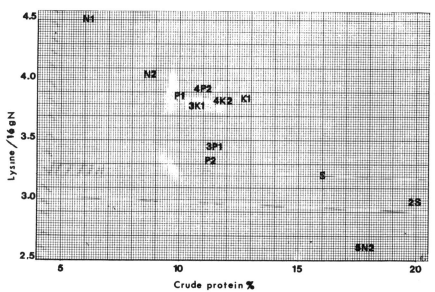

FIG. 11.1. FERTILIZER EFFECTS ON LYSINE (GRAMS PER 16 GM N) AND CRUDE
PROTEIN CONTENT PERCENTAGE IN SEEDS FROM BARLEY VARIETY Sv 59132

S = standard dose (2N2, 2K2, 2P2). The standard dose is varied with higher and lower
amounts of N, P, and K, e.g., N1, K1, P1 and 5N2, 4K2, 4P2.

Genetic Control

A Search for High-lysine Varieties.—It is thus no wonder that Bishop and others in their classical studies on the "principle of regularity" of barley proteins placed emphasis on the environmental effects, although well-established varietal effects were found. This opinion was also founded on the fact that the commercial varieties at that time were (and still are) founded on a limited number of cultivars of mainly European origin. Therefore, in 1967, we decided at Svalof to screen the World Barley Collection, supplied by USDA, Beltsville, for variation in crude protein and lysine content. The collection comprises about 15,000 accessions from different parts of the world. From the outset, we were faced with the problem of developing a suitable, inexpensive, and efficient screening method for lysine content. Combining studies of heat damage of lysine (Munck 1964B, 1968) with a dye-binding (DBC) method (Mossberg 1969), it was shown that Acilane Orange G (monosodium salt of 1-phenylazo-2-naphtol-6-sulfonic acid) binding to meal at pH 2.6 was very well correlated with the basic amino acids, and among them lysine (Munck 1968; Mossberg 1969). It was thus possible to relate the DBC value in mMol-bound dye per kilogram sample with the crude protein content according to Kjeldahl and obtain an estimate of lysine content of protein. The DBC method is further discussed by Munck (1970).

In the first 1000 lines that were screened from the World Barley Collection, we found (in 1968) a high-lysine, high-protein variety (called Hiproly), CI 3947, of Ethiopian origin (Hagberg and Karlsson 1968; Munck *et al.* 1971). Analyses of F_2 progenies with DBC in 4 crosses indicated a 1:3 segregation in 2 combinations, whereas in the other 2 crosses the ratio differed significantly from the 1:3 ratio (Munck *et al.* 1970; Hagberg *et al.* 1970). The F_1 was low in DBC. The high lysine (Hily) character was not related to the crude protein level of the seed. The usual negative regression line between protein level and lysine (grams per 16gm N) in the normal material was thus elevated to a 20–30% higher level of lysine (grams per 16gm N) in the Hily material (Munck 1970). A similar line, CI 4362, with an identical plant habit and origin as Hiproly was found to have the same high-protein content and likewise the naked character, but was, on the other hand, low in lysine. Hiproly was significantly lower in yield and had shrunken, smaller seeds when compared with CI 4362. In the F_2 generation in four crosses—Hiproly X high-yielding, low-protein varieties—the Hily types, on an average, were lower in yield on a single plant basis. A large variation with respect to yield was, however, found and it was possible to select high-yielding Hily lines with good seed quality (Hagberg *et al.* 1970). Thus, CI 4362 is not completely isogenic with Hiproly. In Table 11.3, the means of 35 normal plants and 24 Hily plants segregating in the F_2 are compared with respect to amino acid composition. Lysine, asparagine, methionine, and alanine are significantly increased, while NH_3, glutamine, proline, and cysteine are decreased in the Hily

TABLE 11.3

AMINO ACID COMPOSITION (PER 16GM N) OF HILY AND
NORMAL BARLEY SEGREGANTS (WHOLE SEED) AND OF ENDOSPERMS
FROM HIPROLY BARLEY AND *OPAQUE-2* AND *FLOURY-2* MAIZE

| | Whole Seed from F_2 Segregants | | Barley Hiproly | Endosperm Maize (Nelson 1969) | |
	35 Normal Plants	27 Hily Plants		W64-O *Opaque-2*	Fl-2 *Floury-2*
Lysine	3.4 ± 0.26	4.1 ± 0.18	4.0	3.7	3.3
Histidine	2.1 ± 0.09	2.2 ± 0.10	2.1	3.2	2.2
NH$_3$	3.6 ± 0.23	3.2 ± 0.15	3.3	–	–
Arginine	4.6 ± 0.29	4.9 ± 0.24	4.4	5.2	4.5
Asparagine	6.0 ± 0.57	6.8 ± 0.31	6.2	10.8	8.1
Threonine	3.4 ± 0.34	3.6 ± 0.13	3.5	3.7	3.3
Serine	4.3 ± 0.15	4.4 ± 0.16	4.6	4.8	4.8
Glutamine	26.8 ± 1.33	23.9 ± 0.77	24.0	19.8	19.1
Proline	12.6 ± 1.08	11.3 ± 0.48	12.0	8.6	8.3
Glycine	3.6 ± 0.25	3.9 ± 0.15	3.7	4.7	3.7
Alanine	3.8 ± 0.22	4.4 ± 0.16	4.2	7.2	8.0
Cysteine	1.1* ± 0.12	0.9* ± 0.18	1.7	–	1.8
Valine	4.8 ± 0.18	5.2 ± 0.18	5.3	5.3	5.2
Methionine	1.2* ± 0.18	1.5* ± 0.17	2.0	1.8	3.2
Isoleucine	3.7 ± 0.28	3.9 ± 0.13	3.9	3.9	4.0
Leucine	6.7 ± 0.68	7.1 ± 0.25	7.0	11.6	13.3
Tyrosine	2.8 ± 0.10	2.8 ± 0.09	2.9	3.9	4.5
Phenylalanine	5.9 ± 0.49	5.9 ± 0.20	6.0	4.9	5.1
DBC/protein	1.06 ± 0.06	1.14 ± 0.05	–	–	–
Protein	15.7 ± 2.86	16.8 ± 2.43	17.2	11.1	13.6

Source: According to Eaker (1970).
*Absolute figures unsafe due to nonoxidative analysis.

material. The Hilys have a significantly higher DBC of protein than the normal plants. In the F_4 material grown at Svalof in 1970, plants (Hiproly X Sv 66350) from families high in lysine in the F_2 generation were significantly associated with two different levels of DBC. This was verified with amino acid analyses (Fig. 11.2) in which a high lysine level at about 4.4 gm differed from a medium level at about 4.1 gm as compared to a normal level of approximately 3.6 gm per 16gm N. All lysine levels were negatively correlated with the crude protein level. Lysine:arginine, methionine:cysteine, and aspartic acid:glutamic acid ratios were in the two highest lysine levels typical for a Hily material as described by Munck (1970). Information is thus pointing towards the possibility of additional genetic factors that could act in this case with the recessive "major" gene as a background. The likelihood of detecting several minor genes for protein regulation in barley seeds is strengthened by the recent, independent results of two Danish scientists. Vieuf (1970), in a barley collection, and Doll (1970), in an EMS-treated barley material, employing the DBC method, found 1 and 2 lines, respectively, with increased lysine content. In Fig. 11.2 these lines are

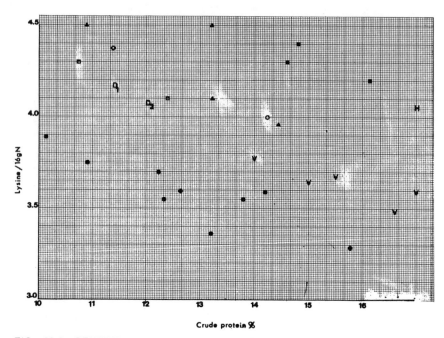

FIG. 11.2. GENETICAL VARIATION IN LYSINE (GRAMS PER 16 GM N) AT
DIFFERENT CRUDE PROTEIN PERCENTAGE LEVELS IN BARLEY

Normal barley (black dots). Hily segregants from the cross Hiproly X Sv 66350 plants from
different families in F_4. (High lysine level: black squares and triangles; moderate lysine
level: open circles, squares, and triangles.) V = Wish. 35-7-2-1-3 (Vieuf 1970) grown in the
same environment as the Hily and the normal material. D_1 and D_2 are mutants from
Carlsberg II according to data from Doll (1970).

marked V, D_1, and D_2, V being grown in Svalof under identical conditions as
the original material. D_1 and D_2 values are from Doll (1970), and are about 15%
higher in lysine per 16gm N as compared with the basic variety Carlsberg II. It is
obvious that the D_1 and D_2 lines fall at the lower level of the Hily material,
while the V line is intermediate between this level and the lysine content of
normal barley protein. The increased genetic variability thus demonstrated
showing minor and major effects could be used as a worthwhile tool in studying
the mechanisms of seed-protein synthesis, particularly genetic implications.

Morphology and Seed Protein Quantity and Quality.—Sections of Hiproly and
the similar normal line, CI 4362, were studied under the microscope and
compared with standard varieties. Acilane Orange G was used for staining. A
large variation in gross morphology was found, with Hiproly and CI 4362 having
a marked, densely colored, starch-free subaleurone layer resembling that of hard
wheats (Kent 1970). In Fig. 11.3 Hiproly (B) is compared with the high-protein,
naked variety CI 5195(A)—the latter almost lacking a subaleurone layer. The

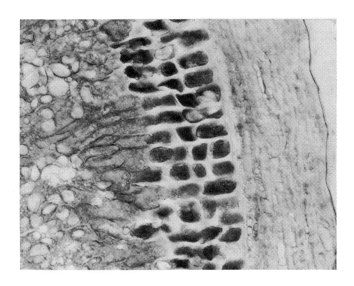

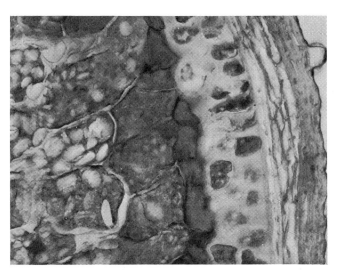

FIG. 11.3. "SOFT" BARLEY CI 5195 (TOP) AND "HARD" BARLEY
HIPROLY CI 3947 (BOTTOM) (NOTE SUBALEURONE LAYER)

Acilane Orange G dye pH 2.6. The varieties have similar protein content.

significance of these structures with respect to feeding value and malting quality are obscure.

In plant breeding, convenient screening procedures on single seed bases are very important. Meal was scratched free with a needle from an incision in the endosperm without affecting viability. The removed material was stained with Acilane Orange G and studied under the microscope. Hiproly was harder and had a darker and "flinty" cut surface when compared with CI 4362. In Fig. 11.4, meal preparation from Hiproly (A) and normal barley (B) as well as F_5 segregants (C, D) are compared, with the high lysine varieties having starch grains strongly adhering to protein tissues. Hily segregants were found with better separation between the starch grains in meal preparations, but they often displayed decreased birfringence of the outer parts of the starch grains. Large starch grains (Munck $et\ al.$ 1970) were more closely associated with the Hily character than small ones. One hundred seeds from a heterozygotic F_1 plant (F_2 seeds) were studied with the scratch technique in a blind experiment together with 20 seeds of each of the parental lines (Munck 1970). The starch granule adhering character showed a significant 1:3 segregation—39 of the 40 parental seeds being correctly classified. Thirty-one of the F_2 seeds of the different morphological classes were checked by amino acid analyses, 30 of which agreed with the morphological classification. Thus, it is shown (see Fig. 11.5) that the Hily gene complex segregates in the F_2 showing dominance for the normal allele in the triploid endosperm. Note the negative correlation between lysine (grams per 16gm N) and crude protein both on the Hily and normal levels from seeds from the same plant. The amino acid analysed F_2 plant material (Table 11.3) was followed up as regards F_3 single-seed morphology and in the F_4 on yields from F_3 plants with respect to both DBC and morphology. Thus, 235 lines related to Hily F_2 plants should be homozygous in the F_4 generation and show high DBC and strong starch-protein adhering properties. Twelve of these lines were low in DBC (but with the "adhering" phenotype) and 17 were normal in morphology (but high in DBC). The number of deviating lines was thus reasonably low with regard to the possibility of spontaneous cross-fertilization and other errors. These lines, however, were significantly associated with certain families, indicating other explanations for the deviation. Hence, 221 lines that were related to normal F_2 plants should display a 1:3 segregation as regards the F_4 yields from F_3 seeds. DBC segregated 1:3.8 and the morphological trait 1:2.7. The efficiency of DBC was verified by amino acid analyses in backcrosses to high-yielding, normal varieties. The preliminary morphological screening technique, however, was difficult to use with this material, giving rise to a wide variation of intermediates and new types. Two normal barley varieties, J5 and Monte Christo, were also difficult to differentiate from Hiproly by microscopic examination of meal preparations. Thus, it is likely that the adhering trait of the starch grains in meal preparations is closely linked with, but not pleiotropic to, the major Hily gene. This does not exclude that a pleiotropic relationship

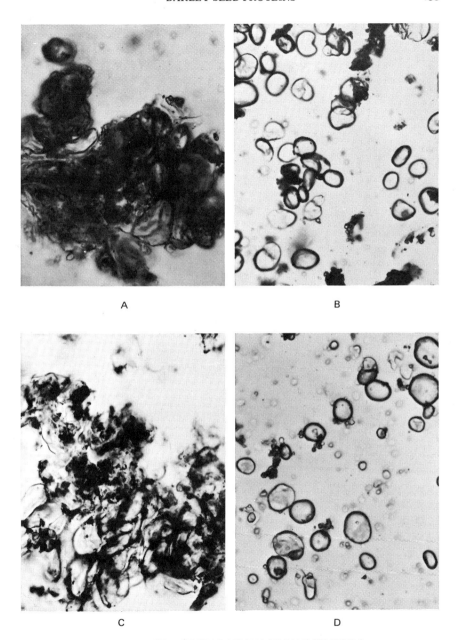

FIG. 11.4. MEAL PREPARATIONS FROM DRY SEEDS

A—Hiproly. B—Normal variety Sv 66303. C—Hily segregant. D—Normal segregant. Segregants in F_5 seeds.

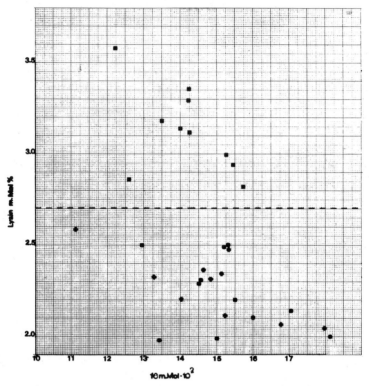

FIG. 11.5. SEEDS FROM A SINGLE PLANT (HIPROLY X Sv 66303)
SEGREGATING FOR LYSINE CONTENT (MOL %) AT DIFFERENT CRUDE
PROTEIN LEVELS

Seeds selected by morphological screening according to Munck (1970).

between the high-lysine trait and a more defined morphological character can
exist. It is also possible that the study of the starch-protein adhering trait could
be considered as an indirect way of studying the hard-grain character.

As far as we can judge from studying seeds at various stages, the protein
bodies containing prolamins are intact in the Hily grain. More basic research as
regards seed ultrastructure is needed both to explain effects of new genes for
changed protein synthesis, but also for screening purposes and to translate the
morphological observations into terms of functional factors of importance for
the use of the barley seed.

It is clear that an uncritical screening for crude protein content or a specific
amino acid composition could, in fact, lead to a selection for, e.g., small seeds,
low number of seeds per spike, low level of starch synthesis, and changed ratios
between embryo and endosperm. Mutants for high protein content (Doll 1970)
often show these effects and are consequently lower in yield than the variety of

origin. This does not seem to be the case with regard to the two high-lysine mutants, D_1 and D_2 (Fig. 11.2). Favret *et al.* (1969, 1970) discussed these effects with respect to barley. A smaller seed size can, however, to some extent, be compensated for by a high number of seeds per spike and by increased numbers of tillers. In breeding for high-protein and high-lysine yield, all of these factors must be considered. The Hily lines have on the average a lower seed weight than normal segregants. But large, positive variation exists (Hagberg *et al.* 1970). In barley, the embryo plus the scutellum make up 3.5–4.5% of the dry seed weight, 5.4–11.5% of the seed protein, and 12.0–20.7% of the seed lysine content (Munck 1970). The effect of the embryo plus the scutellum on the whole seed, amino acid composition is much smaller in barley than in maize. Whole seeds of Hiproly are thus about 20% less in lysine per 16gm N as compared with some lines of *opaque-2* and *floury-2* (Nelson 1969B). In Table 11.3 the endosperms are compared—with Hiproly having higher lysine, glutamine, proline, and phenylalanine content and lower asparagine, alanine, and leucine than *opaque-2* and *floury-2*. It is obvious that comparisons between cereals regarding similarities in amino acid pattern should preferably be made on the endosperm level.

The Action of the Hily Gene Complex on Protein Synthesis.—It is a complicated problem to analyze endosperm protein because of the difficulties in obtaining true solutions. Osborne extraction techniques combined with salting-out and dialysis procedures and final examination with ultracentrifuging and electrophoresis have been used by different workers (Harris 1962). The problems associated with extraction problems are further discussed by Dimler (1966). Preliminary studies (Munck *et al.* 1971) with normal varieties of barley showed that the high-protein lines were, as expected, lower in water-soluble and higher in alcohol-soluble proteins when compared with low-protein lines. Although high in protein, Hiproly resembled the low-protein normal varieties in this respect, which also was valid in the comparison with amino acids such as lysine, asparagine, and glutamine. Methionine is, however, higher and cysteine is significantly decreased in the total seed protein of Hiproly as compared with both the high and the low normal protein lines. The latter effect prevailed through almost all extraction fractions in a slightly modified Osborne technique. Although most of these analyses were performed with unoxidized hydrolysates, and therefore the absolute value of cysteine is too low, the relative effects are reproducible in numerous analyses. The lysine content of the crude, water-soluble fraction was also found to be slightly more lysine-rich in the Hiproly barley, especially when compared with normal varieties with the same protein level. To obtain a more detailed picture for clarifying these effects, polyacrylamide electrophoresis was performed on the crude extracts from single endosperms at different stages of development and water contents (Fig. 11.6, 11.7, and 11.8). The embryo plus scutellum was discarded. The cut seed was homogenized in a small centrifuge tube with a glass pestle and successive extractions with (1) 300μl water, (2)

$100\mu l$ 0.2 M barbituric buffer pH 8.6, (3) $100\mu l$ 0.1 M formic acid. Each extraction was made at 5°C and left to stand overnight. Between each step, washings were made with water. For the water and salt extracts, $10\mu l$ was pipetted onto a 15% polyacrylamide gel containing 0.1–0.05 M tris buffer at pH 8.6. Runs were made both towards the anode and the cathode. Gels were pre-flushed for 60 min before the sample was applied, which was made without a stacking gel. For the formic acid-soluble proteins, $10\mu l$ of the extract was run towards the cathode in a 7.5% gel in alanine-HAc buffer at pH 5.0. Coomassie Brilliant Blue R-250 was the dye used.

In Fig. 11.6, the electrophoretic patterns from the water extracts from ripe seeds (10% H_2O) exhibited less successful extraction (faint bands) than at 40% H_2O. Sampling of seeds before complete dryness should be preferred. The most striking difference between Hiproly and CI 4362 is the much stronger bands especially in the water-soluble fractions run towards the anode. It is also apparent that one of these bands is synthesized in large amounts earlier (65% H_2O) in development than in CI 4362. When comparing the water-soluble and salt-soluble fractions, homologies are indicated, illustrating the lack of selectiveness of the two solvents. The proteins separated toward the cathode are

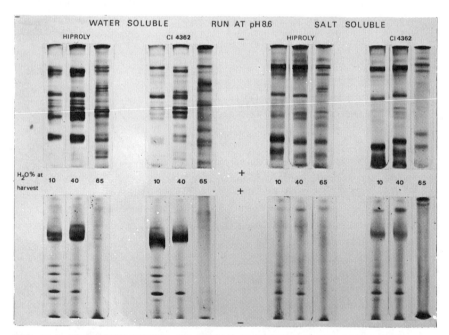

FIG. 11.6. ELECTROPHORESIS OF SINGLE ENDOSPERMS OF HIPROLY (CI 3947) AND NORMAL CI 4362

Water and salt extracts at different stages of maturity. The varieties have similar protein content.

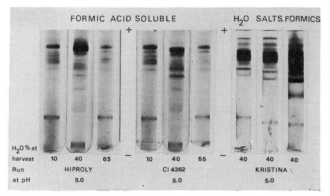

FIG. 11.7. ELECTROPHORESIS OF FORMIC ACID EXTRACT
FOLLOWED AFTER THE EXTRACTIONS IN FIG. 11.6 FOR HIPROLY
AND CI 4362

Direct water, salt, and formic acid extracts from normal variety Kristina.

apparently synthesized at a later stage in development, as almost no bands are
seen from seeds with a 65% water content. Hiproly and CI 4362 are rather
similar at the later stages, except the diffuse part with low mobility, which is
much stronger for CI 3462 both regarding the water and the salt fractions. In
Fig. 11.7, the formic acid-soluble fractions after water- and salt-solution
extractions are shown run at pH 5.0 at the different developmental stages for
Hiproly and CI 4362. Only a few bands are seen at the early stage of seed
formation. There are only small differences between the two barley types. To
demonstrate that the formic acid-soluble proteins also are extracted by water
and buffer solutions, the normal, low protein barley Kristina, at 40% water
content, was extracted with these solvents, including formic acid, without
previous treatments. It is shown in Fig. 11.7 that the electrophoretic patterns at
pH 5.0 are essentially alike. The highest amount of extracted protein was
extracted with the formic acid solution. Some of the proteins that were
extracted with water or buffer and run at pH 5.0 should be homologous with
proteins that were run toward the cathode at pH 8.6 (see Fig. 11.6). The storage
character of these proteins is demonstrated because they appear late in seed
development. Only one slow band was detected when the direct water, buffer,
and formic acid extracts of variety Kristina were run toward the anode at pH
5.0. To demonstrate further the effect of the Hily gene(s), electrophoresis was
carried out on the crude water and salt extracts from single seeds that were
obtained from a cross (Hiproly X Sv 66449) that segregated in the F_4
generation. The electrophoretic patterns (slightly modified method) of a Hily
and a normal segregant (40% water content) with comparable protein levels in
the seed are shown together with amino acid analyses (Fig. 11.8). It is obvious
that at least six water-soluble protein bands are strongly increased in the Hily

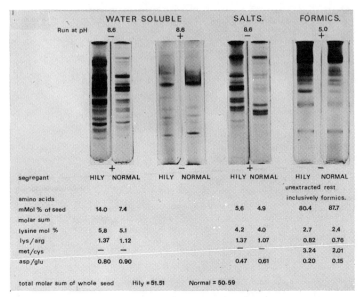

FIG. 11.8. ELECTROPHORESIS PATTERNS FROM TWO SEGREGANTS FROM THE CROSS HIPROLY X Sv 66449

Amino acids analyses are made from the same extracts from the individual endosperms except for formic acid extracts. The latter extracts come from endosperms from the same plant having identical globulin fraction patterns with those in the figure.

segregant. The electrophoretic patterns of the water extract of the normal seed towards the anode were very like that of the salt extract from the Hily seed, again indicating the limitations with the extraction techniques. The diffuse, slow-moving proteins toward the cathode are more abundant in the normal seed. Comparing other segregants from the same cross related to F_2 plants that were determined as Hilys by amino acid analyses, it is shown that the major part of the Hily gene complex is associated with: (1) increased synthesis of certain water- and salt-soluble proteins; (2) decreased synthesis of certain cathodic components soluble in water and buffer and with a diffuse mobility at pH 8.6; (3) no certain qualitative traits associated with the gene(s) possibly with exception to the extra fast-moving component from the anodic part of the salt extract; (4) variation within the Hily and normal classes were found with respect to the formic acid-soluble fraction at pH 5.0, but this variation did not seem to be associated with the Hily character.

Nine extreme segregants were analyzed for amino acid composition of the endosperm protein fractions and compared with the electrophoretic patterns. Two segregants, one Hily and one normal, are presented in Fig. 11.8. The water-soluble protein fraction was doubled in the Hily endosperm when compared with the normal variety, which contained a higher proportion of unextracted residual proteins (not salt-soluble). Lysine (mo1%) of the fractions was slightly higher for the Hily endosperm in all fractions. The lysine:arginine ratio differed considerably both in the water- and salt-soluble fractions, whereas the difference was small in the remaining residues. Cysteine was too low in the water and salt extracts to be estimated by the nonoxidative amino acid procedure that was used in this experiment. The aspartic acid:glutamic acid ratio was lower for the Hily type in the water and salt fractions, while the opposite tendency prevailed for the residue. The higher aspartic acid and lower glutamic acid contents for the whole Hily seed protein thus were caused by differences in the unextracted proteins in this experiment. To date, only segregants from the lower level of lysine (grams per 16gm N) among the Hily types (defined by the increased molar ratio lysine:arginine) (Munck 1970) have been studied. The electrophoretic method could be developed into a potent screening method when the effect of the different Hily types presented in Fig. 11.2 have been characterized.

The nature of the barley salt-soluble proteins has been carefully studied by Djurtoft (1961). The extraction is very complex, being dependent not only on protein components but also on carbohydrates and phytic acid. Proteolytic enzymes from the seed increase the salt-soluble protein extract yield. Djurtoft inhibited this hydrolysis by adding KBr. Using a refined ultracentrifuge technique, Djurtoft found the following amounts of different components in 1 kg of barley (protein content 10.4%): α-globulin, 3.38 gm; β-globulin, 0.85 gm; γ-globulin, 0.47 gm; albumins, ca 0.1 gm; and others, 1.3 gm.

Totally, 6.1 gm of these components were extracted together with a similar amount of low-molecular nitrogen substances. The salt-soluble proteins of Djurtoft (1961) should be only about 6%. Omitting KBr, Djurtoft reported greatly increased yields, which were presumably due to the effect of proteolytic enzymes. It has been shown that the embryo of barley contains γ-globulin exclusively, while α-, β-, and a small amount of γ-globulins are present in the endosperm. The β- and γ-globulins were found only in barley. Enari and Mikola (1962), Enari et al. (1963), and Mikola (1965) fractionated basic albumins employing DEAE cellulose chromatography. These proteins migrate toward the anode in polyacrylamide electrophoresis at pH 8.6. Finally, a globulin that is very rich in lysine and soluble in buffer solution and petroleum ether has been isolated in small amounts (Redman and Fisher 1969).

The roles of these different proteins are still very obscure. Some of them are enzymes such as esterases (Nielsen and Frydenberg 1971); others are β- and α-amylases (Frydenberg et al. 1970). Both esterase and amylase isoenzymes display Mendelian inheritance and could be used as markers for variety characterization and for chromosome mapping. Some of the salt-soluble proteins might have connections with the ultrastructures used for protein storage. Dimler (1966) found in maize that reduced glutelins migrating in starch electrophoresis show similar bands compared to reduced zein and reduced globulin. The number of S-S bridges should be reduced for the Hily endosperm compared with normal varieties because of low cysteine content.

The following model could tentatively be promulgated as a working hypothesis: (1) Prolamins are stored in protein bodies in the endosperm. They are probably comparatively easy to extract. (2) The matrix proteins consist of (a) small amounts of proteins rich in lysine and poor in cysteine which have a *potentially* high solubility; (b) prolamins more difficult to extract than (1). These molecules (a) and (b) are trapped in (c): a network of S-S bridges from the high-molecular weight glutelins. (3) Membrane proteins surround the starch granules.

According to this model, the proteins of 2 (a) or 3 type should be stimulated and those of 2 (c) type relatively depressed owing to the Hily gene(s) leading to a less-bridged glutelin network and increasing lysine in the total seed. When the gene effects are more precisely correlated to certain defined proteins, the nomenclature should be revised to represent adequate symbols for the proteins or ultrastructures that are involved.

PROPERTIES AND PROCESSING

While the extraction difficulties of the barley proteins from the chemical point of view may be overcome by the use of strong chemicals, they constitute a difficult extractive and digestive problem for the animal digestive system. The nutritive value of the barley proteins are not only associated with their amino acid composition and solubilities in more or less pure form, but are also dependent on their localization on the structural level. Munck (1964A) fractionated barley meal from different varieties that had received different nitrogen dressings and determined digestibility for pepsin *in vitro*. The coarse fraction contained more protein in the aleurone layer (Fig. 11.3) than the fine meal (below 120 mesh). There were significant differences in digestibility owing to variety and nitrogen-dressing effects in the whole meal. These relative differences were retained in the meal fractions, in which the finest meal from the endosperm region was highest in digestibility when compared with the coarse meal from the aleurone region. In both fractions and in the whole meal, the digestibility was increased when protein content was elevated by variety or by nitrogen fertilizer. Eggum (1970) confirmed in nitrogen balance tests with rats that the "true digestibility" (TD) of nitrogen increases with rising nitrogen content in the barley seed. At the same time, the "biological value" (BV) decreased owing to the negative relationship between lysine and crude protein level. The product between TD and BV—the net protein utilization (NPU), however, increased as BV fell less than TD when protein content was elevated. It seems that the prolamins are more easily digested than many of the more suitable proteins in the seed from the point of view of amino acid composition (Munck 1964 B). Eggum (1970) demonstrated that lysine and methionine have the lowest TD of the corresponding amino acids for rats.

Comparing the TD of amino acids for Hiproly with a reference, high-protein, normal sample Munck *et al.* (1970) demonstrated increases for Hiproly. The least available amino acids (lysine, methionine, alanine, asparagine, glycine, threonine, and tyrosine) are on a higher level in TD for Hiproly as compared with the reference sample, while the differences for the other amino acids are small. The effects described are likely to correspond to the higher content of proteins in Hiproly that are rich in lysine in the water and salt extracts, as demonstrated in the electrophoretic patterns in Fig. 11.8. To stress further the importance of availability studies of barley proteins, a comparative experiment between mice and rats for coarse and whole meal barley flour is shown in Table 11.4B. The coarse fraction (over 120 mesh) constitutes 50% of the whole meal and is 4.1 gm per 16 gm N in lysine as compared with 3.5 gm for whole meal (barley variety Kristina). The coarse meal fraction had a TD of nitrogen for rats of 75.4 compared with 82.2 for the whole meal. The BV's were completely opposite in tendency: 80.5 and 75.2, respectively. The NPU's were thus about the same, the whole meal being slightly higher. The animals, therefore, seem to have no use for the higher amounts of essential amino acids (lysine) in the coarse meal. Experiments with mice with carcass analyses at two protein levels point in the same direction that the whole meal protein is more utilizable than the coarse meal protein, as measured by nitrogen retention. The fact that about 25% of the important amino acids lysine and methionine are unavailable in the barley seed protein points to the possibility to improve the nutritional value with a change towards more available proteins without altering the total amino acid composition. Table 11.4A demonstrates the nutritional value of Hiproly, the normal reference sample, and CI 4362. The TD for nitrogen seems to be higher in the two closely related Ethiopian varieties Hiproly and CI 4362 as compared with normal barleys of high- and low-protein contents (Table 11.4B). It is possible that the "availability" problem is only partly associated with the Hily complex, and therefore needs attention per se in further screening procedures. The BV (Table 11.4A) was increased in Hiproly because the changed lysine composition led to a high NPU in the rat trial and a correspondingly increased nitrogen retention in the mice experiment. It is quite clear that a deeper knowledge of cereal ultrastructure and protein chemistry will be gained from studies of deviating genotypes in the same spirit as the classical biochemists studied mutants of *Neurospora* and *Escherichia* to trace the fundamentals of cell metabolism. At the same time, it will be possible to define the concept of the nutritional value more clearly with the aid of these new barley types. If high-yielding, nutritious barley varieties can be produced, this will have effects on the feed industry and lead to a decreased need for protein additives and stimulate amino acid supplementation. On the other hand, there are about 100 million people, mostly in the underdeveloped agrarian countries—with concomitant harsh climatic conditions and poor soils—in the Middle East and South America who are dependent on the drought-resistant barley varieties as their major cereal crop.

TABLE 11.4

FEEDING TRIALS WITH RATS AND MICE ON DIFFERENT BARLEY DIETS

Variety	Seed Protein (%)	Lysine (Gm/16Gm N)	Dietary Protein (%)	9 Days Restricted Rat Feeding			14 Days Ad Lib Mice Feeding		
				TD for N (%)	BV (%)	NPU (%)	Gain (Gm)	Nitrogen Retention (%)	Protein Consumed (%)
A. Comparing Hiproly with Normal High Protein Varieties									
Hiproly Sweden Reference[1]	16.8	4.08	9.4	85.2	76.0	64.8	1.70	32.7	5.2
Sweden Hiproly USA	17.3	3.13	9.4	83.2	71.1	59.2	1.43	28.4	5.0
Hiproly USA CI 4362 USA	19.9	4.10	9.4	86.0	75.9	65.3	–	–	–
USA	18.3	3.03	9.4	88.0	68.2	60.0	–	–	–
B. Comparing Normal Commercial Variety Kristina with a Coarse Fraction (Over 120 Mesh) from Meal from the Same Variety									
Kristina	11.4	3.46	9.4	82.2	75.2	61.7	1.66	32.9	5.0
Coarse	11.8	4.07	9.4	75.4	80.5	60.7	1.54	29.1	5.0
Kristina	11.4	3.46	11.4	–	–	–	1.83	28.8	6.7
Coarse	11.8	4.07	11.8	–	–	–	1.72	27.3	6.3

[1] Mixture of four naked, high protein, normal lysine varieties.

Hiproly barley at 16.8% protein and 4.1 gm lysine per 16 gm N is theoretically used as the sole source of protein in a diet for restrictively fed pigs and is compared with an isocaloric barley skim milk ration according to Clausen *et al.* (1963). The differences in amino acid consumption (grams per day) + or − relative the skim milk—barley diet is calculated for the different amino acids from 20 to 90 kg. It seems possible that an addition of 4 gm lysine and 1 gm methionine per kilogram of barley at 20–40 kg and 2 gm lysine from 40 to 90 kg could be sufficient for a reasonable feed efficiency and slaughter quality. The new lines (Fig. 11.2) at 15% protein content and 4.4 gm lysine per 16gm N could be still more useful if amino acid availability could be kept at a high level.

In birds, feeding barley is associated with special problems regarding viscosity giving rise to decreased utilization of protein. Because of the low water content in the upper part of the digestive system, birds are dependent on viscosity-decreasing enzymes from the barley seed itself or from additives, especially when the crop has been grown under dry conditions, as in western United States. Burnett (1966) found that this effect could be correlated with the β-glucane-β glucanase system, which is one of the factors that influence the viscosity of barley meal extracts.

BIBLIOGRAPHY

BURNETT, G. S. 1966. Studies of viscosity as the probable factor involved in the improvement of certain barleys for chickens by enzyme supplementation. Brit. Poultry Sci. *7,* 55-75.

CLAUSEN, H., MADSEN, A., and HANSEN, V. 1963. Protein intake related to slaughter products. Rept. Autumn Meeting Natl. Inst. Animal Sci., Copenhagen, Denmark. (Danish)

COOK, A. H. (Editor) 1962. Barley and Malt. Academic Press, New York, London.

DIMLER, R. J. 1966. Alcohol-insoluble proteins of cereal grains. Federation Proc. *25,* 1670-1675.

DJURTOFT, R. 1961. Salt-soluble proteins of barley (*Hordeum Distichum*). Dansk Videnskabs Forlag A/S, Copenhagen, Denmark.

DOLL, H. 1970. Variation in protein quantity and quality induced in barley by EMS treatment. Manuscript. Danish AEC Research, Riso, DK-4000 Roskilde, Denmark.

EAKER, D. 1970. Determination of free and protein bound amino acids. Symp. Evaluation Novel Protein Prod. Stockholm. Pergamon Press.

EGGUM, O. E. 1970. Relationship between protein quality and nitrogen content of barley. Z. Tierphysiol. Tierernahr. Futtermittelk. *26,* 65-71. (German)

ENARI, T-M., and MIKOLA, J. 1962. Fractionation of barley albumins by chromatography on DEAE-cellulose. European Brewery Conv. Proc. (1961).

ENARI, T-M., MIKOLA, J., and NUMMI, M. 1963. Preparation and starch gel electrophoresis of basic, water soluble proteins from barley and other kinds of cereals. Brauwissenschaft *16,* 189-193. (German)

FAVRET, E. A., SOLARI, R., MANGHERS, L., and AVILA, A. 1969. Genetic control of the qualitative and quantitative production of endosperm proteins in wheat and barley. Symp. New Approaches to Breeding for Improved Plant Protein. IAEA STI Publ. *212,* Vienna.

FAVRET, E. A. *et al.* 1970. Gene control of protein production in cereal seeds. Symp. Improving Plant Protein by Nuclear Techniques. IAEA STI Publ. *258,* Vienna.

FRYDENBERG, O., NIELSEN, G., and SANDFAER, J. 1970. The inheritance and distribution of α-amylase types and DDT responses in barley. Z. Pflanzenzucht. *61,* 201-215.

GUSTAFSSON, Å. 1969. A study on induced mutations in plants. Symp. Induced Mutations in Plants. IAEA STI Publ. *231*, Vienna.

HAGBERG, A., and KARLSSON, K-E. 1968. Breeding for high protein content and quality in barley. Symp. New Approaches to Breeding for Improved Plant Protein. IAEA STI Publ. *212*, Vienna (1969).

HAGBERG, A., KARLSSON, K-E., and MUNCK, L. 1970. Use of Hiproly in barley breeding. Symp. Improving Plant Protein by Nuclear Techniques. IAEA STI Publ. *258*, Vienna.

HARRIS, G. 1962. The structural chemistry of barley and malt. *In* Barley and Malt, A. H. Cook (Editor). Academic Press, New York, London.

KENT, N. L. 1970. Structural and nutritional properties of cereal proteins. *In* Proteins as Human Food, R. A. Lawrie (Editor). In United States and Canada available from Avi Publishing Co., Westport, Conn.; otherwise available from Butterworths, London.

MIKOLA, J. 1965. Disc electrophoresis studies on the basic proteins of barley grain. Ann. Acad. Sci. Fennicae. Ser. *A II* Chemica, 130.

MOSSBERG, R. 1969. Evaluation of protein quality and quantity by dye-binding capacity: a tool in plant breeding. Symp. New Approaches to Breeding for Improved Plant Protein. IAEA STI Publ. *212*, Vienna.

MUNCK, L. 1964A. The variation of nutritive value in barley. I. Variety and nitrogen fertilizer effects on chemical composition and laboratory feeding experiments. Hereditas *52*, 1-34.

MUNCK, L. 1964B. The variation of nutritive value in barley. III. Favourable and unfavourable effects of damaging and processing to the nutritional value considered in relation to plant breeding. Hereditas *52*, 59-76.

MUNCK, L. 1964C. The variation of nutritive value in barley. IV. Plant breeding and nutritional value in cereals. Hereditas *52*, 151-165.

MUNCK, L. 1968. Cereals for feed—quality and utilization. J. Swedish Seed Assoc. *LXXVIII*, 137-201.

MUNCK, L. 1970. Increasing the nutritional value in cereal protein: basic research on the hily character. Symp. Improving Plant Protein by Nuclear Techniques. IAEA STI Publ. *258*, Vienna.

MUNCK, L., KARLSSON, K-E., HAGBERG, A., and EGGUM, O. E. 1970. Gene for improved nutritional value in barley seed protein. Science *168*, 985-987.

MUNCK, L., KARLSSON, K-E., and HAGBERG, A. 1971. Selection and characterization of a high-protein, high-lysine variety from the world collection. Symp. Barley Genetics, Wash. Univ. Press, Pullman, Wash. (1969).

NELSON, O. E. 1969A. Genetic modification of protein quality in plants. Advan. Agr. *21*, 171-193.

NELSON, O. E. 1969B. The modification by mutation of protein quality in maize. Symp. New Approaches to Breeding for Improved Plant Protein. IAEA STI Publ. *212*, Vienna.

NILAN, R. A. 1964. The cytology and genetics of barley, 1951-1962. Wash. State Univ. Monogram Suppl. *3*, No. 1, 32.

NIELSEN, G., and FRYDENBERG, O. 1971. The inheritance and distribution of esterase isozymes in barley. Symp. Barley Genetics II, Wash. Univ. Press. Pullman, Wash. (1969).

REDMAN, D. G., and FISHER, N. 1969. Purothionin analogues from barley flour. J. Sci. Food Agr. *20*, 427-432.

SMITH, L. 1951. Cytology and genetics of barley. Botan. Rev. *17*, 1,3,5.

STOY, V. 1966. Photosynthetic production after ear emergence as a yield-limiting factor in the culture of cereals. Acta Agr. Scand. Suppl. 178-182.

VIEUF, B. T. 1970. Examination of various barley varieties for protein quality and nitrogen content. Ph.D. Thesis. Royal Veterinary Agricultural Highschool, Copenhagen. (Danish)

WRAY, J. L., RIES, S. K., and FILNER, P. 1970. Protein accumulation and the regulation of nitrate reduction in higher plants. Symp. Improving Plant Protein by Nuclear Techniques. IAEA STI Publ. *258*, Vienna.

CHAPTER 12

Walter L. Clark
and
George C. Potter

The Composition and Nutritional Properties of Protein in Selected Oat Varieties

INTRODUCTION

The origin of the oat (*Avena sp.*) is not clear but it is thought to have first appeared as a weed in other cultivated cereals. The term oat is said to be derived from the Middle English term "Ote." It is curious that no other early Germanic language has a term for the oat.

As a cultivated crop the oat is a newcomer as compared to rice, wheat, and barley. Whether this is due to the oat being a crop of the cooler temperate zone while the earliest civilizations arose in hotter climes must remain a matter of conjecture. The oat was probably not cultivated as a crop before the fifth century A.D. and then only in northern Europe. The reader is referred to Coffman (1961) for further information as to the early history of the oat.

The oat belongs to the cereal grasses (*Graminae*) and is of the genus *Avena*. It occurs in spring- and fall-sown types with approximately 80% of the world's oat production of the spring-sown type (*Avena sativa*) while most of the remainder is fall-sown (*Avena byzantina*) (Coffman 1961). The *A. sativa* oats are commonly called white oats and *A. byzantina* red oats but such distinctions are dubious as there has been extensive inbreeding of the two types. *A. nuda* or naked oats, a type having no hull, has enjoyed brief periods of popularity but due to such factors as lessened disease resistance and harvesting difficulties is not commonly encountered.

IMPORTANCE AS FEED AND FOOD INGREDIENT

At the present time the oat is ranked fifth in total production following wheat, corn, rice, and barley, and is mainly utilized for animal feed. Webster *et al.* (1970) have stated that 75–80% of the oat crop is utilized for animal feeds at the farm level. In view of the world's present need for protein this is regrettable, as the oat contains 13–18% protein of good nutritional quality. The protein efficiency ratio (PER) of rolled oats has been given as 2.2 by Campbell and Chapman (1959) when casein is arbitrarily set at a PER of 3.0. Unpublished data accumulated over a period of years by The Quaker Oats Company agree with this estimate.

Whole oats are generally used for animal feeds although small amounts of reject oat groats (dehulled oats from oat middling operations) are sometimes utilized in animal feeds. The oats to be utilized in rolled and other edible oat products are processed through extensive cleaning, sorting, drying, and dehulling operations. The final product (before rolling or processing by other procedures) is commonly referred to as a dried groat. The technology of these processing steps is not within the scope of this review, and interested readers are referred to Coffman (1961) for further information.

VARIATIONS IN PROTEIN CONTENT

In a collaborative study in 1965 undertaken by USDA and The Quaker Oats Company, samples of oats from various Near Eastern areas were analyzed for amino acid profiles and crude protein content. A crude protein content of 14–25% was found in the groats (dehulled oats). The fact that the protein content of these oats is much higher than that of the usually cultivated domestic oats indicates that the oat has genetic traits which regulate its protein content. It should therefore be possible to develop a higher protein oat for use in human foods.

A protein range of 11–15% in oat groats utilized for food is more probable. The groats used in rolled oat production are blends of a number of varieties, and quality control analyses by The Quaker Oats Company show a remarkable uniformity of protein content in any year's receipts of the blended oats. As is true of other cereal grains, the protein content of oats will decrease as the yield of crop is increased by modern agricultural practices.

An extensive literature survey uncovered a surprisingly small amount of information on whether different varieties of the oat differed in the nutritional quality of their proteins. This was true not only of oats for inclusion in the human diet but also of those destined for use in animal feeds. As has been noted above, a major portion of the oat crop is consumed at the farm level for livestock feeding, and the portion that is processed for human consumption is a blend of a number of varieties. Another contributing factor is that oats are not sold on the basis of protein content, so very little research has been done as to the relative value of the protein of one variety as compared to another. Agronomists have traditionally looked for improved yields, disease resistance, and harvesting characteristics, and have had neither the training nor inclination to consider the nutritional characteristics of cereals. It would not be surprising to see the work of agronomists and nutritionists more intimately linked in the crucial years ahead.

NUTRITIONAL ASPECTS OF OAT PROTEIN

The pioneer studies of McCollum *et al.* (1921) on the nutritional qualities of oat protein were based on rolled oats of commercial origin. It is somewhat difficult to make comparisons of the data resulting from these early studies on protein quality as no standard experimental methods were used. However, it is safe to say that the pioneering studies on the cereal proteins uncovered several rather important facts. All cereal proteins were inferior to animal proteins in promoting growth of young animals, and proper supplementation gave an appreciable boost to the quality of cereal protein as measured by the growth of young animals. These early workers also found the proteins of oat, rice, and rye were superior to those of corn, wheat, and barley, regardless of the total protein level in the diet. Jones et al. (1948) further verified these conclusions as to the relative biological value of cereal proteins.

By the end of World War I, McCollum and colleagues had realized that the amino acid lysine was the first limiting factor in cereal proteins (McCollum et al. 1939). Since then, other amino acids have been found to be secondary limiting factors. Mitchell and Smuts (1932) noted that the addition of lysine to a diet utilizing oat protein boosted growth slightly, but it was not until the classical study of Rose et al. (1943) that threonine was discovered to be the final essential amino acid both in synthetic diets and in such materials as oat protein. The position of threonine in the oat is somewhat ambiguous. Chemical analysis indicates an adequate level in oat protein as compared to FAO standards, but feeding trials do not verify this, as the supplementation of oat protein with threonine results in improved growth. Tang and co-workers (1958) suggested that threonine was not completely available to the animal while the lysine and methionine contained in oat protein had normal availability. This aspect of oat protein has not been clarified up to the present time.

The cereals are consumed as dormant seeds and so have at harvest time an appreciable amount of what is called "storage proteins." Zein in corn is typical of the cereal storage proteins in that its lysine content is negligible. The greater the content of storage proteins in a cereal, the poorer it becomes as a source of protein. Any upgrading of cereal proteins must come about in 1 of 2 ways—by increasing the amount of lysine in the storage proteins, or decreasing the level of storage proteins. The first approach does not appear very feasible but the second shows promise, as exemplified by opaque-2 corn where the higher lysine levels are a result of a lowered level of the storage protein. The pioneering work of Mertz et al. (1964) has shown that cereal protein quality can be improved by genetic selection.

QUALITY IMPROVEMENT OF THE OAT

The oat is a self-pollinating plant, which means that oat breeders do not have the advantage that corn breeders have of crossing sterile and nonsterile plants to produce hybrids with improved characteristics. In the small grains such as the oat, the manpower needed to remove the required sex-parts of the plant under field conditions would be beyond the financial capabilities of seed producers.

Coffman (1961) has indicated that mass selection was probably the first attempt by man to improve oat quality. In this technique, oats were selected for desirability in a crop area at a given time. Somewhat later, the selection of individual plants for propagation was utilized by oat breeders. Both methods require considerable time and effort to produce new varieties.

The latter technique of hybridization, where the pollination of the plants is controlled, is of major interest now, but it is also a slow process beset with many pitfalls. It is unfortunate that plant diseases and insect pests seem to be able to adjust to new plants at a rate about equal to the production of new resistant varieties.

Another and very real difficulty in breeding oat varieties, or for that matter any small grain, for improved protein quality is the relatively large amount of material needed to perform the usual animal feeding studies. When one realizes that only a few grams of seed are recovered in the original hybridization step and quantities of 10 kg and more are needed for feeding trials, it is little wonder that the potential nutritional quality of new varieties is ignored at this stage.

The oat has long been regarded as a grain more fit for animal feed than for consumption by the human. However, the northern European races and especially the Scottish have long considered the oat to have equally good nutritional properties for man and horse. An example of this is the exchange said to have taken place between Samuel Johnson, an Englishman, and Boswell, a Scotsman. Johnson defined the oat as a grain eaten by men in Scotland, but in England considered fit only for horses. Boswell's reply was, "and pray, where do we see such men as in Scotland and such horses as in England?" As McCollum *et al.* (1939) stated, this viewpoint is not completely supported by experimental evidence.

There are many factors that enter into the overall nutritional quality of a cereal such as the oat. In this review only the nutritional quality of the protein of oat is to be considered, but this is not to be inferred to mean that other constituents are not of nutritional significance. At the present time the caloric value of the oil and carbohydrate portions of oats are of more practical interest to animal feeders than the quality and quantity of protein.

The fact that the oat has not been considered for its protein contribution is reflected in the small number of research papers presented on differences in nutritional quality of proteins of different oat varieties. Most nutritional studies on oat protein have utilized commercially produced rolled oats. Rolled oats are produced from blends of oats as received by the oat processors from grain market storage facilities, thus making it impossible to state what impact any specific variety had on the overall nutritional quality of the oat product.

Frey (1951) noted that the alcohol-soluble fraction of oat protein did not increase with an increase in overall protein level as was found to be the case for corn zein. Zein proteins are severely deficient in lysine as the major limiting amino acid, with other secondary limiting acids such as tryptophan in corn and methionine for oats appearing if the diet is supplemented with lysine. The alcohol-soluble portion of oat protein remains constant at about 18% of the total protein content in varieties that ranged from 1.49 to 2.53% total nitrogen, or 9.4 to 15.6% protein using a 6.25 conversion factor. The above observation of Frey (1951) is of primary importance, as the oat breeder wishing to improve the protein quality of the oat has only to select higher protein oats that also measure up to other standards such as yield, resistance to disease, and weather damage.

Weber *et al*. (1957), in an experiment using a restricted feeding technique, found that while winter oats appeared to be somewhat superior to spring varieties, the differences were not statistically significant.

There were some varieties of both winter and spring oats which produced consistently poorer weight gains in the feeding studies. In the case of individual varieties, the differences were found to be statistically significant. Microbiological assays for the lysine and methionine failed to show any correlation between amino acid content and weight gain of the groups receiving the various oat samples. The weight gains of the rats on oat-based diets were approximately 73% of those on casein-based diets at a 10.9% protein level. This compares very well with the data presented by Hischke *et al*. (1968) and Webster *et al*. (1970) where the PER of diets containing 10% oat protein was about 73% of that of a diet based on casein at a 10% protein level.

In further studies, Weber *et al*. (1957) decided that environmental factors were of lesser importance than genetic factors in determining the nutritive characteristics of oat protein. By planting the same varieties at various locations and times, they established that such practices did not appreciably alter the nutritive quality of the oat varieties. The availability rather than the total amount of essential amino acids was considered by Weber *et al*. (1957) to be of primary importance in the nutritive quality of oat protein.

With the emergence of the Spackman *et al*. (1958) system for analysis of amino acid from protein hydrolysates, a new approach to nutritive quality of proteins was made available. Instead of kilogram quantities for feeding studies, only milligram quantities were needed for a chemical analysis of amino acid content of various proteins. However, it must be stressed that this technique for evaluating protein quality must be verified by feeding studies. The unavailability of certain amino acids or the presence of factors deleterious to animal growth will not be obvious for amino acid composition studies. Nevertheless, with these limitations in mind, the rapid automated analysis of the amino acids of cereal protein is a valuable tool in protein improvement studies.

Hischke *et al*. (1968) attempted to correlate the amino acid content of seven oat varieties of known purity with the results of rat growth feeding trials. In Table 12.1 the diets fed are presented. It must be noted that all diets were based on dehulled oats commonly referred to as oat groats. The oat hull is commonly found to comprise 25–30% of the whole oat and is primarily composed of inedible roughage.

In Table 12.2 the total protein and amino acid contents of the varieties utilized by Hischke *et al*. (1968) are presented. The lysine content of these oats ranges from 83 to 88% of that of the FAO provisional standard while threonine ranges from 103 to 106% of the FAO value.

The data on methionine and cystine content must be evaluated with caution

TABLE 12.1

COMPOSITION OF OAT DIETS FORMULATED TO A 10% PROTEIN CONTENT
AND A 10% FAT CONTENT

Ration	1	2	3	4	5	6	7
Corn Starch(%)	27.8	25.2	35.3	28.4	34.2	27.1	33.1
Corn Oil (%)	7.5	6.4	7.2	6.6	7.1	4.5	7.7
Sample (%)							
Garland	56.5						
Clintland		60.2					
Bonkee			49.3				
Newton				56.8			
Beedee					50.5		
Lodi						60.2	
Nemaha							51.0
Premix[1]	8.2	8.2	8.2	8.2	8.2	8.2	8.2

[1] Premix includes 4.0% salt mix, 2.2% vitamin mix and 2.0% celluflour.

TABLE 12.2

PROTEIN CONTENT AND AMINO ACID COMPOSITION[1]
OF SEVEN OAT VARIETIES

Amino Acids	Garland (%)	Clintland (%)	Bonkee (%)	Newton (%)	Beedee (%)	Lodi (%)	Nemaha (%)
Glycine	5.03	4.90	4.81	4.98	4.83	5.19	4.93
Alanine	4.83	4.78	4.59	4.79	4.66	4.73	4.71
Half cystine	1.03	1.41	1.42	1.36	1.38	1.75	1.24
Valine	5.48	5.60	5.29	5.40	5.36	5.23	5.47
Methionine	1.51	1.60	1.44	1.43	1.13	1.42	1.56
Isoleucine	4.07	4.09	4.02	4.00	4.02	3.88	4.02
Leucine	7.85	7.87	7.83	7.92	7.79	7.78	7.84
Tyrosine	3.34	3.24	3.42	3.21	3.48	3.45	3.53
Phenylalanine	5.55	5.61	5.50	5.57	5.52	5.49	5.59

[1] The amino acids are expressed as a percentage of the total amino acids plus ammonia recovered. Each value is an average of four replicates. Protein values are expressed on a moisture-free basis.

as these results were obtained from acid hydrolysis which results in some destruction of these amino acids, so no attempt will be made to compare these data with the FAO standards.

Lysine, glutamic acid, glycine and alanine were found to have significant differences between mean values. Regression lines were fitted to the data of the four amino acids and the results are presented in Fig. 12.1, 12.2, 12.3, and 12.4.

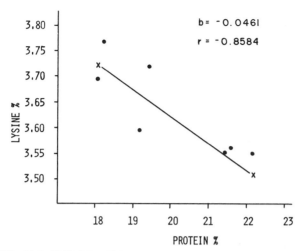

FIG. 12.1. THE RELATIONSHIP BETWEEN THE LYSINE AND
PROTEIN CONTENT OF SEVEN VARIETIES OF OATS

Lysine is expressed as a percentage of the total amino acids plus
ammonia recovered.

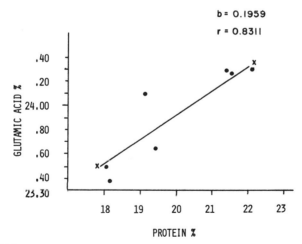

FIG. 12.2. THE RELATIONSHIP BETWEEN THE GLUTAMIC
ACID AND PROTEIN CONTENT OF SEVEN VARIETIES OF
OATS

Glutamic acid is expressed as a percentage of the total amino acids
plus ammonia recovered.

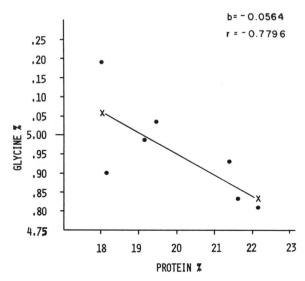

FIG. 12.3. THE RELATIONSHIP BETWEEN THE GLYCINE AND
PROTEIN CONTENT OF SEVEN VARIETIES OF OATS

Glycine is expressed as a percentage of the total amino acids plus
ammonia recovered.

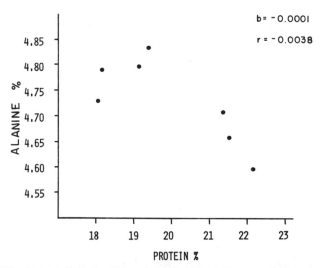

FIG. 12.4. THE RELATIONSHIP BETWEEN THE ALANINE AND
PROTEIN CONTENT OF SEVEN VARIETIES OF OATS

Alanine is expressed as a percentage of the total amino acids plus
ammonia recovered.

Lysine and glycine were negatively related to protein content while glutamic acid had a positive correlation. Alanine exhibited a nonsignificant negative correlation to protein level.

As can be seen from Table 12.3, the results of feeding the pure oat varieties analyzed by Hischke *et al.* (1968) indicated that any differences in amino acid content were not great enough to influence growth response.

The use of automated amino acid analyzers and computers to screen large numbers of cultivars for a specific characteristic such as higher lysine contents is exemplified in the paper by Robbins *et al.* (1971). The authors analyzed 289 samples of oat groats for crude protein content and their amino acid profiles. Their results were essentially the same as those reported by Hischke *et al.* (1968) but the correlation between lysine and crude protein was found to be less negative. A significant negative correlation was found between threonine and protein content. Since the range of threonine content is small, this negative relationship would not be expected to exert a drastic effect on the protein quality of oats. This indicates that it may be possible to increase the protein content of the oat without decreasing the lysine levels. This is an important fact, since as the authors point out, the oat is highest in protein and lysine content of the cereals commonly grown in the United States.

Tkachuk and Irvine (1969) analyzed Canadian oat groats and found the amino acid profile agreed well with those found by Hischke *et al.* (1968). The authors pointed out that discrepancies between various samples should not be strictly interpreted as being significant as they may merely reflect "differing environmental conditions."

One may very well ask, how well are the results of chemical and statistical analyses correlated with feeding studies? Table 12.3 is a presentation of results of oat feeding trials, and the only conclusion to be reached is that the

TABLE 12.3

TWENTY-ONE DAY FEEDING[1] DATA

Ration	Weight Gain (Gm)	Feed Intake (Gm)	PER[2] (Gm)
Garland	46.5	194.6	2.37
Clintland	48.6	203.9	2.38
Bonkee	42.9	188.0	2.28
Newton	42.9	190.3	2.25
Beedee	44.3	194.1	2.27
Lodi	43.1	189.9	2.25
Nemaha	43.1	188.9	2.28

[1] Each value is an average of ten rats.
[2] PER is grams gained in body weight of protein intake.

differences in amino acid composition found by chemical analysis were not detectable by animal feeding studies. This further substantiates the conclusions of Weber *et al.* (1957) that differences of biological availability of specific amino acids may be of more importance than the amino acid content of these materials as found by chemical analysis.

SUMMARY

The oat is a major cereal crop in terms of total acreage grown in the world. Only a minor portion of the total crop appears in the form of edible products for use in the human diet. This is an unfortunate situation as the oat has been shown to have protein of a quality equivalent to or superior to other common cereals. Normal processing of the oat does not result in major protein losses as is the case with rice, wheat, and corn. The oat is consumed as a whole grain and is not degerminated with the resulting loss of protein having nutritive values superior to the protein commonly found in the endosperm of cereal products.

The selection of oat varieties for their biological protein value has not received much attention as most developmental work has been involved in such practical matters as yield, insect and disease resistance, and ease of harvesting by means of mechanical reapers. However, some oat types apparently have the unusual property of not suffering a loss in biological value of the protein as the selection processes used increases the total protein content. With our precarious situation in regard to the world's protein supply, this property alone would merit the oat receiving serious consideration as a potential protein source.

BIBLIOGRAPHY

CAMPBELL, J. A., and CHAPMAN, D. G. 1959. Evaluation of protein in foods—criteria for describing protein value. J. Can. Dietetic Assoc. *21*, 51.

COFFMAN, F. A. 1961. Origin and history. *In* Oats and Oat Improvement, F. A. Coffman (Editor). Am. Soc. Agron., Madison, Wisc.

FREY, K. J. 1951. The relation between alcohol-soluble and total nitrogen content of oats. Cereal Chem. *28*, 506-509.

HISCHKE, H. H., Jr., POTTER, G. C., and GRAHAM, W. R. 1968. The nutritive value of oat protein. I. Varietial differences as measured by amino acid analysis and rat growth response. Cereal Chem. *45*, 374-378.

JONES, D. B., CALDWELL, A., and WIDENESS K. D. 1948. Comparative growth-promoting values of the proteins of cereal grains. J. Nutr. *35*, 639.

McCOLLUM, E. V., SIMMONS, N., and PARSONS, H. T. 1921. IV. The supplementary relations of cereal grain with cereal grain. J. Biol. Chem. *47*, 175-206.

McCOLLUM, E. V., ORENT-KEILES, E., and DAY, H. G. 1939. The Newer Knowledge of Nutrition. Macmillan Co., New York.

MERTZ, E. T., BATES, T. S., and NELSON, O. E. 1964. Mutant gene that changes protein composition and increases lysine content of maize endosperm. Science *145*, 279-280.

MITCHELL, H. H., and SMUTS, D. B. 1932. The amino acid deficiencies of beef, wheat, corn, oats and soya beans for growth in the white rat. J. Biol. Chem. *95*, 263.

ROBBINS, G. S., POMERANZ, Y., and BRIGGLE, L. W. 1971. Amino acid composition of oat groats. Agr. Food Chem. *19*, 536-539.

ROSE, W. C., HAINES, W. J., JOHNSON, J. E., and WARNER, D. T. 1943. Further experiments on the role of the amino acids in human nutrition. J. Biol. Chem. *148*, 457-458.

SPACKMAN, D. H., STEIN, W. H., and MOORE, S. 1958. Automatic recording apparatus for use in the chromatography of amino acids. Anal. Chem. *30*, 1190-1206.

TANG, J. J. N., LAUDICK, L. L., and BENTON, D. A. 1958. Studies of amino acid availability with oats. J. Nutr. *66*, 533-543.

TKACHUK, R., and IRVINE, G. N. 1969. Amino acid composition of cereals and oilseed meals. Cereal Chem. *46*, 206-218.

WEBER, E. B. *et al.* 1957. Protein quality of oat varieties. J. Agr. Food Chem. *5*, 926-928.

WEBSTER, F. H. *et al.* 1970. Oats: a potential protein source. *In* SOS/70 Proc. 3rd Intern. Food Sci. Technol. Conf., Washington, D.C., August. Inst. Food Technologists.

George E. Inglett | **Corn Proteins Related to Grain Processing and Nutritional Value of Products**

The number one cereal in U.S. agriculture is corn, whose major constituent is starch. The corn wet-milling industry was established to supply starch for food and industrial uses. Another type of industrial processing of corn, dry-milling, evolved from early corn grinding operations. Corn grits and flour from today's dry-milling industry are utilized primarily for their carbohydrate value. Corn has traditionally been regarded as an energy source for animal feed or carbohydrate source for industrial products.

In the United States, corn has evolved from an indispensable food staple during early colonial and pioneer days to the most important feed grain. As the United States evolved from a rural society to an industrial and urban one, more corn went into feeds and commercial outlets. In 1967, 4.7 billion bushels were produced in the United States. Feed and seed uses on the farm account for 2.2 billion bushels while much of the 2.5 billion bushels sold from the farm went to feed, other uses being for food processing and export. A total of 360 million bushels (nearly 9% of production) is used annually by the wet millers, dry millers, and fermentation processors. Wet millers use slightly more than 200 million bushels (5%) and the dry corn millers process more than 130 million bushels (3.5%). In addition, approximately 30 million bushels (1%) are fermented to neutral spirits and alcoholic beverages (Inglett 1970).

Principal components of whole kernel corn besides starch are proteins and lipids, with lesser quantities of fiber, sugars, minerals, and miscellaneous organic substances including vitamins.

CORN PROTEINS

Proteins constitute nearly 10% of whole kernel corn. These substances are classified according to their solubility into albumins, globulins, prolamins, and glutelins. It is important to understand the properties of corn proteins because they influence the nutritional composition of food products, the processing of the grain, and the stability of corn products. Only in this decade has the corn breeder given his attention to the protein quality and quantity of the corn kernel. The increased interest in corn protein is the realization that the nutritive value of corn could be improved by manipulating its genes.

The proteins of ordinary dent corn are deficient in the essential amino acids, lysine and tryptophan, particularly in the endosperm which represents 80% of the kernel.

One of the most significant plant breeding studies was made by Mertz and co-workers (1964) at Purdue. They discovered that proteins in the endosperm of certain floury varieties were significantly higher in biological quality than those in ordinary corns. *Opaque-2* maize has an unusually high lysine content, about 3.6% based on endosperm protein, which is about twice the amount of lysine found in ordinary corn hybrids. *Floury-2* maize has a lysine content of 3.2%. Since *opaque-2* maize is a floury variety, its physical characteristics are distinctly different from ordinary corn. Protein is distributed more uniformly throughout the endosperm than in ordinary dent corns. *Opaque-2* lacks the densely packed, translucent, endosperm cells common in dent corns, which are sometimes referred to as horny or vitreous in character. Almost all cells in the endosperm are of the floury type; that is, the starch and protein are loosely arranged within the cells with air spaces between the starch and the protein. Intermingling air spaces and cell solids impart the opaque appearance to floury kernels in transmitted light. The air spaces also are responsible for the lower test weight of floury corns. Milling floury corns, such as *opaque-2*, gives a high proportion of flour, as might be expected. This increase in flour is contrary to the accustomed processing of ordinary dent corn. Ordinary dent corn tends to hold together in milling and to give large groups of intact cells called grits—an important product to the corn dry miller. The horny sections of the kernel are responsible for the grits obtained from ordinary corn.

Endosperm protein in ordinary corn contains about 40–50% zein and 20–30% glutelin. A marked deviation from these proportions was noted by Mertz and his co-workers in *opaque-2*. They found that in *opaque-2* the proportions of zein and glutelin were reversed. Zein in this corn was down to about 15% of the protein and glutelin rose to more than 40%. The nutritional importance of this finding becomes obvious from a study of the amino acid composition of corn glutelin and zein, particularly as they relate to lysine. Zein contains virtually no lysine, whereas corn glutelin contains about 4.7 gm of lysine per 100 gm of protein. Consequently, any increase in the proportion of glutelin at the expense of zein increases the nutritional value of the corn protein.

MICROSCOPIC STUDIES ON CORN PROTEINS

A change in protein composition as great as that found by Mertz and his colleagues results in a corresponding change in protein structure at the subcellular level. The subcellular structure of storage proteins and endosperm proteins of ordinary and high lysine corn has been carefully examined by Wolf and co-workers (1969). Their studies provide a structural basis for the altered protein composition and give some information concerning the milling characteristics of the new varieties of corn.

Starchy endosperm structures of an ordinary dent corn and of a flour corn are compared in Fig. 13.1. Horny endosperm of a dent corn, Iowa 306, is shown on the left and the typical floury endosperm of *opaque-2* is seen on the right. Because of the overlying starch, protein is difficult to see in the sections. After setting the protein with formaldehyde or other fixative to preserve the protein structure from breaking down on subsequent treatment and digesting the starch with α-amylase, the protein distribution can be readily visualized. Sections destarched in this manner are reproduced in Fig. 13.2. These are low-power photomicrographs. Ordinary dent corn is on the left. On the right, the gross structural pattern of the floury mutant *opaque-2* is typical of endosperm of other flour corns. Except for cell walls, most of the material that can be seen is protein. Judging from the depth of staining, ordinary corn shows a high protein density in the outer endosperm (inside the aleurone layer) and a decreasing gradient of protein concentration toward the center of the endosperm. In the flour corn, *opaque-2,* the protein distribution, evaluated similarly, is more uniform throughout the endosperm than in ordinary corn. This relative uniformity is typical of flour corns.

In each cell the protein forms a network. Before enzymatic digestion each opening in the network is occupied by a starch granule. In cells of the horny endosperm (outer endosperm in ordinary corn) the network is well developed so

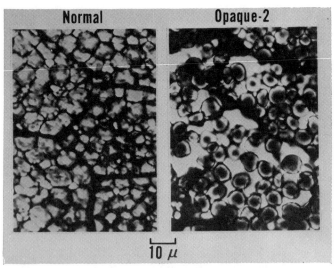

FIG. 13.1. COMPARISON OF HORNY AND FLOURY ENDOSPERM STRUCTURES OF CORN

Horny endosperm (left) of ordinary Iowa 306 dent corn; densely packed starch granules and protein leave little room for air space. Floury endosperm (right) of *opaque-2*; loosely arranged starch and protein leave numerous air spaces in cells.

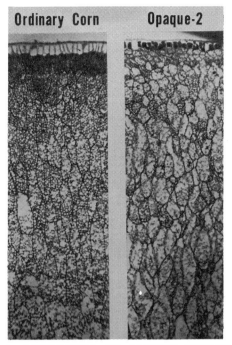

FIG. 13.2. COMPARISON OF PROTEIN STRUCTURE AND PROTEIN DISTRIBUTION IN DENT CORN (U.S. 13) AND FLOURY CORN (*opaque-2*)

Destarched with α-amylase to reveal subcellular protein structure.

that all starch granules of the cell are held firmly in a protein matrix. This arrangement explains why little loose starch is formed from horny endosperm in milling. On the other hand, in floury cells (inner endosperm in ordinary corn or virtually the entire endosperm of flour corn) a protein network is only fragmentary. Starch is loosely held in such cells as those observed in Fig. 13.1. Consequently, milling floury tissues (of ordinary corn or flour corn) yields large proportions of flour.

Even in flour corns certain endosperm cells have a relatively high content of protein compared with adjacent cells. The endosperm cells appear opaque or nearly opaque in photomicrographs. There may be more or less continuous layers of such cells just inside the aleurone. In addition, there is a random scattering of individual high-protein cells in the outer endosperm. In Fig. 13.3, protein subaleurone cells can be readily observed in the flour corn variety Mandan. As in the previous microscopic sections (Fig. 13.2), the starch granules have been digested. The subaleurone cells had virtually no starch, only protein. In some cells this protein is granular; in others it is amorphous. This subaleurone layer was first observed in the high-lysine mutant, *opaque-2,* and it was thought

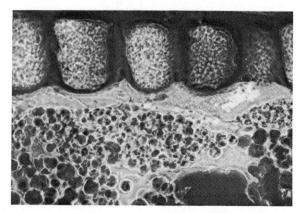

FIG. 13.3. OUTER ENDOSPERM OF MANDAN VARIETY FLOUR
CORN SHOWING THE HIGH PROTEIN SUBALEURONE LAYER

Destarched with α-amylase to expose subcellular protein structure.

at the time that this layer, with its high protein content, was unique to *opaque-2* and may have something to do with the high lysine content of that mutant. Subsequently, the same structure was detected in other flour corns and even in some ordinary dent corn inbreds and hybrids. All these corns have a normal amino acid pattern. Therefore, it was concluded from this finding that the high lysine character of *opaque-2* was not related specifically to the presence of this structure.

In the ordinary dent hybrid we can see embedded in the matrix protein small granules averaging about 2 μ in diameter. These granules can be observed in the photomicrograph (Fig. 13.4). This figure compares sections of an ordinary corn and the high-lysine mutant, *opaque-2*. As in the other photomicrographs, the microscopic sections are starch free. Protein granules have been observed in ordinary corn endosperm for many years. It was not until 1961, however, that Duvick (1961) suggested these granules might be the site of zein deposition. When the subcellular structure of endosperm protein of *opaque-2* was examined, no protein granules were observed under a light microscope. The apparent absence of granular protein in *opaque-2* suggests that the basis for the improved protein composition in this mutant could have been a failure to deposit zein granules; however, Mertz' data were necessary for the correct interpretation. The deficiency in zein protein is apparently compensated by increased matrix protein formation. It was known that *opaque-2* is not completely deficient in alcohol-soluble protein. Therefore, the fine structure of *opaque-2* was compared with ordinary corn endosperm by an electron microscope. The electron photomicrographs are shown in Fig. 13.5. The micron scale at the bottom of the photomicrographs applies to both pictures. Relatively large granules ranging down in diameter to well below 1 μ can be seen in the dent hybrid. In *opaque-2*,

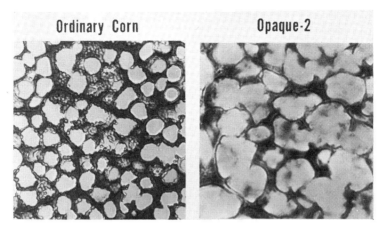

FIG. 13.4. COMPARISON OF SUBCELLULAR PROTEIN STRUCTURE OF ORDINARY CORN (W64A) AND *opaque-2* MUTANT (W64Ao$_2$)

Ordinary corn protein has two components: granular bodies and matrix protein. *Opaque-2* protein has no granular protein bodies.

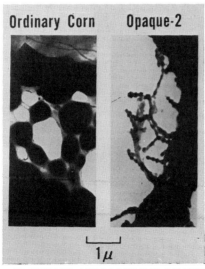

FIG. 13.5. ELECTRON MICROGRAPHS OF SUBCELLULAR ENDOSPERM PROTEIN IN ORDINARY CORN (LEFT) AND *opaque-2* MUTANT (RIGHT)

Ordinary corn has large protein bodies compared to *opaque-2*.

obviously, granules are also present; however, they are only about 1/10 μ in diameter. This size is well below the limits of resolution of the light microscope. The electron photomicrograph also reveals the matrix protein, which is the glutelin. A vivid illustration of protein bodies in ordinary corn and their apparent absence in high-lysine corn was observed in a scanning electron micrograph (Fig. 13.6).

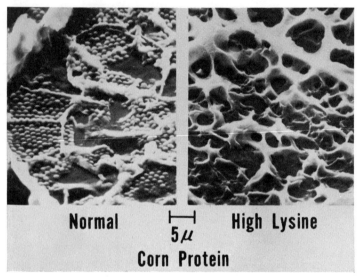

FIG. 13.6. SCANNING ELECTRON MICROGRAPHS OF SUBCELLULAR ENDOSPERM PROTEIN IN ORDINARY CORN (LEFT) AND *opaque-2* MUTANT (RIGHT)

PROTEIN SUBCELLULAR PARTICULATES

Isolation of subcellular particulates was undertaken at the Northern Regional Research Laboratory (Christianson *et al.* 1969). Zone sedimentation was used as a means for isolating and purifying protein bodies, matrix proteins, and soluble proteins from corn.

In immature endosperm cells, starch granules and protein structures appear to be unbound in the cell fluid. Therefore, particulates from homogenates of immature endosperm cells were separated from cell debris and starch granules by sedimentation under gravity. Almost 90% of the total endosperm protein remained in the clarified homogenate.

A typical view of the centrifuge tube after zonal centrifugation of the clarified homogenate from ordinary corn on a 15–17% sucrose gradient is illustrated in Fig. 13.7. After centrifugation, the larger and more dense particles will sediment further into the gradient than will the smaller particles of different size and density. Bright areas of scattered light are due to suspended particulates

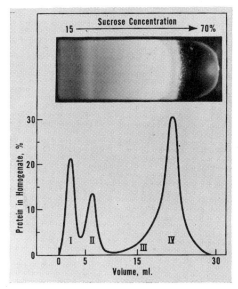

FIG. 13.7. ZONAL CENTRIFUGATION ILLUSTRATION

from the endosperm homogenate. As we progress downward into the tube, the light-scattering is more pronounced with the highest concentration near the bottom. Sequential fractions were taken from different depths of the gradient tube and analyzed for protein. Protein concentrations were assessed by α-amino nitrogen analysis of alkaline-hydrolyzed fractions. The lower portion of the slide shows a plot of the protein concentrate versus position in the tube. Three peaks were located. Two of these protein-rich regions correspond directly to the light-scattering zones within the tube. An uppermost zone (I) contained soluble albumins and globulins which did not exhibit light-scattering. Zone II contained an insoluble protein which scattered light. The protein-containing zone sedimenting most rapidly was skewed on the ascending side enough to indicate the presence of two components. Therefore, this peak was divided into two zones, III and IV.

Insoluble components of each of the zones were collected by ultra-centrifugation and examined directly by scanning election microscopy. Photo-micrographs of the particulates in zones III and IV are shown in Fig. 13.8. The circular bodies as seen in the left photograph are defined as free protein bodies by their shape, size, and structure when compared to the endosperm tissue. These clearly-defined spherical bodies are less distinct in the other micrograph. The relatively larger protein bodies appear clumped together with an amorphous substance. This substance is correlated with the matrix protein, glutelin.

Proteins contained in the particulates were characterized by their amino acid composition and their electrophoretic patterns on starch gels. Zones II and III

FIG. 13.8. SCANNING ELECTRON PHOTOMICRO-
GRAPHS OF PROTEIN PARTICULATES

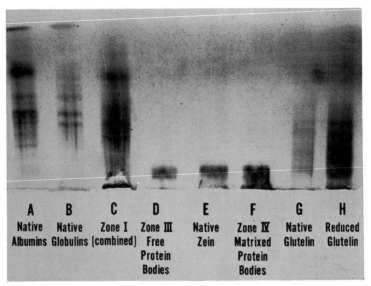

FIG. 13.9. GEL ELECTROPHORESIS OF PROTEIN BODIES

were analyzed directly, whereas in zone IV, the matrixed protein bodies were
first extracted with ethanol to remove zein thereby isolating matrix protein. The
amino acid composition of zone III and the ethanol-soluble portion of zone IV
compare closely with zein isolated from the endosperm meal. Therefore, we have
located zein in the free bodies and in the matrix protein bodies. The structural
proteins representing zone II and the ethanol-insoluble portion of zone IV have
unique amino acid composition but represent in composite glutelin proteins as
isolated by extraction from corn meal.

The mobilities of the proteins in zone I(C), zone III, free protein bodies (D), and zone IV, matrixed protein bodies (F) isolated by extraction are compared in Fig. 13.9. The mobilities of the proteins contained in the free and matrixed protein bodies resemble those of the native zein.

Since we are able to isolate proteins from ordinary corn by these procedures, the method was applied to the mutants of *opaque-2* and *floury-2* endosperms. Similar separation of particulates was evidenced with these endosperms. Confirming earlier studies, zein proteins decreased owing to the reduced protein body size. The amounts of proteins differed greatly in the two mutants.

In *opaque-2* mutant, the protein glutelin in zone II appears to be preferentially synthesized over both the matrix glutelin in zone IV and the zein protein bodies. These changes in the amounts of glutelin proteins contribute to the improved lysine concentrations. In *floury-2* the increased lysine is also dependent on the increased amounts of glutelin proteins.

CHARACTERIZATION OF GLUTELIN

Interest in the structure of corn glutelin protein exists because of its function in maintaining the endosperm matrix and its higher level in high-lysine corn. Because of its insoluble nature, little information previously had been obtained about this protein. The classical method of isolating glutelin from corn endosperm involved preliminary removal of water-solubles and zeins followed by extraction of the residue with 0.1 N sodium hydroxide. As shown in Fig. 13.10, if conditions of extraction are mild, O°, and prompt neutralization follows, the

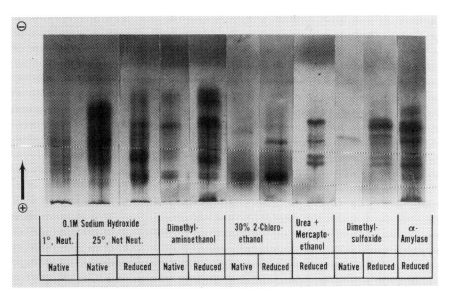

FIG. 13.10. CORN GLUTELIN PREPARED BY VARIOUS METHODS — STARCH-GEL
ELECTROPHORESIS IN 8 M UREA-ALUMINUM LACTATE

glutelin is of high molecular weight and does not appreciably migrate into starch gel during electrophoresis (Nielsen *et al.* 1970). Room temperature extraction yields more protein which migrates indicating fragmentation of the native structure. Alkali treatment results in considerable disulfide cleavage, deamidation, and other changes. On reducing and alkylating this protein, the derivative migrates electrophoretically but gives an indistinct pattern. Other solvents were tried but gave poor extraction unless reducing agents were present, suggesting that glutelin proteins are extensively cross-linked by disulfide bonds. Urea and mercaptoethanol served as a suitable solvent but the product was contaminated with solubilized starch. The starch was eliminated from the residue by treating with highly purified α-amylase. The remaining insoluble glutelin-rich material was reacted with mercaptoethanol in 8 M urea and alkylated to yield the protein derivative shown in the last pattern. This procedure was used to investigate the structure of glutelins from different sources. Figure 13.11 compares the starch-gel electrophoresis patterns of reduced alkylated proteins from ordinary and high-lysine corns. Only small differences in numbers of bands and intensities of components occur between the proteins extracted from the ordinary and high-lysine corns (Paulis *et al.* 1969). The level of individual protein components rather than difference among protein groups in these grains appears to be responsible for difference in lysine level. These data show that the glutelin appears to be composed of distinctly different polypeptide chains than those of albumins, globulins, and zeins. Most of the glutelin bands are intermediate in mobility between water-solubles and zeins. Possibilities exist of some contamination of glutelin with zein components and of the possible identity of the slowest globulin with the fastest glutelin component.

To test these possibilities the reduced-alkylated glutelin proteins were fractionated by chromatography on Sephadex G-200 and on other gel filtration

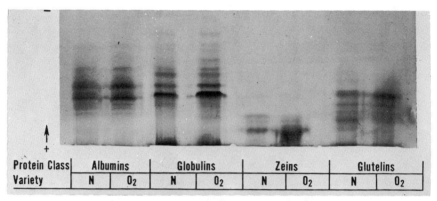

Protein Class	Albumins		Globulins		Zeins		Glutelins	
Variety	N	O₂	N	O₂	N	O₂	N	O₂

FIG. 13.11. STARCH-GEL ELECTROPHORETIC PATTERNS OF REDUCED AND ALKYLATED PROTEINS

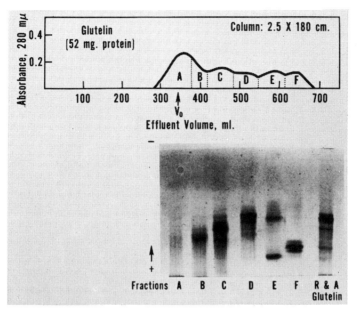

FIG. 13.12. (ABOVE) CHROMATOGRAPHIC SEPARATION OF REDUCED-ALKYLATED CORN GLUTELIN ON SEPHADEX G-200 AND (BELOW) STARCH-GEL ELECTROPHORESIS PATTERNS OF FRACTIONS

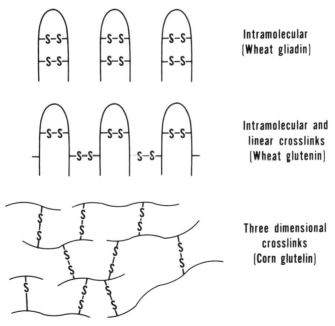

FIG. 13.13. DISULFIDE BONDS IN CEREAL PROTEINS

media. Using 6 M guanidine hydrochloride as solvent gave separation of the unfolded protein chains according to their molecular weight as shown in Fig. 13.12. The separation yields electrophoretically distinct components. Glutelin evidently is composed of a mixture of proteins differing in size and amino acid composition.

As shown in Fig. 13.13, glutenin contains a number of proteins linked by extensive disulfide cross bonds. Wheat gliadin by contrast contains primarily intramolecular disulfide bonds and glutenin contains only limited intermolecular links.

DRY MILLING *Opaque-2* CORN

Since new varieties of high-lysine corn offer an opportunity to produce protein concentrates of excellent nutritional value, a new approach has been taken to disrupt native agglomerates of starch and protein in high-lysine corn flour.

Preconditioning high-lysine corn flour with a buffer solution similar in composition to the original cell fluid releases protein from starch. The flour can either be slurried in a large volume of buffer or conditioned to 26% moisture, equilibrated, and finally dried at room temperature under vacuum. Separation of starch and protein in high-lysine corn flour achieved by this buffer treatment is illustrated photographically (Fig. 13.14). Scanning electron microscopy vividly demonstrates the release of protein from starch. Before treatment, protein adheres to the starch granules in layers. During treatment, these protein layers slide off the starch granules in sheaths and leave the starch granules free and smooth. Thus, the protein fragments and free starch granules can be separated directly be sieving the treated flour (Christianson *et al.* 1971).

Changes that occur in size and composition of agglomerates in the flour during the buffer treatment can also be demonstrated by air classification since

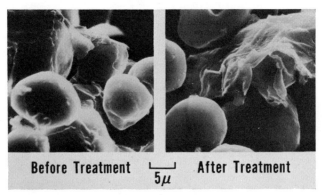

Before Treatment |—| **After Treatment**
5μ

FIG. 13.14. HIGH LYSINE CORN FLOUR BEFORE AND AFTER TREATMENT
WITH ISOTONIC BUFFER

protein and weight shifts differ from those of untreated flour. Air classification of untreated and treated flours pin-milled under similar conditions is plotted in Fig. 13.15. Protein shifts are improved after air classification of buffer-treated high-lysine flour as compared to those for untreated flour. Almost 70% of the protein in the flour can be separated into high-protein fractions as compared to only 33% in untreated flour.

Ordinary corn flour was also buffer treated. In this experiment, the matrix

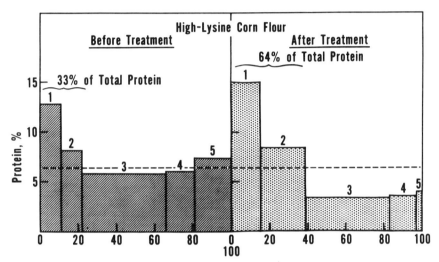

FIG. 13.15. AIR CLASSIFICATION OF BUFFER-TREATED HIGH-LYSINE CORN FLOUR

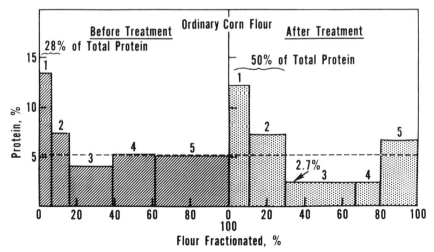

FIG. 13.16. AIR CLASSIFICATION OF BUFFER-TREATED ORDINARY CORN FLOUR

protein is broken into small protein fragments and subjected to pin-milling and air classification. Figure 13.16 compares the results of the separation of protein and starch obtained by pin-milling and air classification of untreated and treated ordinary corn flour. The protein shift is dramatically improved as compared with the shift achieved with the untreated flour. More than 50% of the protein is shifted to the fine fractions 1 and 2 as compared to only 28% in the untreated flour.

High-lysine corn products obtained by air classification may find some food uses. Present varieties of high-lysine corn are especially suitable for milling, fragmenting, and air classifying to gain protein concentrates of good nutritional quality.

Research at the Northern Regional Laboratory on *opaque-2* by conventional corn dry milling has recently been completed (Brekke *et al.* 1971). *Opaque-2* corn was processed into products of acceptable fat content by ordinary dry-milling equipment. A degerminator with a mild abrasive action was required. Gravity table separators could not be used because the germ fraction had a density between those of the germ and grit particles. Several minor changes were made in the flow of the rolls-and-grading system to produce prime products of acceptable fat content. Additional aspiration steps were required. The best degerming results were obtained by corn with moisture contents of 23% and higher.

The prime product spectrum of the *opaque-2* corn was considerably different from that of the ordinary dent corn. No flaking grits were produced while a considerable amount of table grits, meal, and flour were obtained. The protein content of the prime products varied between 6.2 and 10.4% (DB) and the prime product mix contained less protein than the mix prepared from a dent corn of above average protein content.

TABLE 13.1

YIELDS AND PROXIMATE ANALYSES
OF CORN AND HOMINY FEED COMPONENTS

Fraction	Estimated Average Annual Yield % of Grain[1]	Protein (%)	Fat (%)	Fiber (%)	Ash (%)
Yellow dent corn	100	9.0	3.8	2.4	1.7
Fine feed					
(degerminator fines)	14.7	11.0	4.1	2.8	1.6
Oil cake (expeller)	10.4	19	5.5	5.4	8.0
Hull fraction,					
bran, and bran meal	8.1	8.8	7.4	8.5	2.1
Hominy feed	33.2	11.5	5.5	4.7	2.9

[1] Average of three corn dry-milling plants.

NUTRITIONAL VALUE OF CORN DRY-MILLED BY-PRODUCTS

Because corn has traditionally been regarded primarily as a source of calories, its processing by dry milling has been directed to optimum yields of endosperm products — such as grits, meal, and flour — and to good oil recoveries from germ. One-third of the dry-milling product is fractionated to low-cost hominy feed (Table 13.1). The hominy feed fractions, the degerminator fines, and germ oil

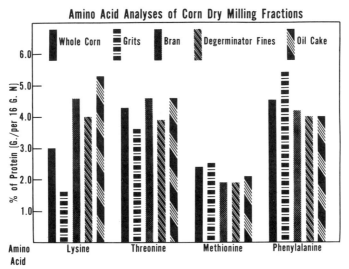

FIG. 13.17. AMINO ACID ANALYSES OF CORN DRY MILLING FRACTIONS

TABLE 13.2

EFFECT OF PROCESSING GERM AND HOMINY FEED
ON PROTEIN EFFICIENCY RATIO (PER)

	Average of 10 Rats for 4 Weeks		
Protein Source	Weight Gain (Gm)	Food[1] Consumption (Gm)	Corrected (PER)
Casein	92	307	2.50
Corn germ		327	2.19
Expeller germ meal cake	49	265	1.54
Solvent-extracted hominy feed	89	413	1.74
Expeller-processed hominy feed	65	341	1.42

[1] 10% protein in diet.

cake are higher in protein than corn and contain high levels of other nutrients including minerals and vitamins. The germ fraction is noteworthy because it contains about 20% protein (Wall *et al*. 1971). Figure 13.17 compares the composition of some essential amino acids in the protein of whole corn and grits to that of the by-product fractions. The level of lysine and threonine in germ and degerminator fines exceeds that of corn and grit protein. Methionine and phenylalanine levels are only slightly lower in germ than in whole corn and grits. While grits and whole corn contain excessive levels of leucine leading to dietary imbalances, leucine levels in germ are satisfactory. The protein of whole germ is readily digested and has good biological value as shown in Table 13.2.

The demand for corn oil has resulted in almost all dry-milled germ being processed to recover the corn oil. Subjecting the corn to high temperature expeller process reduce the nutritional value of the protein as indicated by low protein efficiency ratio values. Oil can be extracted from germ more efficiently by solvent extraction, which yields a germ meal of high nutritional quality. At present, only one dry-milling plant in the United States employs solvent extraction. At the Northern Regional Research Laboratory, we are exploring the production of improved products from corn germ which may find uses in food products (Gardner *et al*. 1971). Preliminary results look promising, but considerably more research is needed.

BIBLIOGRAPHY

BREKKE, O. L., GRIFFIN, E. L., and BROOKS, P. 1971. Dry milling of *opaque-2* (high-lysine) corn. Cereal Chem. *48*, 499-511.

CHRISTIANSON, D. D. *et al*. 1969. Isolation and chemical composition of protein bodies and matrix proteins in corn endosperm. Cereal Chem. *46*, 372-381.

CHRISTIANSON, D. D., STRINGFELLOW, A. C., and BURDICK, D. C. 1971. Endosperm fragmentation of ordinary and high-lysine corn dry-milled products improved by isotonic buffer conditioning. Cereal Chem. *48*, 558-567.

DUVICK, D. N. 1961. Protein granules of maize endosperm cells. Cereal Chem. *38*, 374-385.

GARDNER, H. W. *et al*. 1971. Food products from corn germ: Evaluation as a food supplement after roll-cooking. J. Food Sci. *36*, 640-644.

INGLETT, G. E. 1970. Corn in perspective. *In* Corn: Culture, Processing, Products, G. E. Inglett (Editor). Avi Publishing Co., Westport, Conn.

MERTZ, E. T., BATES, L. S., and NELSON, O. E. 1964. Mutant gene that changes protein composition and increases lysine content of maize endosperm. Science *145*, 279-280.

NIELSEN, H. C., PAULIS, J. W., JAMES, C., and WALL, J. S. 1970. Extraction and structure studies on corn glutelin proteins. Cereal Chem. *47*, 501-512.

PAULIS, J. W., JAMES, C., and WALL, J. S. 1969. Comparison of glutelin proteins in normal and high-lysine corn endosperms. J. Agr. Food Chem. *17*, 1301-1305.

WALL, J. S., CAVINS, J. F., JAMES, C., and INGLETT, G. E. 1971. Economic evaluation of hominy feed. Feedstuffs *43*, 18-19.

WOLF, M. J., KHOO, U., and SECKINGER, H. L. 1969. Distribution and subcellular structure of endosperm protein in varieties of ordinary and high-lysine maize. Cereal Chem. *46*, 253-263.

CHAPTER 14

W. Bushuk | **Ultrastructure Related to Wheat Processing and Products**

In making bread from wheat we are concerned with four products: the wheat kernels themselves; the flour milled from them; the dough prepared from the flour; and the bread baked from the dough. Each of these has a macrostructure apparent to the eye, a microstructure visible under suitable levels of magnification, and an ultrastructure that can be observed in part with more sophisticated forms of microscopy, but which is largely inferred from diverse experimental data.

In this paper we are mainly concerned with the ultrastructure, but must often relate this to the microstructure. Indeed, there is obviously no clear cut separation of microstructure and ultrastructure. Moreover, in this discussion we shall be mainly concerned with the proteins of bread wheats. Our four products, wheat, flour, dough and bread, will be discussed in that order.

WHEAT

Wheat grain is the fruit produced by a member of the grass family *(Gramineae)*. The grain, also called the kernel, consists of three relatively distinct parts: bran, embryo or germ, and endosperm. The bran comprises the seed coat, which encloses the germ and the endosperm, and the outside fruit coat made up of several distinct layers (MacMasters *et al.* 1954). Wheat kernels vary in length from 4 to 10 mm. and in shape from almost spherical to oblong. Their size and shape depend on variety and on location in the spike and spikelet (Bayles and Clark 1954). On one side, the kernel has a longitudinal crease which can extend to varying depths. The germ is at one end and the brush at the other. These gross structural features complicate the milling of wheat into flour. Both the crease and the brush can entrap dirt and dust which must be removed by extensive scouring and washing.

From the utilization viewpoint, most of the attention should be focused on the endosperm of the kernel. Its ultrastructure can, to some extent, be inferred from electron microscopy studies of the developing wheat endosperm by Buttrose (1963) and Jennings *et al.* (1963) but mostly from a large variety of analytical investigations.

The endosperm cells originate from the fusion of two nuclei of the embryo sac and one nucleus of the male sperm after fertilization which occurs immediately after flowering. Rapid outward growth of endosperm cells, by cell division, occurs during the next 12 days and there is essentially no further cell division after 14 days after flowering. From the time when the first endosperm cells are detected to maturity (approximately 30 days after flowering, depending on variety) the endosperm cells continually expand to accommodate the starch and the protein.

Protein synthesis in the kernel commences at about day 5 after flowering. Initially the synthesis is slow but becomes quite rapid after 14 days when the acetic acid soluble or storage proteins are synthesized much faster than the pyrophosphate-buffer-soluble proteins. The storage protein is deposited almost entirely in electron-dense protein bodies varying in size from 1 to 15 μ (Fig. 14.1). Graham *et al.* (1963) found that the small protein bodies contain more gliadin than the large ones. The protein body is enclosed within a parallel array of fine lipoprotein membranes which are presumed to be part of the endoplasmic reticulum of the endosperm cell. Isolated protein bodies contained 75-80% protein, mostly of the acetic acid-soluble (gluten) type as can be shown by starch gel electrophoresis. The nature of the remaining portion of the protein bodies has not been determined although fixation with potassium permanganate produced an electron-dense appearance indicating the presence of lipid.

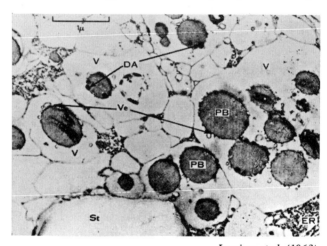

Jennings et al. (1963)

FIG. 14.1. ELECTRON MICROGRAPH OF WHEAT ENDOSPERM CELL AT 14 DAYS AFTER FLOWERING SHOWING PROTEIN BODIES (PB) IN VACUOLAR STRUCTURES (V), STARCH GRAINS (St), AND ENDOPLASMIC RETICULUM (ER)

Protein production in the wheat kernel is limited by the amount of available nitrogen in the soil at different stages of kernel development, and is also a function of soil moisture, mineral nutrients, and climatic factors such as the total and distribution of rainfall and temperature (Barmore 1948). Differences in protein content among similar varieties grown under comparable conditions are small. Under closely controlled greenhouse conditions we have found that the bread wheat variety, Manitou, had a significantly higher protein content at maturity than the durum (Stewart 63) or the soft white winter (Talbot) varieties (Bushuk and Wrigley 1971). In relation to breadmaking quality, both gliadin and glutenin (main proteins of gluten) are synthesized at about the same time and this synthesis continues to maturity (Fig. 14.2); the amount of water-soluble protein, in relation to total protein decreases during maturation. Both gliadin and glutenin are synthesized at about the same rate in the developing endosperm.

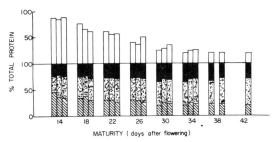

FIG. 14.2. PROPORTIONS OF GRAIN PROTEIN AS GLUTENIN (SOLID COLUMN), GLIADIN (STRIPPLED), ALBUMIN (CROSS-HATCHED), AND SMALL UV-ABSORBING MOLECULES (OPEN) FOR DEVELOPING GRAIN

Manitou, hard red spring wheat (left); Talbot, soft white winter wheat (center), and Stewart 63, durum wheat (right).

Starch synthesis begins at about day 5 after flowering, reaches the maximum rate between day 12 and day 20 and then levels-off to maturity. Initially starch is synthesized faster than protein. The larger ellipsoidal starch granules are formed first and the smaller spherical granules during the later stage of development. Medcalf and Gilles (1965) isolated and characterized the starch from 17 varieties of wheat of four different classes including hard red spring, hard red winter, durum and soft wheat. Differences in granule size among classes were insignificant; however durum starches contained slightly more amylose, had higher water binding capacity, and showed greater rates of iodine absorption than the starches of the other wheat classes.

Synthesis of endosperm lipid after day 14 is much slower than the synthesis of protein and starch. Daftary and Pomeranz (1965) observed both quantitative and qualitative changes in the lipids during maturation. Free lipids decreased from about 2 to 1.7% whereas bound lipids decreased from 1.9 to 1.2%. During maturation, the most marked qualitative change was the consistent decrease in free fatty acids, and in mono- and diglycerides.

From studies of the developing kernel it can be inferred that the mature endosperm cell is tightly filled with adjacent starch granules and protein bodies. However, protein bodies have not yet been detected in mature wheat endosperm (Seckinger and Wolf 1970).

In relation to processing, it is pertinent to comment on the microstructure of the endosperm cell and differences in the structure of different wheats as discerned by microscopy. The physical structure of endosperm cell walls has not been resolved, although it is well established that chemically they are mainly pentosans. Larkin *et al.* (1952) found that in U.S. northwest white wheats, the cell walls near the crease were considerably thicker than the walls in the centre of the kernel (7.3 μ vs. 2.6 μ). Wheats of best milling quality were those that had the thinnest cell walls near the aleurone layer. Seeborg and Barmore (1952) suggested that wheats of good milling quality contained a lower proportion by weight of endosperm cell walls. Sandstedt (1954) postulated that differences in endosperm hardness depends on the protein that fills the space between the starch granules. By ordinary microscopy he showed that, during milling, hard wheat endosperm cells tend to separate as individual cells with breaks following the cell walls. However, essentially all of the cells are partially disrupted, and the starch granules are subjected to severe abrasive damage. The union between the starch and the protein matrix is so strong that hard wheat flour contains very few free starch granules and comprises mainly endosperm cells and cell fragments. In soft wheats, the binding between starch granules is apparently extremely weak so that when these wheats are milled the cells disintegrate easily and release large proportions of undamaged starch granules. Soft wheat flour, therefore, comprises almost entirely free starch granules and particles that are high in protein (Sandstedt and Fleming 1952).

Medcalf *et al.* (1968) found that durum wheat pentosans contained more arabinose suggesting a higher degree of branching. But these workers could not account, on this basis, for the difference in the physical properties of the endosperm of durum and hard red spring wheat. Further studies are needed to delineate the factors that determine grain hardness which, in turn, is directly related to millability.

FLOUR

Flour milling involves the separation of the germ and the bran from the endosperm and the reduction of the endosperm to a fine powder. According to standard definition the flour particles must be smaller than 149 μ. The milling process involves three major operations: breaking, sifting and purifying, and reduction. Because of the awkward shape of the grain, and the firm attachment of the bran to the endosperm, the grain must first be opened up by passing through relatively coarse corrugated break rolls. The initial crushing and shearing of the grain depends on its intrinsic structure and on the modification of the structure by controlled wetting called tempering.

An extremely important property of the wheat kernel at the milling stage is the toughness of the bran, which must not shatter into fine particles but must remain in large fragments so that it can be readily separated from the flour by subsequent sieving and purification. The amount and composition of flour produced at various stages of the milling process depends on the hardness of the wheat endosperm. This affects the physical properties of the flour, such as starch damage, and thereby its subsequent behaviour in the breadmaking process. Within the same type of wheat, hardness depends inversely on protein content. Wheats of different classes, of about the same protein content, can differ markedly in hardness, and therefore in milling properties (Kent and Jones 1952; Sandstedt and Fleming 1952). This is obviously a genetic characteristic but the fundamental chemical and physical properties that give rise to this difference are virtually unknown.

Low extraction flours will have essentially the same chemical composition as the endosperm. On a dry basis, the major components of an average bread flour are: 75% starch, 16% protein, 3% fat, and 3% pentosans. The purity of the flour, which reflects the efficiency of the milling process is usually measured by its ash content. The ash content of endosperm is about 0.48% (dry basis) and about 2% of the combined bran layers. Unusually high ash content and poor color of the flour indicates excessive contamination by bran and an inefficient milling. Milling yield is indirectly correlated with the fibre content of the wheat (Moss and Stenvert 1970).

Flour particles vary in shape, size and composition (Fig. 14.3) as described by Jones *et al.* (1959). Most commercial flours have a bimodal particle-size distribution with the first peak at about 20 μ and the second at about 70 μ (Williams 1970). The composition of flour particles can vary from almost pure protein or starch to that of the endosperm represented by the endosperm chunks. Differences in composition leads to differences in density and hence it is possible to fractionate flour by flotation in air currents. This technique, called air classification, can produce flours differing widely in particle size and protein content. Soft wheat flours give a much wider protein distribution than flours

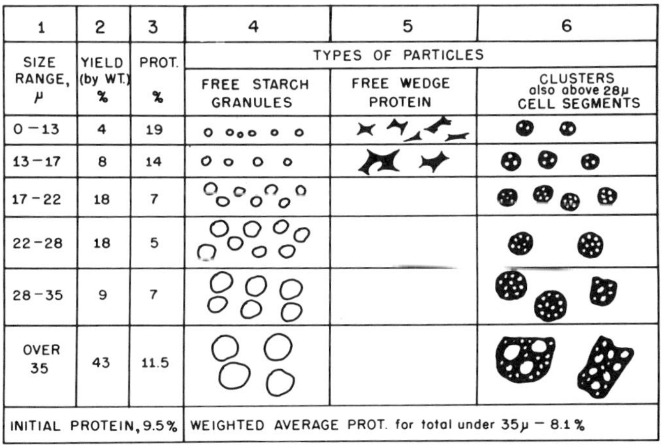

1	2	3	4	5	6
SIZE RANGE, μ	YIELD (by WT.) %	PROT. %	TYPES OF PARTICLES		
			FREE STARCH GRANULES	FREE WEDGE PROTEIN	CLUSTERS also above 28μ CELL SEGMENTS
0 – 13	4	19			
13 – 17	8	14			
17 – 22	18	7			
22 – 28	18	5			
28 – 35	9	7			
OVER 35	43	11.5			
INITIAL PROTEIN, 9.5%			WEIGHTED AVERAGE PROT. for total under 35μ – 8.1%		

Adapted from Jones et al. (1959) with permission

FIG. 14.3. FLOUR PARTICLES IN A SOFT WHEAT FLOUR OF 9.5% PROTEIN

from hard wheats (Mattern *et al.* 1970) confirming the suggestion of Sandstedt (1954) that the separation of the starch and protein in soft wheat endosperm is much more efficient.

On the basis of microscopic observation and density fractionation using non-aqueous media, Hess (1954, 1955) postulated the existence of two relatively different proteins in wheat endosperm: wedge and adhering proteins. The adhering proteins were considered to be strongly cemented to starch granules by phospholipid (Hess and Mahl 1954). Although differences in amino acid composition have been demonstrated between these two proteins (Hess and Hille 1961), they behave similarly. Recent electron microscopic work of Seckinger and Wolf (1970) did not confirm the original hypothesis of Hess that the adhering proteins contained relatively high quantities of bound phospholipid.

The study of Seckinger and Wolf (1970) is significant from another point of view. For the first time, it was clearly demonstrated, by electron microscopy, that there were differences in the ultrastructure of the protein particles from the soft and hard wheat flours. Hard wheat protein particles were smaller and considerably more compact that the particles from soft wheat.

Studies with the scanning electron microscope of (Aranyi and Hawrylewicz 1969) confirmed the cellular structure of wheat endosperm and provided additional new information on the ultrastructure of flour and dough. Fig. 14.4 shows two photographs from that study. On the left are shown chunks of the inner endosperm at 525X magnification. According to the authors, this picture shows both large and small starch granules embedded in proteinaceous material. It also shows evidence of flakes of non-starchy material, presumably interstitial protein. On the right is outer endosperm chunks magnified 3800X. Here, again, we see that the starch granules are covered with films of rather amorphous material.

Since the first comprehensive fractionation study of wheat proteins by Osborne (1907), at the beginning of this century, there have been many

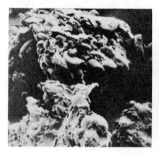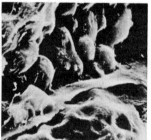

Aranyi and Hawrylewicz (1969)

FIG. 14.4. WHEAT ENDOSPERM PARTICLES BY SCANNING ELEC-
TRON MICROSCOPY (LEFT X 263; RIGHT X 1900)

attempts to relate qualitative and quantitative differences in solubility of flour proteins to differences in breadmaking quality. So far, these studies have provided only a partial answer.

A modified Osborne fractionation technique has been used quite successfully in our laboratory. When applied to widely different cereal grains or wheat types (Chen and Bushuk 1970), this technique gives a clear differentiation of the materials which can, in turn, be used to account for the superior breadmaking potential of the hard red spring wheat of high quality such as the variety Manitou (Fig. 14.5). Differences among common wheats of widely different baking quality (Fig. 14.6) were considerably smaller (Tanaka and Bushuk 1972). The trend that was actually observed with three inferior varieties compared with two better hard red spring wheats showed that the inferior varieties contained somewhat less gliadin and more of the protein that was insoluble in the four solvents used in sequence.

Fractionation of wheat endosperm can also be accomplished on the basis of molecular weight by molecular sieve chromatography using a strongly dissociating solvent (Meredith and Wren 1966). With this technique Bushuk and Wrigley (1971) showed that at an early stage of kernel development (e.g. 14 days after flowing) the proteins of three widely different wheats had essentially the same protein elution profiles (Fig. 14.7). At 14 days, the patterns for the three wheats were the same as the pattern shown for Talbot. As can be seen, there are marked differences at maturity especially in the glutenin component. Whether

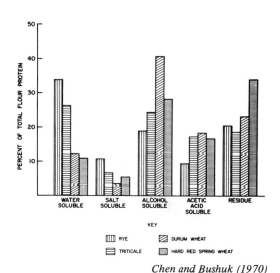

Chen and Bushuk (1970)

FIG. 14.5. DISTRIBUTION OF FLOUR PROTEINS AMONG FIVE SOLUBILITY FRACTIONS FOR FOUR GRAIN SPECIES

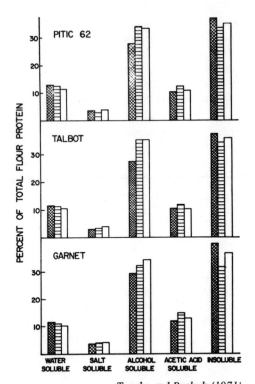

Tanaka and Bushuk (1971)

FIG. 14.6. DISTRIBUTION OF FLOUR PROTEINS
AMONG FIVE SOLUBILITY FRACTIONS

Pitic 62, Talbot, and Garnet compared with values for
Manituo (horizontal hatching) and 11-463A (open).

these analytical differences can be correlated to differences in ultrastructure remains to be investigated.

Detailed chemical and physical structures of the main protein fractions of wheat endosperm (gliadin and glutenin) have not been worked out although some progress has been made in this area. Chromatographic studies of Rohrlich and Schulz (1960) of peptic hydrolysates showed no definite differences between the two proteins. Similar results were obtained by Ewart (1966) with the fingerprinting technique. On the other hand, work by Bietz and Rothfus (1970) showed up a number of significant differences between the peptides obtained from gliadin and glutenin by peptic hydrolysis. These workers postulated, mainly on the basis of the effect of the reduction of disulfide (-S-S-) bonds, that the gliadin molecule is quite compact comprising a single polypeptide chain with intramolecular -S-S- crosslinkages.

It is the opinion of some cereal chemists, including the writer, that the key to

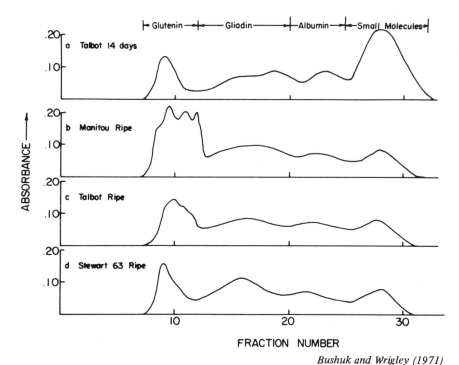

Bushuk and Wrigley (1971)

FIG. 14.7. GEL FILTRATION OF AUC EXTRACTS OF GROUND GRAIN ON SEPHADEX G-150

Elution profiles (absorbance at 280 nm) for (a) Talbot (undried) harvested 14 days after flowering, (b) ripe Manitou, (c) ripe Talbot, and (d) ripe Stewart 63.

varietal differences in baking quality lies in the structure of the glutenin and the insoluble protein of the endosperm. With the advent of new techniques that have been developed for studying the physiochemical structure of protein, the molecular structure of glutenin is being vigorously investigated in many laboratories. A number of different structurers, involving -S-S- crosslinkages in one, two and three dimensions have been proposed (Ewart 1968; Krull and Wall 1969). The linear structure postulated by Ewart (1968) is in essence equivalent to the structure postulated earlier by the Northern Regional Research Laboratory group (Beckwith et al. 1963). According to this latter hypothesis, glutenin was considered to be a polymer of gliadin-like units linked together by -S-S- bonds (Fig. 14.8). The true molecular structure of glutenin and its relation to the ultrastructure of gluten and dough remains to be verified.

The most promising results on the varietal and class differences in glutenin were published by Huebner (1970). Glutenins from different wheat varieties showed widely different responses to precipitation with salt indicating that

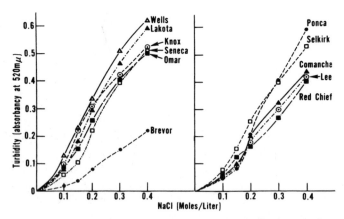

(Upper) Krull and Wall (1969)
(Lower) Ewart (1968)

FIG. 14.8. HYPOTHETICAL MODELS FOR THE STRUC-
TURE OF WHEAT GLUTENIN

rheological behaviour of gluten in dough would be extremely sensitive to changes in ionic strength (Fig. 14.9). Differences in sensitivity to salt of different flours are well known in the baking trade, however at the present times these cannot be explained on the basis of amino acid composition (Tkachuk 1966). Furthermore, reduced and alkylated glutenins showed significant variations in electrophoretic patterns among varieties of the same classes and greater differences among wheats of different classes.

Other components of wheat endosperm that are particularly relevant to processing and products are the enzymes, the lipids and the starch. These will be discussed briefly.

Small amounts of both α and β-amylase are beneficial to the functional quality of flour. They produce the fermentable sugars from starch for proper

Huebner (1970)

FIG. 14.9. SALT PRECIPITATION OF GLUTENIN FROM DIFFERENT
WHEAT FLOURS IN 2 M UREA, 0.03 ACETIC ACID

yeast action. Flour from sound wheat has very little α-amylase and therefore the miller usually adds supplements in the form of malted barley flour or malted wheat flour. Wheat that has suffered sprout damage is excessively high in α-amylase activity and this can lead to lowered water absorption in the dough, and to poor grain and weak crumb in the bread (Tipples *et al.* 1966). There is no practical way of reversing this harmful effect of excessive α-amylase activity. The effect of β-amylase on the physical properties of dough is considerably less drastic and therefore normal fluctuations in its activity are not too significant.

The presence of proteolytic activity in bread flours has been recognized for many years, but it has received relatively little attention in relation to functional properties of normal bread flours. Although proteases may not be involved in the improver reaction of oxidizing agents, as postulated by Jorgenson (1936), they may well have a direct effect on the stability of the gluten during mixing and fermentation (McDonald 1969). Recent studies of Bushuk et al. (1971) indicate that wheats that have weak glutens also have significantly higher proteolytic activities. Furthermore, the activity per grain (kernel) remains essentially constant throughout kernel development and maturation. Accordingly all wheat flours will have a significant proteolytic activity. When flour and water are mixed into a dough, the attack by the native proteases on the insoluble protein of the gluten could be extremely important functionally.

Several lipoxygenases have been detected in flour (Graveland 1970). They catalyze the oxidation of unsaturated fatty acids. The intermediates of this oxidation can be involved in two functionally important reactions. They can oxidize the yellow pigments of the flour and thereby produce a desirable bleaching. Wytase[R], an enzymatically active preparation from soybean, is commercially added to bread dough primarily for this purpose. Secondly, the oxidation products, if present at sufficient level, can oxidize the sulfhydryl groups in the flour proteins (Tsen and Hlynka 1963) and thereby limit the functional role of these groups.

Wheat starch can no longer be considered as an inert filler in dough. Replacement of wheat starch by other starches does not produce satisfactory bread (Rotsch 1953, 1954). There must be specific interactions between the starch and other flour components that are important to breadmaking quality. During milling, starch granules are physically damaged. The damaged starch absorbs at least five times as much water as does undamaged starch (Sandstedt 1954) and can thus have a very important effect on the water absorbing and retaining capacity of dough and bread. The gelatinization properties of damaged starch during baking are also quite different from those of granular starch (Yasunaga *et al.* 1968) and this type of starch thus affects the ultra structure and the quality of the bread.

Damaged starch is more readily attacked by amylases (Sandstedt 1954) — presumably because water, carrying the amylases with it, can penetrate the disrupted granules faster. The digestion, however, is slower than water penetration. A certain level of damaged starch is essential for breadmaking; too much is detrimental. The amount that is considered too high depends on the amylase activity of the flour (Tipples *et al*. 1966). It is obvious that optimum breadmaking quality, by a specific baking process, requires a very critical balance among several interrelated factors.

DOUGH

Dough can be considered as a complex colloidal system comprising solid components of the flour and added ingredients, water, and air. In macrostructure it can be likened to a solid foam. The continuous network of the dough, which determines its rheological properties, is the gluten that is formed when flour is mixed with water. Proper mixing is therefore required not only to incorporate all the ingredients but also to develop the optimum structure of the hydrated gluten. Although protein is the main component of the gluten, other minor components of flour, especially certain lipids (Pomeranz 1967) and pentosans (Myhre 1970) may play an important functional role in the strucure of gluten.

The first fundamental study of the ultrastructure of dough and gluten was published by Traub *et al*. (1957). X-ray diffraction patterns showed that wheat flour was characterized by 47Å spacings which disappeared when the lipid materials were extracted with cold acetone. However, the extracted flour retained its original ability to form gluten, indicating that the extracted lipid was not essential for gluten formation. Hess and Mahl (1954) had postulated earlier that the protein that adheres strongly to starch is held by an adhesive layer of phospholipids. Later X-ray work of Grosskreutz (1960, 1961) led to the lipoprotein model for the structure of gluten in dough (Fig. 14.10). Recent studies of Daniels *et al*. (1970 and others of same series) and Pomeranz (1967) demonstrated a functional role for the lipids in the structure of gluten. Furthermore, Hoseney *et al*. (1970) postulated that the matrix of dough is a gliadin-glycolipid-glutenin complex involving both hydrogen and hydrophobic bonding. They suggested that gliadin-glycolipid interactions involve hydrogen bonds whereas glycolipid-glutenin interactions are mainly hydrophobic. This is an extremely significant observation because it implicates the glycolipids directly and specifically in the structure of gluten. It is somewhat surprising that the interactions between gliadin and glycolipid should turn out to be hydrogen bonds. On the basis of the higher hydrophobicity of gliadin versus glutenin, it would be expected that the interactions between the former protein and glycolipid would be hydrophobic. In addition to hydrogen and hydrophobic bonding, Fullington (1967) from the Western Regional Laboratory has suggested a relatively specific ternary complex between the phospholipids and proteins involving ions such as Ca^{++}, Mg^{++}, etc. In such complexes, trace quantities of these ions may produce marked effects on the physical properties of doughs.

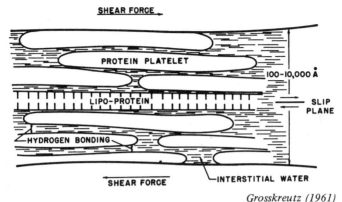

Grosskreutz (1961)

FIG. 14.10. GROSSKREUTZ LIPOPROTEIN MODEL FOR GLUTEN
STRUCTURE

In traditional breadmaking, relatively long periods of fermentation were considered essential for the development of the optimum dough structure. Modern baking technology has shown that vigorous mixing can replace fermentation to a large degree in the development of the optimum gluten structure in dough. This mechanical development may involve conformational and structural changes in the dough proteins, facilitated by -S-S- interchanges, and physical aggregation-dissaggregation. Indirect evidence has been obtained in several laboratories that suggests that -SH groups might be involved in the -S-S- interchanges. The low molecular weight -SH compounds such as glutathione (Sullivan 1948) and cysteinyl glycine (Tkachuk 1969, 1970) could be extremely important in these interchanges because of their higher solubility and mobility in a concentrated system such as dough. Furthermore, -SH groups are known to be directly involved in stabilizing the quaternary structure of a number of enzymes (Klotz et al. 1970). A similar function for -SH groups in the functional structure of gluten can be postulated and should be investigated.

The French workers have recently shown that the transmission electron microscope can be used to observe differences in doughs from different flours at different stages of development by mixing (Crozet et al. 1966). During mechanical development, gluten films showed preferred orientation in the direction of the stretching forces. At the level of resolution used, these workers did not obtain any evidence to support the lipoprotein model postulated by Grosskreutz (1961). However, their work confirmed the results of Sandstedt et al. (1954), obtained much earlier by ordinary microscopy, of the orientation of starch granules in gluten films.

Figure 14.11 shows dough surfaces taken with the scanning electron microscope at two levels of magnification (Aranyi and Hawrylewicz 1969). The lower magnification picture shows a relatively uniform distribution of starch

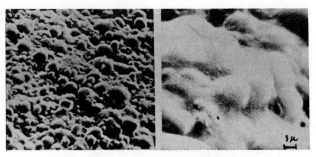

Aranyi and Hawrylewicz (1969)

FIG. 14.11. DOUGH SURFACES BY SCANNING ELECTRON
MICROSCOPY (LEFT X 200; RIGHT X 2000)

granules. At the higher magnification the continuous film, presumed to be gluten, becomes quite evident. However, even at the ultimate level of resolution of this type of microscope (200Å), no regular ultrastructure in the continuous matrix of the dough could be discerned.

The present state of knowledge supports the hypothesis that the viscoelastic properties of dough depend on the critical ratio of gliadin, the viscous component, and glutenin, the elastic component in the dough. The physical properties of these protein components depend on their molecular structure and intermolecular interactions. All other components, and intercomponent interactions, are important, but to a lesser extent. The proper rheological properties must be developed by just the right balance of ingredients and mixing.

During the final stages of the dough phase of the breadmaking process, the volume of the dough expands under the influence of the carbon dioxide produced by fermentation (Sandstedt *et al.* 1954). The process of expansion is hastened later by the rising temperature of the dough in the oven. The structure and strength of the dough must be such that it will expand under this internal pressure without collapsing. In final stages of baking, the dough becomes a rigid loaf by the formation of the hard crust and the relatively rigid internal structure of thermally denatured gluten and deformed, partially gelatinized starch granules.

BREAD

The classic work on the microstructure of a ripe bread dough and bread crumb is the microscopic study of Sandstedt *et al.* (1954). These workers showed that dough has no regular structure at the beginning of the mixing, but during mixing and fermentation, proteinaceous films containing starch granules formed around the gas bubbles (Fig. 14.12).

During baking, the dough expands quite rapidly as a result of thermal expansion of the gas bubbles, and the vaporization of water and the alcohol produced

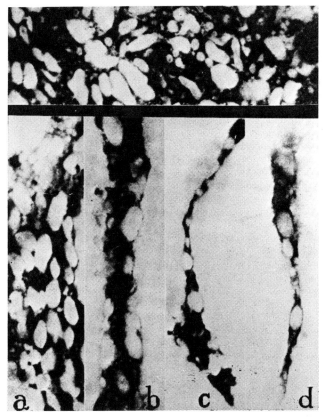

Sandstedt et al. (1954)

FIG. 14.12. NONORIENTED GRANULE ARRANGEMENT IN A FRESHLY MIXED DOUGH WITH NO GAS CELLS (TOP) AND (BOTTOM) ORIENTATION AND DISTRIBUTION OF GRANULES IN DEVELOPED DOUGH

a—thick film at junction of cells. b, c, and d—thin films. X 240.

by the yeast fermentation. As the bubbles increase in size, the cell walls become thinner. As the temperature increases, the starch granules swell, become quite plastic, and take on various shapes to fit around the gas bubbles. Fortunately, the amount of water in the dough is such that the gelatinization of starch is only partial. The deformed starch granules and the thermally denatured proteinaceous matrix form the structure of the crumb of the bread. Figure 14.13, taken from Sandstedt *et al.* (1954), shows how the swollen starch granules are deformed in shape to fit around the gas bubbles and give the desirable crumb grain and texture. The variation in the thickness of the cell walls is readily apparent. It should be emphasized that even after the baking process, there are starch granules in bread crumb that have retained their identity. It is quite obvious that

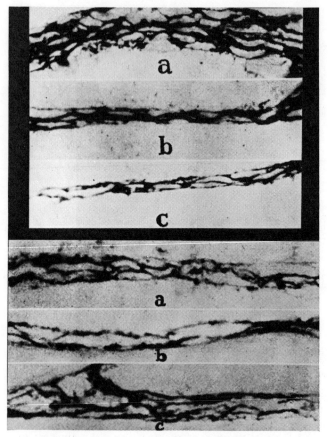

Standstedt et al. (1954)

FIG. 14.13. ORIENTATION AND ELONGATION OF STARCH GRAN-
ULES IN CROSS SECTIONS OF BREAD FILMS (TOP X 250; BOTTOM
X 120)

the ultimate structure of the crumb depends not only on the gluten matrix but also on the starch. If the envelopes that surround the gas bubbles are too weak, they will break, and bubbles will coalesce to form larger bubbles and thereby give a coarse and open grain. If the envelopes are too strong, a crumb grain with small cells and thick walls will result. This is also undesirable. Accordingly, it is easy to appreciate how fine and critical is the balance in the physical properties of the dough for the optimum loaf of bread.

Our interest in the ultrastructure of wheat, flour, dough, and bread, is dynamic rather than static. The differences that exist between one product and the next serve to elucidate the mechanisms involved in the breadmaking process. Eventually, one hopes that better understanding of the processes will serve to improve them, and even to facilitate the breeding of better bread wheat varieties.

BIBLIOGRAPHY

ARANYI, C., and HAWRYLEWICZ, E. J. 1969. Application of scanning electron microscopy to cereal specimens. Cereal Sci. Today *14*, 230-233, 253.

BARMORE, M. A. 1948. Wheat protein. Baker's Dig. *22*, 30-34.

BAYLES, B. B., and CLARK, J. A. 1954. Classification of wheat varieties grown in the United States in 1949. USDA Tech. Bull. *1083*.

BECKWITH, A. C., WALL, J. S., and DIMLER, R. J. 1963. Effect of disulfide bonds on molecular interactions of wheat gluten. Federation Proc. *22*, 348.

BIETZ, J. A., and ROTHFUS, J. A. 1970. Comparison of peptides from wheat gliadin and glutenin. Cereal Chem. *47*, 381-392.

BUSHUK, W., HWANG, P., and WRIGLEY, C. W. 1971. Proteolytic activity of maturing wheat grain. Cereal Chem. *48*, 637-639.

BUSHUK, W., and WRIGLEY, C. W. 1971. Glutenin in developing wheat grain. Cereal Chem. *48*, 448-455.

BUTTROSE, M. S. 1963. Ultrastructure of the developing wheat endosperm. Australian J. Biol. Sci. *16*, 305-317.

CHEN, C. H., and BUSHUK, W. 1970. Nature of proteins in *Triticale* and its parental species. I. Solubility characteristics and amino acid composition of endosperm proteins. Can. J. Plant Sci. *50*, 9-14.

CROZET, N. *et al*. 1966. Ultrastructure microscopy of protein fragments of wheat flour and the gluten separated by leaching. Ind. Aliment. Agr. (Paris) *83*, 559-563. (French)

DAFTARY, R. D., and POMERANZ, Y. 1965. Changes in lipid composition in maturing wheat. J. Food Sci. *30*, 577-582.

DANIELS, N. W. R. *et al*. 1970. Studies on the lipids of flour. V. Effect of air on lipid binding. J. Sci. Food Agr. *21*, 375-384.

EWART, J. A. D. 1966. Fingerprinting of glutenin and gliadin. J. Sci. Food Agr. *17*, 30-33.

EWART, J. A. D. 1968. A hypothesis for the structure and rheology of glutenin. J. Sci. Food Agr. *19*, 617-623.

FULLINGTON, J. G. 1967. Interaction of phospholipid-metal complexes with water-soluble protein. J. Lipid Res. *8*, 609-614.

GRAHAM, J. S. D., MORTON, R. K., and RAISON, J. K. 1963. Isolation and characterization of protein bodies from developing wheat endosperm. Australian J. Biol. Sci. *16*, 375-383.

GRAVELAND, A. 1970. Personal communication. Institute for Cereals, Flour and Bread TNO, Wageningen, The Netherlands.

GROSSKREUTZ, J. C. 1960. The physical structure of wheat protein. Biochim. Biophys. Acta *38*, 400-409.

GROSSKREUTZ, J. C. 1961. A lipoprotein model of wheat gluten structure. Cereal Chem. *38*, 336-349.

HESS, K. 1954. Protein, gluten and lipids in the wheat kernel and flour. Kolloid-Z. *136*, 84-89. (German)

HESS, K. 1955. Importance of wedge protein and adhering protein in wheat flour for dough, bread, and pastry. Kolloid-Z. *141*, 60-79. (German)

HESS, K., and HILLE, E. 1961. Wheat proteins. Z. Lebensm. Untersuch. Forsch. *115*, 211-222. (German)

HESS, K., and MAHL, H. 1954. Electron microscopic observations on wheat flour and flour preparations. Mikroskopie *9*, 81-88. (German)

HOSENEY, R. C., FINNEY, K. F., and POMERANZ, Y. 1970. Functional (breadmaking) and biochemical properties of wheat flour components. VI. Gliadin-lipid-glutenin interaction in wheat gluten. Cereal Chem. *47*, 135-140.

HUEBNER, F. R. 1970. Comparative studies on glutenins from different classes of wheat. J. Agr. Food Chem. *18*, 256-259.

JENNINGS, A. C., MORTON, R. K., and PALK, B. A. 1963. Cytological studies of protein bodies of developing wheat endosperm. Australian J. Biol. Sci. *16*, 366-374.

JONES, C. R., HALTON, P., and STEVENS, D. J. 1959. The separation of flour into fractions of different protein contents by means of air classification. J. Biochem. Microbiol. Technol. Eng. *1*, 77-98.

JORGENSEN, H. 1936. On the existense of powerful but latent proteolytic enzymes in wheat flour. Cereal Chem. *13*, 346-355.

KENT, N. L., and JONES, C. R. 1952. The cellular structure of wheat flour. Cereal Chem. *29*, 383-398.

KLOTZ, I. M., LANGERMAN, N. R., and DARNALL, D. W. 1970. Quarternary structure of proteins. Ann. Rev. Biochem. *39*, 25-62.

KRULL, L. H., and WALL, J. S. 1969. Relationship of amino acid composition and wheat protein properties. Baker's Dig. *43*, 30-39.

LARKIN, R. A., MacMASTERS, M., and RIST, C. E. 1952. Relation of endosperm cell wall thickness to milling quality of seven Pacific northwest wheats. Cereal Chem. *29*, 407-413.

MacMASTERS, M., BRADBURY, D., and HINTON, J. J. C. 1954. Microscopic structure and composition of the wheat kernel. *In* Wheat Chemistry and Technology, I. Hlynka (Editor). Am. Assoc. Cereal Chemists, St. Paul.

MATTERN, P. J., JOHNSON, V. A., and SCHMIDT, J. W. 1970. Air-classification and baking characteristics of high-protein Atlas 66 x Comanche lines of hard red winter wheat. Cereal Chem. *47*, 309-316.

McDONALD, E. C. 1969. Proteolytic enzymes of wheat and their relation to baking quality. Baker's Dig. *43*, 26-28, 30, 72.

MEDCALF, D. G., D'APPOLONIA, B. L., and GILLES, K. A. 1968. Comparison of chemical composition and properties between hard red spring and durum wheat endosperm pentosans. Cereal Chem. *45*, 539-549.

MEDCALF, D. G., and GILLES, K. A. 1965. Wheat starches. I. Comparison of physico-chemical properties. Cereal Chem. *42*, 558-568.

MEREDITH, O. B., and WREN, J. J. 1966. Determination of molecular-weight distribution in wheat flour proteins by extractions and gel filtration in a dissociating medium. Cereal Chem. *43*, 169-186.

MOSS. J., and STENVERT, N. 1970. Personal Communication. Bread Research Institute, Sydney, Australia.

MYHRE, D. V. 1970. Function of carbohydrates in baking. Baker's Dig. *44*, 32-35, 38-39, 60.

OSBORNE, T. B. 1907. The proteins of the wheat kernel. Carnegie Inst. Wash. Publ. *84*.

POMERANZ, Y. 1967. Wheat flour lipids—a minor component of major importance in bread-making. Baker's Dig. *41*, 48-50, 170.

ROHRLICH, M., and SCHULZ, W. B. T. 1960. Investigation on the progress of enzymatic degradation of wheat flour proteins. Naturwissenschaften *47*, 181-187. (German)

ROTSCH, A. 1953. Significance of starch on crumb structure. Brot. Geback *7*, 121-125. (German)

ROTSCH, A. 1954. Chemical and technical investigations on dough. Brot. Geback *8*, 129-130. (German)

SANDSTEDT, R. M. 1954. Photomicrographic studies of wheat starch. III. Enzymatic digestion and granule structure. Cereal Chem. *32*, Suppl. 17-47.

SANDSTEDT, R. M., and FLEMING, J. 1952. Milling properties of wheat. I. Relation to endosperm structure and hardness. Cinemicrograph. Univ. Nebraska.

SANDSTEDT, R. M., SCHAUMBURG, L., and FLEMING, J. 1954. The microscopic structure of bread and dough. Cereal Chem. *31*, 43-49.

SECKINGER, H. L., and WOLF, M. J. 1970. Electron microscopy of endosperm protein from hard and soft wheats. Cereal Chem. *47*, 236-243.

SEEBERG, E. E., and BARMORE, M. A. 1952. The influence of milling moisture on flour yield, flour ash and milling behavior of wheat. Northwest Miller, Milling Prod. *17*, No. 7, 1, 16-19.

SULLIVAN, B. 1948. The mechanism of the oxidation and reduction of flour. Cereal Chem. *25*, No. 6, Suppl.

TANAKA, K., and BUSHUK, W. 1972. Effect of protein content and wheat variety on solubility and electrophoretic properties of flour proteins. Cereal Chem. *49*, 247-257.

TIPPLES, K. H., KILBORN, R. H., and BUSHUK, W. 1966. Effect of malt and sprouted wheat. Cereal Sci. Today *11*, 362-375.

TKACHUK, R. 1966. Amino acid composition of wheat flours. Cereal Chem. *43*, 207-222.

TKACHUK, R. 1969. Involvement of sulfhydryl peptides in the improver reaction. Cereal Chem. *46*, 203-204.

TKACHUK, R. 1970. L-Cysteinylglycine: its occurrence and identification. Can. J. Biochem. *48*, 1029-1036.

TRAUB, N., HUTCHINSON, J. B., and DANIELS, D. G. H. 1957. X-ray studies of the wheat protein complex. Nature *179*, 769-770.

TSEN, C. C., and HLYNKA, I. 1963. Flour lipids and oxidation of sulfhydryl groups in dough. Cereal Chem. *40*, 145-153.

WILLIAMS, P. C. 1970. Particle size analysis of flour with the Coulter counter. Cereal Sci. Today *15*, 102-106, 112.

YASUNAGA, T., BUSHUK, W., and IRVINE, G. N. 1968. Gelatinization of starch during breadbaking. Cereal Chem. *45*, 269-279.

CHAPTER 15

H. L. E. Vix
H. K. Gardner, Jr.
M. G. Lambou
and
M. L. Rollins

Ultrastructure Related to Cottonseed and Peanut Processing and Products

INTRODUCTION

It is well known that both cottonseed and peanuts are important commodities because these crops are valued sources of edible oil and nuts, and nearly all the cake and meal is used as animal feed. Now that the search is on in earnest for new sources of edible protein to feed the population of the United States, as well as of developing nations, seed proteins such as those from cottonseed and peanuts are being evaluated for low-cost, commercially feasible ways of obtaining their proteins. In the 29-yr history of the Southern Marketing and Nutrition Research Division, SMN, a considerable fund of fundamental information and practical processing know-how has been developed on these two commodities. It is from such a wide spectrum of experience that the essential outline of this paper is extracted. It is designed to show that both peanuts and cottonseed are similar ultrastructurally and that this ultrastructure is related to processing and eventually to new products which may be beamed at today's markets. We realize that during commercial solvent extraction, screwpressing, and prepress solvent extraction we are hitting cottonseed with a sledge-hammer. Now we know how important ultrastructure is and are approaching with a microscope and tweezers.

Ultrastructurally, not too many studies have been made on the mature dry seed. Those published from the Southern Division include cotyledonary tissue from *Gossypium hirsutum* L. by Yatsu (1965); lipids from cottonseed by Hensarling *et al.* (1970); isolation of globoids from cottonseed aleurone grains by Lui and Altschul (1967); isolation and characterization of peanut spherosomes by Jacks *et al.* (1967); ontogeny of aleurone grains in cotton embryo by Engleman (1966); and earlier a study of protein bodies during germination of peanut (*Arachis hypogaea* L.) seed by Bagley *et al.* (1963). Other publications in this area can be found in the literature citations of the papers referred to above.

PACKAGING OF CELL CONTENTS

For the purposes of this manuscript, ultrastructure is understood to be the submicroscopic, physicochemical organization of protoplasm. Figure 15.1 shows a spongy mesophyll cell from a cotyledon of dry cotton embryo (top) magnified 7000 times (Yatsu 1965), and a similar peanut cell (bottom). Here one sees that cell contents of both seeds are compartmentalized. Nature has set aside the

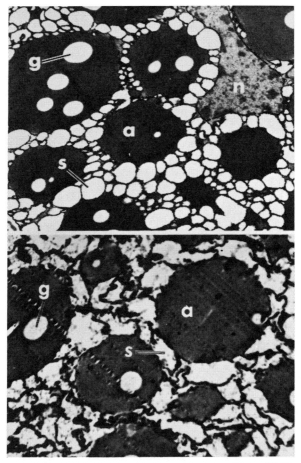

Yatsu (1965)

FIG. 15.1. MESOPHYLL CELL OF DRY COTTON EMBRYO
(TOP) AND SIMILAR CELL FROM PEANUT SEED (BOTTOM)

a—aleurone grains; g—globoids; s—spherosomes; n—nucleus.

various constituents in neat packages. In the cottonseed, the large dark gray bodies (a) are aleurone grains or protein bodies averaging 4–6 μ in diameter (Yatsu 1965) which contain small white globoids (g) (Lui and Altschul 1967). Globoids, which normally stain even darker than the protein bodies with osmium, are sites of phytin storage (Lui and Altschul 1967). The elongated irregular-shaped light gray body (n) is the nucleus. The small, white, mostly round particles ringing the protein bodies are the spherosomes (s) or lipid bodies, generally of the order of 0.3–3 μ in diameter (Hensarling *et al.* 1970; Yatsu 1965). These same packages appear similar in the peanut cell at the bottom.

There are differences in the makeup of each cell but these will be discussed later. Generally speaking, the cell contents from one oilseed resemble closely those from the other.

ULTRASTRUCTURE VERSUS PROCESSING

Four examples, two from each commodity, cottonseed and peanuts, are used to illustrate the relationship between ultrastructure and processing. These include the high-protein products from the Liquid Cyclone Process (LCP) for making a degossypolized cottonseed flour, air classification of glanded cottonseed flour and peanut flour, and the partially defatted peanuts.

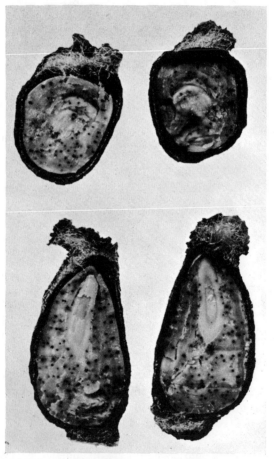

FIG. 15.2. LONGITUDINAL AND CROSS SECTIONS OF GLANDED
COTTONSEED SHOWING PIGMENT GLANDS EMBEDDED
IN THE KERNEL

LCP Process for Making Cottonseed Flour

Practically all cottonseed grown commercially has glands of various sizes containing pigments, predominantly gossypol. Longitudinal and cross sections of glanded cottonseed (Fig. 15.2) show the dark rounded spots which are multicellular secretory structures containing the pigments (Boatner 1948). A workable process for the removal of pigment glands is a prerequisite if food-grade products are to be made from glanded cottonseed.

Figure 15.3 is a scanning electron micrograph of a cross section of hydrated glanded cotton embryo. The large bulbous shape in the center of the figure is a pigment gland. Glands range in size from 50 to 400 μ (Spadaro *et al.* 1948). The small round cells surrounding the pigment gland constitute the spongy tissue of the mesophyll and are similar to the cell shown in Fig. 15.1. These cells range in size from 20 to 40 μ (Yatsu 1971). Some idea of the relative sizes of mesophyll cells to pigment glands can be obtained by comparing ping pong balls to a football. Figure 15.4 is a picture of isolated pigment glands containing pigments.

Boatner *et al.* (1947) showed that the glands are surrounded by platelets (Fig. 15.5) and that the platelets can be separated from each other (Fig. 15.6). Moore and Rollins (1961) found the center of the gland inside the protective platelets to be a complex net-like structure enmeshing globules of gossypol.

Von Bretfeld (1887) and Hanausek (1907) noted that polar solvents ruptured the pigment glands immediately (Fig. 15.7). Pigments stream from a gland as

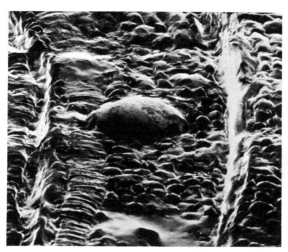

Courtesy of W. R. Goynes, Jr.

FIG. 15.3. SCANNING ELECTRON MICROGRAPH OF CROSS SECTION OF PARTIALLY HYDRATED GLANDED COTTON EMBRYO

Large bulbous shape in center is pigment gland. Small round cells are spongy tissue of the mesophyll.

FIG. 15.4. PIGMENT GLANDS CONTAINING PIGMENTS

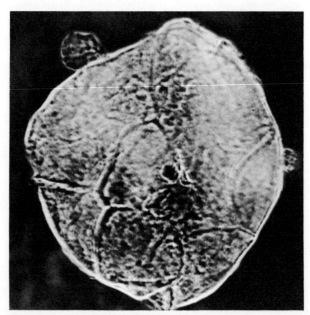

FIG. 15.5. AN EMPTY PIGMENT GLAND (GOSSYPOL RE-MOVED) SHOWING PLATELETS X 180

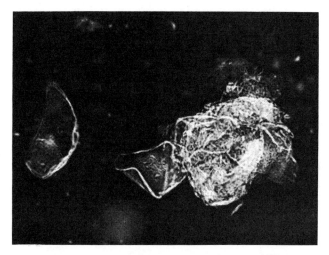

FIG. 15.6. EMPTY PIGMENT GLAND SHOWING SEPARATION
OF PLATELETS X 180

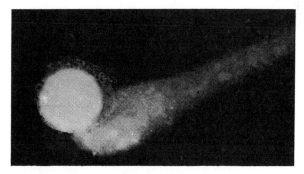

FIG. 15.7. PIGMENTS STREAMING FROM A GLAND ON CON-
TACT WITH WATER X 180

soon as water makes contact (Boatner 1948). This is what happens to pigment glands during commercial processing of cottonseed, especially during the cooking operation.

It was then discovered that nonpolar solvents such as hexane seemingly had no effect on pigment glands (Fig. 15.8) (Boatner and Hall 1946). This information was translated into a gland flotation method on a laboratory scale by Dr. Boatner *et al.* (1949) and then upgraded to a pilot-plant scale by the Engineering and Development Laboratory of the Southern Division (Vix *et al.* 1949). The method was called "differential settling" as illustrated in Fig. 15.9. Note the appearance of the slurry prior to settling (left), after settling for a few minutes (center), and after settling for 12 hr (right). In the center graduate, a layer of pigment glands rests above the settled coarse meal fraction and the major fraction of fine protein particles is suspended in hexane above the pigment layer.

It is the natural packaging of cell constituents as shown in Fig. 15.1 and 15.3 coupled with significant differences in sizes between glands and cells that is

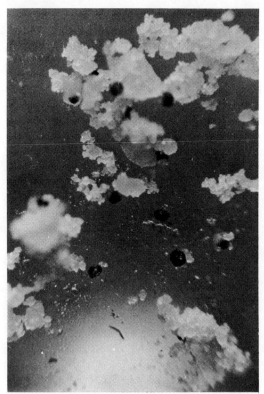

FIG. 15.8. COTTONSEED TISSUE AND PIGMENT GLANDS IN
HEXANE SHOWING INTEGRITY OF GLANDS

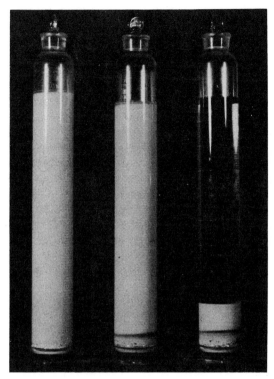

FIG. 15.9. SETTLING OF HEXANE-SLURRIED COTTON-
SEED FLAKES AT DIFFERENT TIME INTERVALS

important to processing glanded cottonseed for edible products. The Liquid Cyclone Process (LCP), developed by Gastrock *et al.* (1969) for making an edible flour from glanded cottonseed, is the outgrowth of the early work outlined here. Since then, the process has been refined by Eaves and Gardner (1970) to yield a product which closely approximates the quality of flour made from glandless cottonseed. The LCP process simultaneously removes pigment glands and lipids while concentrating the protein.

Figure 15.10 is a flow diagram of the refined process upon which has been imposed pictures of whole and cracked cottonseed meats, flakes, overs, unders, and the final flour product. Separation of the overs, an essentially gland-free fraction, is made possible by pumping the milled slurry in hexane through a 3-in. diameter stainless steel Liquid Cyclone (Fig. 15.11). After filtration, washing to remove oil, desolventization, drying, and sterilization, a high-quality cottonseed flour is recovered. Chemical composition of the LCP flour is outlined in Table 15.1.

The high content and high solubility of the protein and low concentration of gossypol contribute to establishing this product as an edible concentrate.

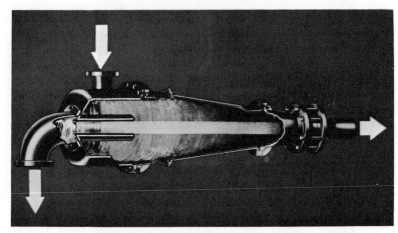

FIG. 15.11. LIQUID CYCLONE

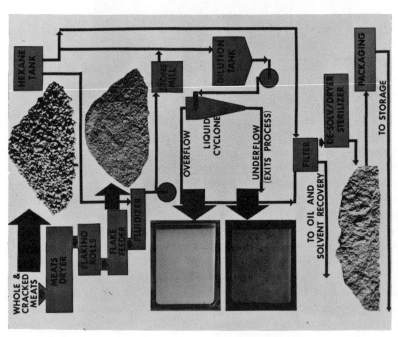

FIG. 15.10. FLOW DIAGRAM OF LIQUID CYCLONE PROCESS
FOR REMOVAL OF PIGMENT GLANDS FROM GLANDED
COTTONSEED

TABLE 15.1

CHEMICAL COMPOSITION OF LCP FLOUR
PREPARED AT THE SOUTHERN REGIONAL
RESEARCH LABORATORY

	(%)
Moisture	3.2
Lipids	1.2
Free gossypol	0.020
Total gossypol	0.065
Nitrogen	10.57
Protein (N × 6.25)	66.1
Protein MFB	68.2
Nitrogen Solubility (0.2N NaOH)	99.5
EAF Lysine	3.96 (gm/16gm N)
Fiber	2.3
Ash	8.9
Flavor	Bland
Color	Light Cream

Air Classification of Hexane-Extracted Cottonseed Flour

The same packaging of constituents of mesophyll cells of cottonseed operates advantageously during air classification. The major particle in the cell is the aleurone grain or protein body which is freed during grinding. Distributed among the cells are the pigment glands.

A large quantity of glanded meal was prepared from high-moisture meats of a commercially purchased lot of 15 tons of seed, crop year 1967, from Greenville, Mississippi, by direct hexane extraction with a minimum amount of heat during desolventization. The meal was ground in a stone mill and then air classified in a pilot-plant model of the Pillsbury[1] classifier. In such a cottonseed flour the principal components are the glands, protein bodies, and a collection of cell wall fragments with adhering cytoplasmic tissue, residual spherosome "membranes," and groups of cells. Air classification separates these particles according to size, shape, and density. A four-cut profile and a final coarse fraction were obtained. Chemical and physical characteristics of the flour fractions are outlined in Table 15.2.

Air classification does accomplish a separation of the flour into two usable products. One product is made up of fractions 1 and 2; the second product is composed of the last 3 fractions. Fractions 1 and 2 contain the smallest particles, highest protein and ash contents, and lowest crude fiber, lipids, and free and total gossypol. They may be combined to yield 53% of a high-protein

---

[1] Use of a company or product name by the USDA does not imply approval or recommendation of the product to the exclusion of others which may also be suitable.

TABLE 15.2

SELECTED PHYSICAL AND CHEMICAL CHARACTERISTICS OF FRACTIONS
OF AN AIR CLASSIFIED COTTON SEED FLOUR[1]

Sample	Yield (%)	Moisture (%)	Nitrogen (%)	Protein (%)	Lipids (%)	Crude Fiber (%)	Ash (%)	Gossypol Free (%)	Gossypol Total (%)
Flour[2]	100	9.1	9.85	61.6	1.3	2.9	7.37	1.7	2.02
Fractions									
1	36.0	8.6	11.24	70.3	1.1	1.9	8.77	0.13	0.33
2	17.1	8.3	10.89	68.1	1.1	2.2	8.47	0.13	0.35
3	13.7	8.0	9.38	58.6	1.5	3.0	8.02	1.20	1.78
4	7.1	7.5	8.97	56.1	2.2	2.9	7.19	2.44	2.54
5	26.1	7.1	7.65	47.8	3.4	4.2	6.18	3.53	3.64

Source: Martinez (1966).
[1] All values except moisture are on a moisture-free basis.
[2] Cottonseed flour prior to air classification.

content flour containing a weighted 69.6% protein on a dry weight basis. For all practical purposes this is considered a protein concentrate. Fractions 3, 4, and 5 may be combined and blended with the feed meal stream at the plant. Fractionation was based on concentrating the protein bodies in fractions 1 and 2, and the pigment glands, crude fiber, and lipids in the final 3 fractions. Air classification of glanded cottonseed does show potential. However, additional research is needed to obtain lower levels of gossypol in the high-protein fractions.

Air Classification of Hexane-Extracted Peanut Flour

Relationship of ultrastructure to air classification of hexane-extracted peanut flour from a Virginia variety (1967) of peanuts is parallel to that outlined for cottonseed in the previous section. Similarly a four-cut profile and a final coarse fraction were obtained. Some chemical and physical characteristics of the fractions are outlined in Table 15.3. Fractions 1 and 2 contain the smallest particles of the highest protein and ash contents and the lowest crude fiber and total sugars. These two fractions can be combined for a 62% yield of an edible, nearly white high-protein content flour containing approximately 72% protein on a dry weight basis. Such a product is considered a protein concentrate. The other three fractions can be combined to produce a flour of lower protein content.

Paired photomicrographs, under transmitted light (on the left) and slightly polarized light (on the right) of the original flour and fractions 1, 3, and 5 at a magnification of X 500 are shown in Fig. 15.12. Particle size increases from (A)

TABLE 15.3

SELECTED PHYSICAL AND CHEMICAL CHARACTERISTICS
OF AN AIR CLASSIFIED PEANUT FLOUR[1]

Sample	Yield (%)	Moisture (%)	Protein (%)	Lipids (%)	Crude Fiber (%)	Ash (%)	Total Sugar[4] (%)
Flour	100	7.8	63.2	0.5	4.1	6.30	3.8
Fractions							
1[2]	37.5	7.3	76.2	0.5	1.9	6.95	3.3
2	24.5	7.3	67.6	0.4	3.3	6.45	3.0
3	26.7	7.6	48.4	0.4	6.4	5.79	3.8
4	5.8	8.0	44.6	0.7	7.0	5.30	3.8
5[3]	5.5	7.7	45.4	0.7	6.6	5.60	5.4

[1] Dry weight basis except for moisture. Mass median diameter: 17–30 μ.
[2] Only fraction containing reducing sugar (0.5% measured as invert).
[3] Coarse residue.
[4] Measured as invert sugars.

to (D) on the left and matches an increase in concentration of starch granules on the right (E) to (H). Air classification in the Pillsbury air classifier effectively sorted the protein bodies from the starch grains. Packaging of the constituents of the cell had conveniently assisted in processing peanut flour to produce a high-protein fraction by dry classification.

Partially Defatted Peanuts

Notable reductions in calorie content of peanuts can be obtained by partially defatting. These have become a popular snack item on the American scene, especially during the cocktail hour. Processing consists of hydraulically pressing deskinned, blanched peanuts to remove oil—up to 60–70% of the oil can be removed at a pressure of 2000 psi for 50 min; expanding in boiling water for 3–8 min; and drying and roasting (Vix *et al.* 1967; Pominski *et al.* 1970 A, B). Figure 15.13 shows the original blanched peanuts (Virginia variety); misshapen pressed peanuts; expanded peanuts; and finally the product after drying and oil roasting, in that order. This process is applicable to all varieties of peanuts, such as Spanish and Runner.

Magnification X 1000 of the peanut cell wall (Fig. 15.14) shows a well-defined pattern of pores visible in 3 walls of what appears to be an 8-sided cell. Minute cytoplasmic threads extend through these openings connecting protoplasts of adjacent living cells. These threads are known as plasmodesms (Wilson and Loomis 1962), a term derived from the Greek *desmos* meaning a bond or connection. Plasmodesms are important as the normal pathway for movement of

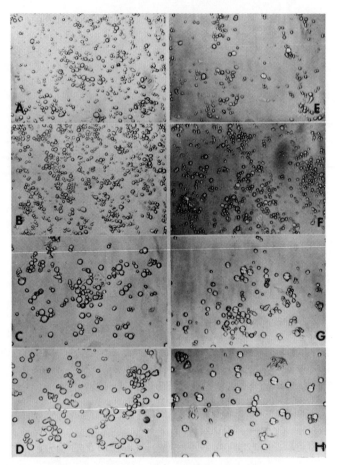

FIG. 15.12. FRACTIONS OF AIR CLASSIFIED PEANUT FLOUR
UNDER TRANSMITTED LIGHT (LEFT) AND SLIGHTLY POLARIZED
LIGHT (RIGHT) X 250

A and E—control flour; B and F—fraction 1; C and G—fraction 3; D and
H—fraction 5.

organic substances between adjoining cells within the seed kernel. It is believed
that the pores housing the plasmodesms play an important role in removal of oil
when pressure is applied and in restoring the original shape when the pressed
nuts are expanded in boiling water.

What occurs when the peanuts are pressed and then expanded is demonstrated in Fig. 15.15. These pictures are electron micrographs (approximately X
4000) of thin sections of normal, pressed, and expanded peanuts embedded in
plastic to retain the structure imposed by the processing steps (Luft 1956; Pease
1964; Spurr 1969). Figure 15.15A (right) is the normal or control from a sample

FIG. 15.13. (LEFT TO RIGHT) BLANCHED VIRGINIA PEA-
NUTS, PRESSED PEANUTS, EXPANDED PEANUTS,
OIL-ROASTED PEANUTS

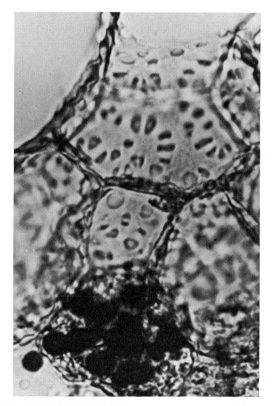

FIG. 15.14. PORES IN PEANUT CELL WALL AT X 667

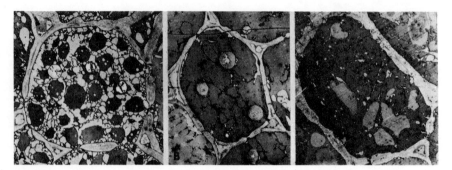

FIG. 15.15. ELECTRON MICROGRAPH OF EFFECTS OF PRESSING AND EXPANSION ON PEANUT CELLS

A—control. B—pressed. C—expanded.

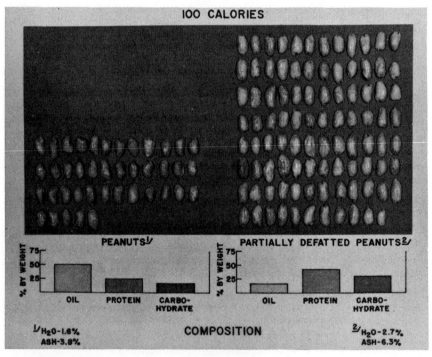

FIG. 15.16. NUMBER OF HALVES AND COMPOSITION OF FULL-FAT AND PARTIALLY DEFATTED PEANUTS (80% OIL REMOVED) REQUIRED TO PRODUCE 100 CAL OF FOOD ENERGY

of medium Virginia, dry-blanched peanuts. Cell structure is normal and resembles that in Fig. 15.1. The protein bodies, globoids, spherosomes, and starch granules are in evidence as expected.

Even the plasmodesms are clearly visible in the compressed areas of the cell wall. These areas are believed to be the sites of pores which were shown in Fig. 15.14. It is noted that the cell wall is completely lined with spherosomes or oil sacs. Figure 15.15B (center) is of the pressed peanuts. Here the spherosomes have been crushed, the "membranes" broken and the oil has escaped. There are now spaces where the spherosomes were just inside the wall. It also appears that the cell wall which has been compressed has lost most of the pore sites. Only one seems to have survived in this picture. Contact between cells via plasmodesms is also disrupted. Pressure caused a coalescence of aleurone grains and the globoid sites are gone. However, the integrity of the starch grains seems to have been preserved.

Figure 15.15C (left) shows a section of the expanded peanut cell. The cell wall has swelled to about twice its original thickness; the cell seems to come back to size. Some of the pores have returned but the plasmodesms are no longer visible. The coalesced protein bodies remain in that state although swelled to some extent. Comparing C with A, one might be tempted to say that the globoids are also present but this is not known. The starch grains have lost their shape, swelled, and may have gelatinized during expansion in boiling water. It is speculated that the coalescence of protein bodies and gelatinization of starch may contribute significantly to the crispness and crunchiness of the new product.

At this point, it would be well to show the calorie difference between regular peanuts and partially defatted peanuts. Figure 15.16 illustrates quite dramatically that 41 full-fat peanut halves are equivalent to 100 cal and are equal to 95 defatted peanut halves. In other words, looking at this both ways; one can eat twice as many partially defatted peanuts as the full-fat peanuts and still consume only 100 cal; or, if one eats the same number of partially defatted peanuts as full-fat peanuts, he will have consumed 43 cal. Should you eat the 95 partially defatted peanuts you would have consumed considerably more protein—approximately 137% more protein than that consumed in 100 cal of full-fat roasted peanuts.

Once again the submicroscopic structure shows pretty clearly what is being accomplished by processing and indicates some explanations for what occurs.

FUTURE PROSPECTS

Through better understanding of the fine structure of cottonseed and peanuts we now have four products which we feel could be potentially important. Many companies in the United States have looked at some of the SMN products, particularly the LCP flour. They are evaluating the utility of these products to upgrade nutritional properties and to impart certain functional qualities to the foods they manufacture. This trend is important to the problem of malnutrition that exists at home as well as in developing countries, particularly because, through it, additional nutrition can be achieved economically.

Examples of some of the products being distributed by U.S.-based industries include CSM which is a dry blend of gelatinized cornmeal, defatted soy flour (toasted), nonfat dry milk solids, minerals, vitamins, and soy oil (Anderson *et al*. 1971). More than one billion pounds have been distributed under the U.S. Food for Peace Program. Beverages, universally popular, are represented by Vitasoy in Hong Kong; Puma which was modeled after Vitasoy and introduced into Guyana by Monsanto; and a caramel-flavored beverage called Saci which has been introduced into Brazil by the Coca-Cola Company. Both LCP cottonseed flour and air classified peanut flour could be used in the above products to increase the protein content in place of the protein source now being used.

The Dorr-Oliver Company is currently operating a Liquid Cyclone processing plant in India on a small commercial scale for the production of an edible high-protein flour from glanded cottonseed. This work is based on the research conducted at the Southern Division to develop the Liquid Cyclone degossypolized cottonseed flour.

The Indian Government has begun fortifying "atta," an unsorted wheat flour or meal, with protein from peanut flour and with essential vitamins and minerals. The cost is absorbed by the Government. This particular product has the advantage of not requiring a change in food habits. Air classified peanut flour such as reported above and LCP-glanded cottonseed flour could be used as sources of protein for the Indian supplementation program involving "atta."

The partially defatted peanut product is an attractive snack item and is currently being manufactured and distributed commercially by three peanut processing plants. Potentially, this product may be used for making many new foods and confections, all more attractive to the enlightened calorie-conscious consumer. Undoubtedly it will help focus attention on peanuts as a good source of available protein domestically as well as throughout the world.

The interested reader is directed to the paper by Altschul and Rosenfield (1970) for an extensive list of examples of the new food technology embracing fortification of staple foods, high-protein foods, beverages, and textured foods. We hope that the products developed at the Southern Division and described above, namely, the LCP flour and air classified flour from cottonseed and the parallel air classified peanut flour and low fat peanuts, will find a number of uses in the new food technology. These, of course, will depend upon ease of availability, knowledge of the technology involved, and cost. Certainly with these developments the Southern Division is well on its way to contributing significantly to the breakthroughs which will precede the passing of hunger and malnutrition in many developing areas as well as in the United States.

SUMMARY

It has been shown that cottonseed and peanuts are ultrastructurally similar. Major cell components are compartmentalized and provide the means of separation by processing. New food products result. Examples cited include: the Liquid Cyclone Process for preparing a degossypolized cottonseed flour of high-protein content; air classification of hexane-extracted cottonseed flour that virtually yields a protein concentrate; air classification of hexane-extracted peanut flour that also produces a protein concentrate; and a process for making defatted peanuts—a new snack item in which 50% of the oil has been removed.

ACKNOWLEDGEMENT

The authors gratefully acknowledge the support and cooperation of E. L. Griffin, Jr., Northern Marketing and Nutrition Research Division; Joseph Pominski, SMN, for furnishing material on the partially defatted peanuts; H. H. Mottern and Wilda Martinez in the preparation of air classified peanut and cottonseed flours; the contributions of Thomas Hensarling, Jarrell H. Carra, Anna M. Cannizzaro, Inez V. deGruy, and Rosalie M. Babin in photo- and electron microscopy; and the photography of Al Fayette and Jack Bergquist.

BIBLIOGRAPHY

ALTSCHUL, A. M., and ROSENFIELD, D. 1970. Protein supplementation: satisfying man's food needs. Progr. (Unilever Quart.) *54*, No. 305, 76-84.

ANDERSON, R. A., PFEIFER, V. F., BOOKWALTER, G. N., and GRIFFIN, E. L., JR. 1971. Instant CSM food blends for world-wide feeding. Cereal Sci. Today *16*, 5-11.

BAGLEY, B. W., CHERRY, J. H., ROLLINS, M. L., and ALTSCHUL, A. M. 1963. A study of protein bodies during germination of peanut (*Arachis hypogaea*) seed. Am. J. Botany *50*, 523-532.

BOATNER, C. H. 1948. Pigments of cottonseed. *In* Cottonseed and Cottonseed Products, A. E. Bailey (Editor). Interscience Publishers, New York.

BOATNER, C. H., and HALL, C. M. 1946. The pigment glands of cottonseed. I. Behavior of the glands toward organic solvents. Oil & Soap *23*, 123-128.

BOATNER, C. H., HALL, C. M., ROLLINS, M. L., and CASTILLON, L. E. 1947. Pigment glands of cottonseed. II. Nature and properties of gland walls. Botan. Gaz. *108*, 484-494.

BOATNER, C. H., HALL, C. M., and MERRIFIELD, A. L. 1949. Heavy gravity liquid separation of cottonseed. U.S. Pat. 2,482,141. Sept. 20.

EAVES, P. H., and GARDNER, H. K., JR. 1970. Glanded cottonseed processing. Presented at the 19th Cottonseed Clinic, New Orleans, La., Feb. 16-17.

ENGLEMAN, M. E. 1966. Ontogeny of aleurone grains in cotton embryo. Am. J. Botany *53*, 231-237.

GASTROCK, E. A., D'AQUIN, E. L., EAVES, P. H., and CROSS, D. E. 1969. Edible flour from cottonseed by liquid classification using hexane. Cereal Sci. Today *14*, 8-11.

HANAUSEK, T. F. 1907. Microscopy of Technical Products. (Translation from German by A. L. Winton.) John Wiley & Sons, New York.

HENSARLING, T. P., YATSU, L. Y., and JACKS, T. J. 1970. Extraction of lipids from cottonseed tissue: II. Ultrastructural effects of lipid extraction. J. Am. Oil Chemists' Soc. *47*, 224-225.

JACKS, T. J., YATSU, L. Y., and ALTSCHUL, A. M. 1967. Isolation and characterization of peanut spherosomes. Plant Physiol. *42*, 585-597.

LUFT, J. H. 1956. Permanganate—a new fixative for electron microscopy. J. Biophys. Biochem. Cytol. 2, 799-802.

LUI, N. S. T., and ALTSCHUL, A. M. 1967. Isolation of globoids from cottonseed aleurone grain. Arch. Biochem. Biophys. 121, 678-684.

MARTINEZ, W. H. 1966. Private communication. Southern Marketing and Nutrition Research Division, Agricultural Research Service, USDA, New Orleans.

MOORE, A. T., and ROLLINS, M. L. 1961. New information on the morphology of the gossypol pigment gland of cottonseed. J. Am. Oil Chemists' Soc. 38, 156-160.

PEASE, D. C. 1964. Histological Techniques for Electron Microscopy. Academic Press, New York.

POMINSKI, J., PEARCE, H. M., JR., and SPADARO, J. J. 1970A. Partially defatted peanuts—factors affecting oil removal during pressing. Food Technol. 24, No. 6, 92-94.

POMINSKI, J., PEARCE, H. M., JR., and SPADARO, J. J. 1970B. Factors affecting water solubles of partially defatted peanuts. Food Technol. 24, No. 9, 76-79.

SPADARO, J. J. et al. 1948. Pilot-plant fractionation of cottonseed. I. Disintegration of cottonseed meats. J. Am. Oil Chemists' Soc. 25, 345-353.

SPURR, A. R. 1969. A low viscosity epoxy resin embedding medium for electron microscopy. J. Ultrastruct. Res. 26, 31-43.

VIX, H. L. E. et al. 1949. Pilot-plant fractionation of cottonseed. II. Differential settling. J. Am. Oil Chemists' Soc. 26, 526-530.

VIX, H. L. E., POMINSKI, J., PEARCE, H. M., JR., and SPADARO, J. J. 1967. Development and potential of partially defatted peanuts. Peanut J. & Nut World 46, No. 3, 10-11; No. 4, 10-11, 18; No. 6, 10-11.

VON BRETFELD, J. 1887. Anatomy of cottonseeds and kapokseeds. Examination for construction, identification and quality. J. Landw. 35, 29-56. (German)

WILSON, C. L., and LOOMIS, W. E. 1962. The matter and mechanics of cells. In Botany, 3rd Edition. Holt, Rinehart & Winston, New York.

YATSU, L. W. 1965. The ultrastructure of cotyledonary tissue from Gossypium hirsutum L. seeds. J. Cell Biol. 25, No. 2, Part 1, 193-199.

YATSU, L. W. 1971. Private communication. SMN, New Orleans, La.

CHAPTER 16

Walter J. Wolf

Soybean Ultrastructure and Its Relationship to Processing

INTRODUCTION

Studies on the ultrastructure of soybeans lag considerably behind similar research on other seeds including wheat, corn, cottonseed, and peanuts. Although the outlines of soybean ultrastructure based on optical microscopy were described more than 50 yr ago, little additional work was done until the early 1960's when more sophisticated tools and techniques, such as the electron microscope, became readily available. Since then, a few investigators have filled in some of the details of soybean structure, but much more research is needed. The relationship of soybean ultrastructure to processing is sketchy, and little of the known information has been applied to developing new processing methods. This chapter summarizes present knowledge of soybean ultrastructure and attempts to relate this information to changes in cellular structure during processing of soybean products.

SOYBEAN ULTRASTRUCTURE

Microscopic Studies

The major anatomical parts of the soybean are hypocotyl and plumule (2%), seed coat or hull (8%), and cotyledons (90%). The structures of these entities as discerned with the optical microscope have been known for more than 50 yr (Kondo 1913; Wallis 1913). Proceeding inward the hull consists of the following layers: an epidermis of palisade cells, large (30–70 μ long) hourglass cells, 6–8 layers of spongy parenchyma cells, aleurone cells, and finally, compressed endosperm cells (Fig. 16.1). The cotyledon is covered with an epidermis of small cubical cells followed by a layer of larger cells. The remainder of the cotyledon is made up of polygonal cells 30–50 μ long and 15–25 μ wide. Wallis (1913) described these cells as being filled with "aleurone grains and numerous droplets of fixed oil." These subcellular structures were also known to exist in many other oilseeds but received little attention until the 1960's when biochemical separation techniques and electron microscopy were applied to isolate and characterize them. The electron microscope shows that the aleurone grains or protein bodies are large (2–10 μ in diameter) nearly spherical particles while the oil is located in much smaller (approximately 0.2–0.5 μ in diameter) particles designated spherosomes (Fig. 16.2). The protein bodies are surrounded by a membrane probably consisting of phospholipids; the membrane is stable to diethyl ether and to hexane (Tombs 1967).

Although not shown in Fig. 16.2, cotyledon cells also contain ribosomes, the sites of protein synthesis. Ribosomes from soybean cotyledons were isolated by Matsushita *et al.* (1966) and characterized as 80S particles when analyzed in an

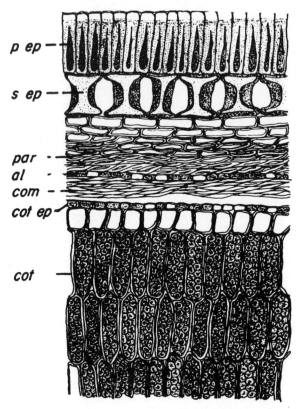

Williams (1950)

FIG. 16.1. CROSS SECTION OF SOYBEAN HULL AND COTYLEDON

The hull consists of p ep, palisade cells; s ep, hourglass cells; par, parenchyma; al, aleurone cells; and com, compressed cells. The cotyledon is covered with cot ep, epidermis and contains mainly cot, aleurone cells.

ultracentrifuge. Hypocotyl ribosomes were also isolated and were a mixture of 80S and 115S particles.

Isolation of Protein Bodies

Three procedures have been developed to isolate and characterize protein bodies of soybeans, whereas the spherosomes have not been studied. Saio and Watanabe (1966) homogenized soybeans in cottonseed oil and separated the protein bodies by differential centrifugation in cottonseed oil-carbon tetrachloride mixture of varying densities.

Tombs (1967) isolated protein bodies by homogenizing hexane-defatted flour (350 mesh) in 20% sucrose-pH 5 citrate buffer and centrifuging in a 70–90%

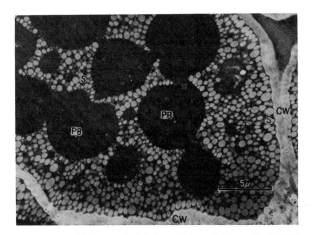

With permission from Saio and Watanabe (1968)

FIG. 16.2. TRANSMISSION ELECTRON MICROGRAPH OF SOY-
BEAN COTYLEDON AFTER SOAKING IN WATER OVERNIGHT

Fixed with osmium tetroxide, and stained with uranium acetate and
lead citrate. Structures are PB, protein bodies; S, spherosomes; and
CW, cell wall.

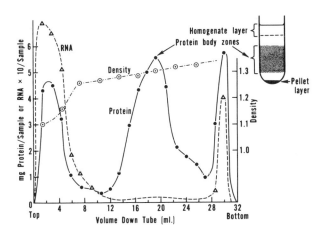

From Tombs (1967) courtesy of Plant Physiology

FIG. 16.3. DISTRIBUTION OF PROTEIN AND RIBONUCLEIC ACID
IN SUCROSE DENSITY GRADIENT

Inset is diagram of centrifuge tube.

sucrose gradient (Fig. 16.3). The protein bodies were usually obtained as two bands: a light fraction (density <1.30) and a heavy fraction ($1.30<$density<1.32). Soluble (cytoplasmic) proteins remained at the top of the centrifuge tube, and proteins associated with cell wall debris and intact cells sedimented to the bottom (pellet layer). Applying a buffer at the isoelectric point (pH 5) of the globulins minimized disruption of the protein bodies.

The third technique for isolating protein bodies was described by Komoda *et al.* (1968). After soaking soybeans overnight in water they homogenized them in 0.5 M sucrose, removed the cellular debris, diluted the homogenate to 0.25 M sucrose, and centrifuged. The precipitate was believed to contain the protein bodies.

Analytical data for protein bodies isolated by the three procedures are summarized in Table 16.1. Included are Tombs' (1967) analyses for his total protein body preparation, the light and the heavy fractions plus the starting soybean flour. Protein contents for the different preparations vary widely. The protein bodies isolated by Komoda *et al.* (1968) are low in protein but high in lipid (23–60%). Because no electron microscopic results were reported, it is not known whether the lipids were present as spherosomes or were associated with the proteins within the protein bodies. Because of the known instability of the protein bodies in aqueous solutions at the pH of soybeans (approximately 6.5), it is questionable whether the protein bodies remained intact during this isolation procedure. In washing their protein bodies with carbon tetrachloride, Saio and Watanabe (1966) probably removed most lipids, but about 17% of their preparations are unaccounted for. Even though Tombs' heavy fraction contained cell wall fragments and cytoplasmic attachments, it was nearly 80% protein. His light protein-body preparation was almost pure protein; protein bodies of such high protein content appear unique to soybeans.

Protein bodies isolated by Saio and Watanabe were 2–10 μ in diameter and Tombs reports similar results. We have prepared protein bodies by Tombs' method and found them to be quite uniform in size but only about 1–3 μ in diameter (Fig. 16.4a). Examination of the pellet layer likewise revealed particles of 1–3 μ diameter (Fig. 16.4b). A 5-μ particle was seen in the pellet layer (Fig. 16.5a), but was obviously an aggregate of the smaller sized particles. Because larger particles were present in the starting soybean flour (Fig. 16.5b), disintegration of the large particles probably occurred during isolation.

The question of protein body stability needs further study; but with this possible limitation, Tombs' procedure is the present method of choice for preparing these materials from soybeans. Protein body yields are high (58–62% of the total protein is recovered as protein bodies) and the mixture of light and heavy fractions is obtained by a single density-gradient centrifugation.

Distribution of Cellular Components

Information on distribution of constituents between the subcellular struc-

TABLE 16.1

COMPOSITION OF SOYBEAN PROTEIN BODIES[1]

Constituent	Total Preparation[3] (%)	Preparation[4] Total (%)	Preparation[4] Light (%)	Preparation[4] Heavy (%)	Defatted Soybean[4] (%)	Preparation No.[5] 1 (%)	2 (%)	3 (%)	4 (%)
Protein (N × 5.8)[2]	65.0	82.5	97.5	78.5	50.0	35	52	39	34
Total phosphorus	0.94	0.84	0.84	0.90	—	—	—	—	—
Ribonucleic acid	0.53	1.29	0.43	2.04	1.66	—	—	—	—
Phospholipid	—	1.0	—	—	2.25	29	44	23	60
Total lipid	—	11.3	1.5	5.6	8.6	—	—	—	—
Phytic acid	—	1.35	2.6	2.3	2.24	—	—	—	—
Carbohydrate	8.5	3.0	—	—	—	—	—	—	—
Ash	7.7	—	—	—	.[1]	—	—	—	—
Total	82.7	99.3	100.8	87.4	64.8	—	—	—	—

[1] Calculated on a moisture-free basis.
[2] Nitrogen-to-protein conversion factor based on amino acid composition data (Tombs 1967).
[3] Saio and Watanabe (1966).
[4] Tombs (1967).
[5] Komoda et al. (1968).

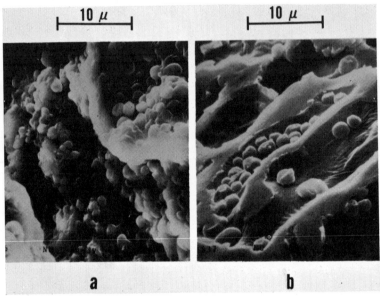

Wolf (1970)

FIG. 16.4. SCANNING ELECTRON MICROGRAPH OF (a) SOYBEAN PROTEIN BODIES
AND (b) PELLET FRACTION OBTAINED BY SUCROSE DENSITY GRADIENT
CENTRIFUGATION

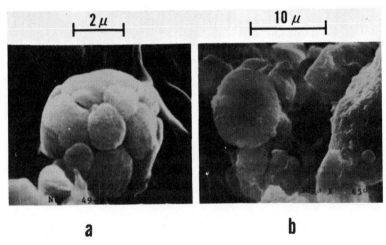

Wolf (1970)

FIG. 16.5. SCANNING ELECTRON MICROGRAPH OF (a) AGGREGATE OF PROTEIN
BODIES IN PELLET FRACTION AND (b) PROTEIN BODIES IN UNDEFATTED
GROUND SOYBEANS

tures and the cytoplasm is sparse. Tombs (1967) estimated that at least 70% of the protein in a cell resides in protein bodies. This finding is consistent with the conclusions of Saio and Watanabe (1966) that proteins from isolated protein bodies and water-extractable proteins of defatted meal are nearly identical when measured by gel filtration or polyacrylamide gel electrophoresis. Catsimpoolas *et al.* (1968) prepared protein bodies by Tombs' technique and found six immuno-chemically different components by disc immunoelectrophoresis. The disc gel electrophoretic pattern for protein body components was essentially the same as the pattern for crude globulins isolated from defatted cotyledons by extraction with salt solution followed by dialysis against water. Ultracentrifugal patterns for extracts of the protein bodies are similar to patterns for water-extractable proteins of defatted meal, except for reduced amounts of 2S fraction (Wolf 1970; Koshiyama and Fukushima 1970). Gel electrophoresis indicates that trypsin inhibitor, which is in the 2S fraction of water-extractable proteins, is located in the cytoplasm rather than in protein bodies (Tombs 1967). None of the other cytoplasmic proteins have been identified so far. In summary, it appears that only some of the minor proteins, such as trypsin inhibitors and possibly enzymes, are found outside of protein bodies.

Tombs found ribonucleic acids (RNA) primarily in the cytoplasmic fraction (homogenate layer in Fig. 16.3); RNA in the pellet fraction was attributed to intact cells. Tombs reported a lower content of phytate in his protein body preparations than for his original meal; consequently, phytate may not be located exclusively in protein bodies. He found no evidence for localization of phytate in globoids within the protein bodies as noted with other seeds (Lui and Altschul 1967). Tombs suggested that phytate is combined with proteins rather than compartmentalized.

Komoda *et al.* (1968) isolated protein bodies as described earlier, but the purpose of their study was to determine the intracellular distribution of tocopherol in soybean cotyledons. They found that cellular particles (mitochondria, microsomes, etc.), cytoplasmic solution, and protein bodies contained 10, 15, and 60%, respectively, of the total tocopherol in soybeans.

Cytological location of enzymes, including α-amylase, lipoxygenase, and urease of interest in food and feed processing, is still unknown. Origin and intracellular distribution of the compounds responsible for the characteristic beany and bitter flavors of raw soybeans likewise remain undetermined.

Ultrastructure During Development and Germination

Prominent structures of the mature cotyledon—spherosomes and protein bodies—are absent during early stages of development. Fifteen days after flowering the structures observed are the nucleus, ribonucleoprotein particles, and endoplasmic reticulum. After 18 days chloroplasts, spherosomes, protein bodies, mitochondria, and starch grains appear. As development continues the protein bodies and spherosomes become more numerous and increase in size while chloroplasts and mitochondria decrease in number. Starch grains are present until less than a week before maturation but then disappear almost completely (Bils and Howell 1963).

During early stages of germination protein bodies become granular, their membrane disappears, and their shape becomes irregular. In subsequent stages, protein bodies disrupt and the proteins coalesce into large masses. At the same time, spherosomes decrease in number and immature plastids plus starch grains become detectable (Tombs 1967). A study of the proteins of soybeans during germination, as followed by disc electrophoresis and disc immuno-electrophoresis, revealed differences in rates of breakdown for the various proteins. The 11S protein, the major component of the protein bodies, was detectable up to 16 days after germination. One of the 7S components, however, decreased to low levels after only 9 days of germination (Catsimpoolas et al. 1968).

ULTRASTRUCTURE AND PROCESSING

In the United States the major form of processing soybeans is an integrated operation consisting of extracting the oil and then desolventizing-toasting the defatted flakes for use as animal feeds. Small but increasing amounts of defatted flakes are also being converted into food products where the flake toasting operation is not used (Wolf and Cowan 1971). Recently, extrusion cooking of soybeans has received attention as a means for converting soybeans into edible products (Mustakas et al 1970). In Japan and other parts of the Orient, certain food products are made from soybeans by extraction and fermentation techniques. The relationship between ultrastructure and these various processes has received only limited attention, but available information is reviewed here.

Oil Extraction and Desolventizing-Toasting

If whole or half soybeans are extracted with hexane for a week, respective percentages of the original oil extracted are only 0.08 and 0.19% (Othmer and Agarwal 1955). Such results clearly demonstrate that to extract oil successfully the seed structure needs to be disrupted. Also, such results do not answer the question: Is the oil unextractable because the cell walls are impermeable to hexane or because the spherosomes are? It has long been known that to extract oil, soybeans and other oilseeds must be decreased in particle size so that at least 1 dimension is about 0.010-in. (Karnofsky 1949A). Almost universally, soybeans are reduced to this size by flaking. After preliminary cracking and dehulling, the broken beans are heated to 140°–170°F, adjusted to 10–11% moisture, and passed through smooth rolls to produce flakes about 0.010 in. (254 μ) in thickness. If the cotyledon cells are 20–40 μ in diameter, this flake thickness corresponds to 6–12 layers of cells. Since the solvent penetrates the flakes from both surfaces, extraction occurs through a thickness of 3–6 cells. The extent of cell disruption during flaking of soybeans does not appear to have been investigated (Karnofsky 1949B). Moreover, there is as yet no agreement whether grinding or flaking of other oilseeds disrupts a large fraction of the oil-bearing cells (Norris 1964).

The effect of desolventizing-toasting on cellular structure of hexane-defatted soybean flakes has been reported (Sipos and Witte 1961). In this operation hexane-laden flakes from an extractor are treated with live steam to volatilize the hexane and to cook the flakes to inactivate antinutritional factors found in raw flakes. Toasting destroys cell walls and agglomerates and coalesces the protein bodies into large masses. In a properly toasted flake the cellular structure is completely obliterated.

Japanese Food Products

Saio and Watanabe (1968) investigated structural changes that occur during conversion of soybeans into several Japanese foods. No changes were noted on soaking beans in water overnight, but on steaming at 115°C for 30 min the protein bodies ruptured, and the proteins appeared coagulated. Steaming also broke up the spherosomes and the released oil appeared as droplets 1–3 μ in diameter. After incubation of the steamed beans with *Bacillus natto* to produce natto, the cell walls were partly destroyed and some of the oil droplets appeared meshlike in structure.

Fresh tofu (bean curd) is prepared by soaking beans in water, grinding, heating, and filtering to yield soybean milk. Addition of calcium salts to soy milk coagulates the protein-oil complex and produces tofu. Electron microscopy of tofu reveals a meshy network of coagulated protein particles (0.1–0.2 μ in diameter) interspersed with oil droplets up to 1 μ in diameter. No protein bodies or spherosomes are apparent; likely, these structures are destroyed during the hot extraction used to prepare soy milk. Also examined was kori-tofu (dried or frozen tofu) made by freezing a specially prepared fresh tofu at −10°C, aging at −1 to −2°C, thawing, and finally drying. Kori-tofu in the wet curd stage has a denser network of coagulated protein than ordinary tofu. On freezing, the protein-oil network becomes more compact and, on thawing and drying, a fibrous structure develops.

Air Classification of Defatted Soy Flours

Separation of a high-protein fraction from defatted soy flour by air classification has been tried, but increases in protein content were small (Pfeifer *et al.* 1960). The starting flour with a protein content of 55.5% was separated into 6 fractions with maximum protein contents of only 59.9 and 59.3% in 2 trials. Pfeifer *et al.* used a 200-mesh flour ground from commercial hexane-defatted flakes. Possibly alternate methods for preparing the flour are needed so that grinding frees the intact protein bodies from each other and remaining cellular material. Flaking may disrupt many of the protein bodies as well as squeezing them into a compacted mass in which the other cellular structures are also embedded. Studies on the effects of flaking and grinding processes on the ultrastructure of soybeans might provide the information needed for the successful development of an air classification process for increasing protein content of soy flours. Theoretically, protein bodies should be capable of separation by this technique. Indeed, fractions rich in protein bodies have been separated from defatted cottonseed flour by air classification (Martinez 1969). This approach is especially attractive because Tombs (1967) demonstrated that protein bodies are nearly pure protein and because it is a dry process that avoids problems of water pollution associated with wet fractionation methods.

BIBLIOGRAPHY

BILS, R. F., and HOWELL, R. W. 1963. Biochemical and cytological changes in developing soybean cotyledons. Crop Sci. *3,* 304-308.

CATSIMPOOLAS, N., CAMPBELL, T. G., and MEYER, E. W. 1968. Immunochemical study of changes in reserve proteins of germinating soybean seeds. Plant Physiol. *43,* 799-805.

KARNOFSKY, G. 1949A. The mechanics of solvent extraction. J. Am. Oil Chemists' Soc. *26,* 570-574.

KARNOFSKY, G. 1949B. The theory of solvent extraction. J. Am. Oil Chemists' Soc. *26,* 564-569.

KOMODA, M., MATSUSHITA, S., and HARADA, I. 1968. Intracellular distribution of tocopherol in soybean cotyledons. Cereal Chem. *45,* 581-588.

KONDO, M. 1913. The anatomical structure of some commercial foreign legumes. Z. Nahr. Genussm. *25,* 1-56.

KOSHIYAMA, I., and FUKUSHIMA, D. 1970. Recent advances in soybean globulin research. Eiyo to Shokuryo (J. Jap. Soc. Food Nutr.) *23,* 297-311.

LUI, N. S. T., and ALTSCHUL, A. M. 1967. Isolation of globoids from cottonseed aleurone grain. Arch. Biochem. Biophys. *121,* 678-684.

MARTINEZ, W. H. 1969. Cottonseed protein concentrates by air classification. *In* Conference on Protein-Rich Food Products from Oilseeds, New Orleans. USDA, ARS *72-71,* 33-39.

MATSUSHITA, S., MORI, T., and HATA, T. 1966. Enzyme activities associated with ribosomes from soybean seedlings. Plant Cell Physiol. (Tokyo) *7,* 533-545.

MUSTAKAS, G. C. *et al.* 1970. Extruder-processing to improve nutritional quality, flavor, and keeping quality of full-fat soy flour. Food Technol. *24,* 1290-1296.

NORRIS, F. A. 1964. Extraction of fats and oils. *In* Bailey's Industrial Oil and Fat Products, 3rd Edition, D. Swern (Editor). Interscience Publishers, New York.

OTHMER, D. F., and AGARWAL, J. C. 1955. Extraction of soybeans: Theory and mechanism. Chem. Eng. Progr. *51,* 372-378.

PFEIFER, V. F., STRINGFELLOW, A. C., and GRIFFIN, E. L., Jr. 1960. Fractionating corn, sorghum and soy flours by fine grinding and air classification. Am. Miller *88,* No. 8, 11-13, 24.

SAIO, K., and WATANABE, T. 1966. Preliminary investigation on protein bodies of soybean seeds. Agr. Biol. Chem. (Tokyo) *30,* 1133-1138.

SAIO, K., and WATANABE, T. 1968. Observations of soybean foods under electron microscope. Nippon Shokuhin Kogyo Gakkaishi (J. Food Sci. Technol. (Tokyo) *15,* 290-296.

SIPOS, E., and WITTE, N. H. 1961. The desolventizer-toaster process for soybean oil meal. J. Am. Oil Chemists' Soc. *38,* No. 3, 11-12, 17-19.

TOMBS, M. P. 1967. Protein bodies of the soybean. Plant Physiol. *42,* 797-813.

WALLIS, T. E. 1913. The structure of the soya bean. Pharm. J. *37,* 120-123.

WILLIAMS, L. F. 1950. Structure and genetic characteristics of the soybean. *In* Soybeans and Soybean Products, K. S. Markley (Editor). Interscience Publishers, New York.

WOLF, W. J. 1970. Scanning electron microscopy of soybean protein bodies. J. Am. Oil Chemists' Soc. *47,* 107-108.

WOLF, W. J., and COWAN, J. C. 1971. Soybeans as a food source. Crit. Rev. Food Technol. *2,* 81-158.

CHAPTER 17

Sidney J. Circle
and
Allan K. Smith

Functional Properties of
Commercial Edible Soybean
Protein Products

INTRODUCTION

The steadily increasing cost of food grade animal proteins derived from the meat, poultry and egg, dairy, and fishing industries has compelled the food industry to look to the low-cost vegetable sources of protein, principally the oilseeds, as an alternate source. At present in the United States, the least expensive and most readily available high quality food grade protein comes from the soybean crop in the form of various derivatives. These soy protein products may be identified as fractions of the soybean or combinations of fractions (Circle and Smith 1972), and may be grouped according to protein content dry basis in the following categories: <50% protein (full-fat and refatted soy grits and flours), 50% protein (defatted soy grits and flours, and extruded-expanded textured soy meals), 70% protein (soy protein concentrates), and 90% protein (soy protein isolates). Another category is based on enzyme-hydrolyzed soy protein, and still others on fatted or defatted soybean "milks," "curds," or spun fibers (the latter three marketed in Japan). In the United States, prices range as follows: defatted soy flour, 6½–7¢ per lb; extruded-textured soy meal, 33–80¢ per lb; soy protein concentrate, 20–28¢ per lb; soy protein isolate, 38–47¢ per lb; and enzyme-modified soy protein, 80¢–$1.20 per lb. (Eley 1968).

FUNCTIONAL PROPERTIES OF SOY PROTEIN PRODUCTS

The processing of soybeans into various edible soy protein products has developed into viable commercial enterprises only because of the acceptance by the food industry of these products. It is emphasized that these commercial edible soy protein products are not consumed as such, but are used as ingredients in food formulations. Except for some consumer-ready textured foods (meat analogs), the soy protein products have little gustatory appeal or palatability in themselves, being sold for the most part in the form of unflavored dry powders, grits, granules or chunks. The current interest in these products on the part of food manufacturers lies not in their appearance, but in those properties which enable their use as ingredients to impart favorable, desirable, and acceptable changes in structure, texture, and composition to the finished foods, at an attractive price.

In many cases they are employed to supplement, extend, or replace more costly ingredients derived from animal protein sources including meat, eggs, and milk; in other cases they compete with grain or other plant derivatives. In any event, the choice of soy protein products as food ingredients depends on their providing economical performance on a par with what they replace, without detracting from finished product quality. These soy products (regular and extruded flour products, concentrates, isolates, and modified isolates) are normally marketed in the United States conforming to various specifications of color, particle size, bulk density, solubility, and other characteristics. These are usually classified as physical properties, although solubility, involving solvation, may also be considered as a chemical property. Physical and chemical properties of soy protein fractions in relation to their functional properties have been discussed by several workers (Wolf and Smith 1961; Wolf 1969, 1970; Briskey 1970). Rather arbitrarily, a tacit understanding has evolved among workers in this field to categorize separately the physical, chemical, sensory (flavor), nutritional, and functional properties of these soy protein products without attempting to define these precisely. Although the nutritional value of these products as proteins does approach closely that of animal-derived proteins, and generally is better than that of most other vegetable proteins, the chief interest at present is focussed on their functional properties, that is, those which contribute some performance aspect, especially on manipulation in aqueous dispersion, to affect structure and texture of the formulation favorably (in order to incorporate soy protein products into finished foods, at some point in the processing hydration is required; this is true even for dry blends such as prepared mixes, which usually call for addition of water in order to put them to use).

Delineation of Functional Properties

These were first listed in a separate category by Circle and Johnson (1958) for soy protein isolate aqueous dispersions, to encompass physicochemical properties such as dough-forming, moisture-holding, fat-binding, emulsifying, foaming, film-forming, thickening, stabilizing, gelling, cohesiveness, adhesiveness, and others (see also Johnson and Circle 1959), but the appellation "functional" was coined later (Johnson and Circle 1962). Smith (1958) referred to several similar properties of protein isolates intended for industrial use. Anson (1958) mentioned combinations of oilseed protein and oilseed fat for simulating dairy products, and discussed meatlike textures achieved by suitable manipulation of oilseed protein, namely by heat gelation and fiber-spinning of gel precursors.

Tests for Functional Properties

The desirability of having available relatively simple physical or chemical tests which could be used to predict how soy protein products will affect food systems in which they are incorporated has occurred to several workers (Circle *et al.* 1964; Eldridge *et al.* 1963A,B; Kelley and Pressey 1966; Catsimpoolas and Meyer 1970; Johnson 1970A,B; Meyer 1970; Wolf 1970; Mattil 1971). Quoting from Mattil (1971) concerning "the problems of those who would evaluate the functional properties of various proteins. The questions must always be asked, 'In what system; and what is the effect of the other components of the system?' How do the data obtained from an emulsifiability test run in a beaker relate to the complex ionic environment in a frankfurter? Do thickening properties measured by traditional viscosity tests really predict how a protein will act in a meat system? It has been demonstrated that solubility profiles can be used to detect protein denaturation, and to predict modes of preparation of protein concentrates and isolates, but to what extent can they be used to predict functionality in real life systems?" Johnson (1970A,B) states that physical and chemical tests for functionality of a protein product will give useful information, but that in the final analysis it is necessary to incorporate the product as an ingredient into the food formulation and produce the finished product in order to determine its performance beyond doubt. Nevertheless, it is considered worthwhile to enumerate several tests for functional performance which have been described, and which have found use value in the food industry. It should be pointed out that there must exist a sizable number of unpublished tests developed over the years, and reserved as proprietary information for in-house use by the larger food firms, especially in the baked goods, dairy, and meat fields.

Water Absorption and NSI.—The degree of water retention by soy flour and grits and soy protein concentrate is considered to be useful as an indication of performance in several food formulations, especially some involving dough handling. One such test is given by the National Soybean Processors Association (1946). The relationship of water-absorption to nitrogen solubility index (NSI) was indicated by Lemancik and Ziemba (1962) and Johnson (1970A). The NSI itself is used as a predictor for selection and performance of heat-treated soy flour products for use in certain food formulations (Johnson, 1970A). Extruder-expanded textured soy flour chunks are said to reach 3½ times their original dry weight in the fully-hydrated, drained condition (Martin and LeClair 1967). Paulsen (1961) described a water absorption test for macaroni, and applied it to compare macaroni products with and without incorporated soy flour.

Aerating Capability.—Whipability or foaming action is a property desirable for whip toppings, whipped desserts, and frozen desserts, among others. Eldridge et al. (1963A,B) described foam preparation from soy protein isolate dispersions, along with methods of measuring foam volume and stability.

Emulsifying Capacity of Soy Protein Isolate.—The use of soy protein isolate as a minor additive in sausage emulsions for its moisture-binding, fat-emulsifying and emulsion-stabilizing properties has been discussed by several workers (Meyer 1970; Rakosky 1970; Rock et al. 1966). Although Pearson et al. (1965) claimed that soy protein isolate was not as good an emulsifier as milk proteins in their model test procedure, this was refuted in actual sausage tests reported by Inklaar and Fortuin (1969). Further work on the model test method would appear to be warranted.

Rheological Properties of Soy Protein Isolate.—These include viscosity and gelation. Circle *et al.* (1964) studied the rheological behavior of relatively concentrated aqueous dispersions of a commercially available soy protein isolate. The viscosity was found to rise exponentially with increase of concentration in the absence of heat. Heat caused thickening and then gelation, in concentrations above 7% by weight, with a temperature threshold of 65°C. Rate of gelling and gel firmness were found to be dependent on temperature, time of heating, and protein concentration. At 8–14% protein concentration, gels formed within 10–30 min at 70°–100°C, but were disrupted if overheated at 125°C. The gels were firm, resilient, and self-supporting at protein concentrations above 16–17%, and less susceptible to disruption by over-heating. Figure 17.1 shows the effect

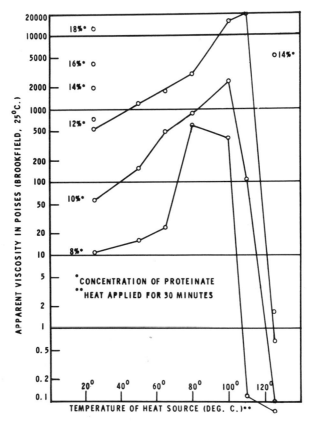

FIG. 17.1. EFFECT OF TEMPERATURE ON APPARENT VIS-
COSITY OF 8, 10, 12, 14, 16, and 18% SODIUM SOY
PROTEINATE AQUEOUS DISPERSIONS HELD FOR 30 MIN AT
TEMPERATURE SHOWN

of temperature on apparent viscosity of dispersions containing 8–18% soy protein isolate.

Circle *et al.* (1964) also investigated the influence of additives on the heat-gelation of soy protein isolate dispersions. They found that the viscosities of both unheated and heated 10% dispersions were raised by added soy oil, soy lecithin, wheat starch, carboxymethylcellulose or carrageenan. The addition of salts to 10% dispersions lowered the viscosity of the unheated dispersion, but raised that of the heated. The specific disulfide reducing agents, sodium sulfite and cysteine, profoundly lowered the viscosity of both unheated and heated dispersions, and prevented gelation.

Catsimpoolas and Meyer (1970) proposed the concept of conversion of a protein sol to a "progel" state on heating, with formation of a gel on cooling. The gel-progel transition was considered reversible, with the sol-progel change being irreversible. Both progel and gel are subject to disruption on excessive heating to what they term a "metasol" which no longer gels on cooling. Further discussion on gelation is given by Wolf (1972) and Circle and Smith (1972).

FUNCTIONAL PROPERTIES IN FOOD APPLICATIONS

Comminuted Meat Products and Meat Analogs

Soy protein products are used in processing frankfurters, bologna, nonspecific meat loaves, meat balls, meat patties, salisbury steak, chili con carne, luncheon meats, and similar items.

Rock *et al.* (1966) have reported that many of the problems of sausage making are related to the wide variation in the raw material or meat trimmings used in the process. These variations are the results of the age, diet, weight, and sex of the animal, anatomical origin of the trimmings, and whether the latter are fresh or frozen. They state that the proper use of soy protein products in the formulations as binders compensates for these variations.

Rakosky (1967) has described the processing of frankfurters and bologna as a problem of finely comminuting the meat materials and additives by chopping, flaking, or grinding into a continuous matrix, and encapsulating the finely-divided fat particles. In this emulsion the salt-soluble muscle-proteins, myosin and actomyosin, function as a heat coagulable continuous phase. The function of the soy protein is to supplement the myosin and actomyosin as an emulsifying and encapsulating agent to prevent fat separation and to hold the meat juices, especially during cooking. In federally-inspected plants, soy flour and protein concentrate are permitted in frankfurter to the extent of 3.5%; soy protein isolate to the extent of 2%. The latter must be tagged with titanium dioxide at the level of 1000 ppm Ti for analytical detection purposes.

The functional effectiveness of soy flour, concentrate, and isolate in comminuted meats is dependent to a degree on the content of water-dispersible protein, but not entirely. Soy protein isolate, which is almost completely soluble in water at its pH of 6.9–7.2, and which contains more than 90% protein, is accordingly the most effective of the several soy products used in comminuted meats, but also the most expensive. There exists the possibility in sausage making that the emulsion in the finished cooked frankfurter will break and the separated white fat particles will create an undesirable appearance requiring that the product be reprocessed. Thus, it is important to use additional emulsifier-binder as insurance against a processing failure.

A good example of both the functional and nutritional use of soy protein is in making meat patties where soy flour, grits, or the concentrate may be used as described by Rakosky (1970). The amount of these products which can be used is designated by the Meat Inspection Division of the USDA as "sufficient for the purpose." In making patties it is necessary to add water at 2–3 times the weight of the soy protein. The primary function of the protein is to improve dimensional stability of the pattie while cooking and to decrease cooking losses. When properly used the patties will be tastier, will have higher protein and lower fat, and thus be better balanced nutritionally. Huffman and Powell (1970) have reported that patties made with as little as 2% soy grits had higher tenderness scores than without the soy product.

Rakosky (1970) has discussed, also, the use of soy products in preparation of chili con carne. In the usual method of making chili, much of the fat is rendered out of the meat, and this may be retained by the addition of corn meal or of soy protein products. When a soy product is used it increases the protein content more than by the addition of corn meal; and if soy grits are used, they give a pleasing grainy texture. Soy flour, grits, concentrate, and isolate are permitted at a level of 8% in chili products in federally-inspected plants. Other foods related to the meat industry which can use soy protein products at a level sufficient for the purpose are stews, scrapple, tamales, meat pies, and beef and pork barbecue sauce.

Meat analogs, also called textured protein products (TPP) are products processed from vegetable proteins to resemble meat in chewiness and flavor. Presently the two most important methods are (1) by spinning soy protein isolate into a "tow" of fibers, and shaping and flavoring into the desired meat-like product; and (2) by processing a formulated soy flour mass through an extruder into different sizes and shapes to give a chewy product of the desired flavor. A third method is to form a chewy gel by heat gelation of a soy protein isolate dispersion under the proper conditions. A fourth method of achieving textures uses graded coarsely ground granular particles of isoelectric soy protein isolate which are soaked in water containing calcium hydroxide or trisodium phosphate to raise the pH from 4.5 to the range 5.5–7.5, depending on the type of "bite" desired (Circle and Johnson 1958).

Spun Fiber Products.—Considerable literature is now extant on spun fiber type meat analogs (reviewed by Smith and Circle 1972). Kelley and Pressey (1966) reported their studies of several factors affecting viscosity of the "spinning dope" and fiber formation. Boyer and Saewert (1956) stated that the toughness of the fibers is controlled by the pH, salt concentration, and temperature of the coagulating bath.

According to Odell (1967), the spun fiber foods, on a dry basis, contain 1/3–2/3 of their weight in monofilament fiber. Although the composition of the final product can be made to vary widely, a typical example reported by Thulin and Kuramoto (1967) contains approximately 40% protein fiber, 10% binder, 20% fat, and 30% flavor, color, and supplemental nutrients. Another analysis on a dry basis was reported as 60% protein, 20% fat, 17% carbohydrates, and 3% ash. As initially manufactured they have a moisture range of 50–70%. The final products have flavors resembling ham, bacon, chicken, turkey, beef, and other meat products and usually are more tender than the natural product. They can be refrigerated, canned, frozen, or dried.

In comparing the cost of analogs with meat, Odell (1967) has estimated the cost of unflavored spun fibers as 50¢ per lb. He points out that the analogs are boneless, cooked, and completely in an edible form, and that price comparisons with meat should be made only on that basis. With this type of comparison the cost of the analog will be much below the average price of meat. Also, this price advantage is expected to improve, since the price of meat will probably increase, whereas the price of soy protein isolate may be expected to decrease from its present level.

Extrusion-Cooked Products.—Several companies have developed meat analogs by the method of continuous extrusion under heat and pressure of a prepared soy protein product formulation (Atkinson 1969, 1970; Anon. 1967A, 1967B, 1970). The process makes small chunks which, when hydrated, have a chewy texture and taste like meat. The principal advantage of this method is the use of low-cost soy flour in the formulation rather than the more expensive protein isolate.

A premix of soy flour with added fat, flavoring, carbohydrates, coloring, and other desirable components is made, preconditioned, and passed through an extruder under predetermined conditions of moisture, temperature, time, and pressure. When the product comes from the extruder to atmospheric pressure it usually will expand. The shape of the chunk will be determined by the shape of the die at the extruder outlet, and the speed of the revolving knife which cuts the extruded product to the desired length.

According to Martin and LeClair (1967) an unflavored, dried product made from defatted soy flour, at a moisture content of 6–8%, has a composition of protein 50–53%, fat 1.0%, fiber 3.0%, and ash 5–6%, which is the same as the original soy flour. The composition of the product can be modified by the addition of fat up to 30% and of carbohydrates and other food components.

Mustakas *et al*. (1970) have described the operation of an extruder for making full-fat soy flour, and the technics described can be useful in making extruded meat analogs. It should be possible to use the full-fat soy flour as well as the defatted for making meat analogs.

Heat-gelled Products.—A food product developed by Anson and Pader (1958, 1959) comprised a protein system in the form of a chewy protein gel, and a process entailing three steps: (1) adjusting concentration and pH of a protein-water system, (2) shaping of the system, (3) applying heat to the system to set it to a chewy gel.

Frank and Circle (1959) prepared meat analogs simulating bologna and frankfurters in structure, texture, and flavor from isolated soy protein. In their process, the isolated protein (isoelectrically precipitated) was added to a hot solution of trisodium phosphate in a mixer. While mixing, the other ingredients, consisting of fat, emulsifier, soy sauce, smoke flavor, spices, and color were added, and pH was adjusted to about 6.3. The properly stirred mixture was stuffed into cellulose frankfurter casings, linked, tied, and steamed at 10–15 psig for about 10 min. With this heat treatment, the protein dispersion is changed from a viscous sol to a gel which binds the fat, water, and other components together.

Meat Fibers in Heat-gelled Protein Matrix.—A novel approach to forming a dehydrated meat product with improved rehydration characteristics has been patented by Coleman and Creswick (1966). Natural meat tissue (striated skeletal muscle of mammalian, avian, piscine, crustacean, or molluscan origin) is reduced to short, discontinuous rod-like fibers (1/8–1/2 in. in length) by shredding or tearing, with or without prior cooking. These are mixed with an aqueous paste of edible heat-coagulable protein of vegetable or animal origin (neutral soy protein isolate or egg albumin, for example; or mixtures of the two). A dough formed by thorough mixing is forced through an extruder as a ribbon (aligning the fibers in parallel form), and the ribbon cut into shaped pieces, which are cooked and dried.

Other Food Applications

Soy flour and grits and soy protein concentrate contain insoluble carbohydrate in addition to protein. Although the protein is considered the component most responsible for the functional value of soy protein products, the insoluble carbohydrate may also have influence, both by diluting the protein, and on its own account. This has been pointed out by Johnson (1970B) and Wolf (1970). The insoluble carbohydrate may be detrimental to rheological characteristics (viscosity and heat gelation). It may affect water absorption or moisture-holding capacity. In any event the soy flour and protein concentrate products are less versatile than the protein isolate in their functional properties, but nevertheless find many applications in food formulations.

Soy Flour and Grit Products (Lemancik and Ziemba 1962; Ziemba 1966).—Soy flour is said to make bread and cookie doughs more pliable, easier

TABLE 17.1

FUNCTIONAL PROPERTIES OF SOY PROTEIN PRODUCTS AS FOOD INGREDIENTS

Functional Property	Protein Form Used[1]	Food System	References
Emulsification			
Formation	F,C,I	Frankfurters, bologna, sausages	Rock et al. (1966); Pearson et al. (1965); Inklaar and Fortuin (1969); Wood (1967); Tremple and Meador (1958)
		Breads, cakes, soups	Circle and Johnson (1958)
		Whipped toppings, frozen desserts	
Stabilization	F,C,I	Frankfurters, bologna, sausages	Rock et al (1966); Pearson et al. (1965); Inklaar and Fortuin (1969)
		Soups	Wood (1967)
Fat absorption			
Promotion	F,C,I	Frankfurters, bologna, sausages, meat patties	Rock et al. (1966); Wood (1967)
Prevention	F,I	Doughnuts, pancakes	Ziemba (1966); Eley (1968); Johnson (1970B)
Water absorption			
Uptake	F,C	Breads, cakes	Wood (1967); Tremple and Meador (1958); Turro and Sipos (1968)
		Macaroni	Paulsen (1961)
		Confections	Ziemba (1966)
Retention	F,C	Breads, cakes	Wood (1967); Tremple and Meador (1958)

TABLE 17.1 (Continued)

FUNCTIONAL PROPERTIES OF SOY PROTEIN PRODUCTS AS FOOD INGREDIENTS

Functional Property	Protein Form Used[1]	Food System	References
Texture			
Viscosity	F,C,I	Soups, gravies, chili	Wood (1967); Ziemba (1966)
Gelation	I	Simulated ground meats	Anson and Pader (1958); Circle et al. (1964); Frank and Circle (1959)
Chip and chunk formation	F	Simulated meats	Ziemba (1966, 1969)
Shred formation	F,I	Simulated meats	Rusoff et al. (1962)
Fiber formation	I	Simulated meats	Ziemba (1966, 1969); Thulin and Kuramoto (1967)
Dough formation	F,C,I	Baked goods	Circle and Johnson (1958)
Film formation	I	Frankfurters, bologna	Circle and Johnson (1958); Ziemba (1966)
Adhesion	C,I	Sausages, lunch meats, meat patties, meat loaves and rolls, boned hams	Rock et al. (1966); Ziemba (1966)
		Dehydrated meats	Coleman and Creswick (1966)
Cohesion	F,I	Baked goods	Circle and Johnson (1958)
		Macaroni	Paulsen (1961)
		Simulated meats	Rusoff et al. (1962)
Elasticity	I	Baked goods	Circle and Johnson (1958)
		Simulated meats	Rusoff et al. (1962)
Color control			
Bleaching	F	Breads	Wood (1967)
Browning	F	Breads, pancakes, waffles	Wood (1967); Eley (1968)
Aeration	I	Whipped toppings, chiffon mixes, confections	Ziemba (1966);Eldridge et al. (1963A,B); Circle and Johnson (1958)

Source: Wolf (1970).
[1] F,C, and I represent flours, concentrates, and isolates, respectively.

to handle and machine, and less sticky. After baking, bread has enriched crust color, better eating quality, and increased shelf-life through enhanced moisture retention. With the addition of soy flour piecrust is claimed to be more tender, cookie dough to release more readily from forming dies, and breading mix and doughnuts to have reduced fat absorption on frying and improved browning, all with fewer cripples. Baby cereals are less pasty with better pickup and release on drying rolls, and with improved texture, color, and appearance after drying. Soy flour in fudge retards crystallization and drying out. Caramels handle better on the wrapping machine and do not stick to the wrapper; oil bleeding is checked, and moisture retention improved. Soy flour can be added to soup mixes, gravies, and chili products as a thickening agent. In molasses, spice oils, and garlic powders it has value as an anticaking agent. The enzyme-active flour is useful for its bleaching effect on carotene. Coarse soy grits can be used as nut-like textured particles in toppings and cookies. In macaroni, soy flour imparts a firmer texture (Paulsen 1961).

Soy Protein Concentrates (Ziemba 1966).—The advantage of concentrates over soy flours lies in their blander flavor, lighter color, and higher protein content. In general, they can be used in food formulations in almost the same manner as the soy flours, but their lower flavor profile permits a higher use level without adverse effect on flavor of the final products. However, they do entail a higher cost.

Soy Protein Isolates (Circle and Smith 1972).—In dairy type products soy protein isolates provide emulsifying, suspending, and stabilizing action, and can be used to formulate almost the complete line including margarine, imitation milk and creams, coffee whiteners, imitation cream cheese, frozen desserts, and others. In the formulation of puddings and spreads they are useful as thickening, gelling, and suspending agents. Although competitively attractive for use in dairy-type and meat-type products, their relatively high cost is a drawback to their usage in cereal-based products.

Enzyme-modified Soy Protein Products.—Unmodified soy protein isolates as usually marketed are composed of several different components with molecular weights in the range 200,000–350,000, and are insoluble at pH 4–5 (Wolf 1972). Partial hydrolysis by pepsin treatment (Circle and Smith 1972) reduces the molecular weight to the range 3000–5000, and makes the hydrolysate soluble in water over the entire pH scale including pH 4–5. Such products have been given the misnomer "soy albumen," although they more properly should be classified as "peptones." Their chief applications are in confections (nougat, fudge, caramel, etc.), marshmallow toppings, icings, and dessert mixes as a whipping agent, and in beverages as a foaming agent.

Wolf (1970) has summarized in tabular form the functional properties of soybean protein products in food systems (see Table 17.1).

BIBLIOGRAPHY

ANON. 1970. Improves properties of textured proteins. Food Eng. *42*, No. 7, 161.

ANON. 1967A. New route to texture, flavor, or color control of product uses "tailored proteins." Food Prod. Develop. *1*, No. 1, 35-36.

ANON. 1967B. Soybean protein products new entry in market. Food Technol. *21*, 176.

ANSON, M. L. 1958. Potential uses of isolated oilseed protein in food-stuffs. *In* Processed Plant Protein Foodstuffs, A. M. Altschul (Editor). Academic Press, New York.

ANSON, M. L., and PADER, M. 1958. Protein food products and method of making same. U.S. Pat. 2,830,902. Apr. 15.

ANSON, M. L., and PADER, M. 1959. Method for preparing a meat-like product. U.S. Pat. 2,879,163. Mar. 24.

ATKINSON, W. T. 1969. Process for preparing a high protein snack. U.S. Pat. 3,480,442. Nov. 25.

ATKINSON, W. T. 1970. Meat-like protein food product. U.S. Pat. 3,488,770. Jan. 6.

BRISKEY, E. J. 1970. Functional evaluation of protein in food systems. *In* Evaluation of Novel Protein Products. Pergamon Press, New York.

BOYER, R. A., and SAEWERT, H. E. 1956. Method of preparing imitation meat products. U.S. Pat. 2,730,448. Jan. 10.

CATSIMPOOLAS, N., and MEYER, E. W. 1970. Gelation phenomena of soybean globulins. I. Protein-protein interactions. Cereal Chem. *47*, 559-570.

CIRCLE, S. J., and JOHNSON, D. W. 1958. Edible isolated soybean protein. *In* Processed Plant Protein Foodstuffs, A. M. Altschul (Editor). Academic Press, New York.

CIRCLE, S. J., and SMITH, A. K. 1972. Processing soy flours, protein concentrates and protein isolates. *In* Soybeans: Chemistry and Technology, Vol. 1, Proteins, A. K. Smith and S. J. Circle (Editors). Avi Publishing Co., Westport, Conn.

CIRCLE, S. J., MEYER, E. W., and WHITNEY, R. W. 1964. Rheology of soy protein dispersions. Effect of heat and other factors on gelation. Cereal Chem. *41*, 157-172.

COLEMAN, R. J., and CRESWICK, N. S. 1966. Method of dehydrating meat and product. U.S. Pat. 3,253,931. May 31.

ELDRIDGE, A. C., HALL, P. K., and WOLF, W. J. 1963A. Stable foams from unhydrolyzed soybean protein. Food Technol. *17*, 120-123.

ELDRIDGE, A. C., WOLF, W. J., NASH, A. M., and SMITH, A. K. 1963B. Alcohol washing of soybean protein. J. Agr. Food Chem. *11*, 323-328.

ELEY, C. P. 1968. Food uses of soy protein. *In* Marketing and Transportation Situation. USDA *ERS-388, MTS-170*, 27-30.

FRANK, S. S., and CIRCLE, S. J. 1959. The use of isolated soybean protein for non-meat simulated sausage products. Frankfurter and bologna types. Food Technol. *13*, 307-313.

HUFFMAN, D. L., and POWELL, W. E. 1970. Fat content and soy level effect on tenderness of ground beef patties. Food Technol. *24*, 1418-1419.

INKLAAR, P. A., and FORTUIN, J. 1969. Determining the emulsifying and emulsion-stabilizing capacity of protein meat additives. Food Technol. *23*, 103-107.

JOHNSON, D. W. 1970A. Oilseed proteins-properties and applications. Food Prod. Develop. *3*, No. 8, 78, 80, 84, 87.

JOHNSON, D. W. 1970B. Functional properties of oilseed proteins. J. Am. Oil Chemists' Soc. *47*, 402-407.

JOHNSON, D. W., and CIRCLE, S. J. 1959. Multipurpose quality protein offers 'plus' factors. Food Process. *20*, No. 3, 37-39.

JOHNSON, D. W., and CIRCLE, S. J. 1962. Food uses and properties of protein fractions from soybeans. First Int. Congr. Food Sci. Technol. London, England, Sept.

KELLEY, J. J., and PRESSEY, R. 1966. Studies with soybean protein and fiber formation. Cereal Chem. *43*, 195-206.

LEMANCIK, J. F., and ZIEMBA, J. V. 1962. Versatile soy flours. Food Eng. *34*, No. 7, 90-91.

MARTIN, R. E., and LeCLAIR, D. V. 1967. For creating products—textured proteins. Food Eng. *39*, No. 4, 66-69.

MATTIL, K. F. 1971. Functional requirements of proteins for foods. J. Am. Oil Chemists' Soc. *48*, 477-480.

MEYER, E. W. 1970. Soya protein isolates for food. *In* Proteins as Human Food, R. A. Lawrie (Editor). In United States and Canada available from Avi Publishing Co., Westport, Conn.; otherwise available from Butterworths, London.

MUSTAKAS, G. C. *et al.* 1970. Extruder processing to improve nutritional quality, flavor, and keeping quality of full-fat soy flour. Food Technol. *24*, 1290-1296.

NATIONAL SOYBEAN PROCESSORS ASSOCIATION. 1946. Handbook of Analytical Methods for Soybeans and Soybean Products.

ODELL, A. D. 1967. Meat analogs from modified vegetable tissues. Proc. Int. Conf. Soybean Protein Foods. USDA-ARS. *71-35*, 163-169.

PAULSEN, T. M. 1961. A study of macaroni products containing soy flour. Food Technol. *15*, 118-121.

PEARSON, A. M., SPOONER, M. E., HEGARTY, G. R., and BRATZLER, L. J. 1965. The emulsifying capacity and stability of soy sodium proteinate, potassium caseinate, and nonfat dry milk. Food Technol. *19*, 1841-1845.

RAKOSKY, J. 1967. Soy proteins—their preparation and uses in comminuted meat products. Meat Hygiene, USDA-CMS *8*, No. 6, 1-13.

RAKOSKY, J. 1970. Soy products for the meat industry. J. Agr. Food Chem. *18*, 1005-1009.

ROCK, H., SIPOS, E. F., and MEYER, E. W. 1966. Soy protein—its role in processed meat production. Meat *32*, 52-56.

RUSOFF, I. I., OHAN, W. J., and LONG, C. L. 1962. Protein food product and process. U.S. Pat. 3,047,395. July 31.

SMITH, A. K. 1958. Vegetable protein isolates. *In* Processed Plant Protein Foodstuffs, A. M. Altschul (Editor). Academic Press, New York.

SMITH, A. K., and CIRCLE, S. J. 1972. Protein products as food ingredients. *In* Soybeans: Chemistry and Technology, Vol. 1, Proteins, A. K. Smith and S. J. Circle (Editors). Avi Publishing Co., Westport, Conn.

THULIN, W. W., and KURAMOTO, S. 1967. Bontrae—a new meat-like ingredient for convenience foods. Food Technol. *21*, 168-171.

TREMPLE, L. G., and MEADOR, R. J. 1958. Soy flour in cake baking. Baker's Dig. *32*, No. 4, 32.

TURRO, E. J., and SIPOS, E. 1968. Effect of various soy protein products on bread characteristics. Baker's Dig. *42*, No. 6, 44-50, 61.

WOLF, W. J. 1969. Chemical and physical properties of soybean protein. Baker's Dig. *43*, No. 5, 30-37.

WOLF, W. J. 1970. Soybean proteins. Their functional, chemical and physical properties. J. Agr. Food Chem. *18*, 969-976.

WOLF, W. J. 1972. Purification and properties of the proteins. *In* Soybeans: Chemistry and Technology, Vol. 1, Proteins, A. K. Smith and S. J. Circle (Editors). Avi Publishing Co., Westport, Conn.

WOLF, W. J., and SMITH, A. K. 1961. Food uses and properties of soybean protein. II. Physical and chemical properties of soybean protein. Food Technol. *15*, No. 5, 12-13, 16, 18, 21, 23, 26, 28, 31, 33.

WOOD, J. C. 1967. Soy flour in food products. Ingredient Survey. Food Manuf. *42*, 11, 12-15.

ZIEMBA, J. V. 1966. Let soy proteins work wonders for you. Food Eng. *38*, No. 5, 82-89.

ZIEMBA, J. V. 1969. Simulated meats—how they're made. Food Eng. *41*, No. 11, 72-75.

Constance Kies

Evaluation of the Protein Value of Cereal/Plant/Oilseed Products by Human Biological Assay Techniques

The use of humans as biological models in the evaluation of protein nutritive value of any food presents problems in philosophy, in morality, in legality, in theory, and in practice which are unique to this organism. Although final laboratory evaluation of any food or feed is obviously "best" done using the specie for which the product is intended, nutritional evaluations of food products using humans under controlled conditions is usually omitted.

The use of cereal/plant/oilseed products in human diets can hardly be considered new; however, their importance as current and future primary protein resources for human populations has been commonly recognized only within the last decade. These sources are generally believed to be of poorer protein nutritional value than are animal origin products. This has generated much interest and effort toward theoretical and practical, and quantitative and qualitative improvement in protein value of cereal/plant/oilseed products for humans in research and action programs. Human biological evaluation can be extremely important in setting guidelines and in final evaluation.

In theory, the protein nutritional value of a food product could be defined as the ability of the protein content of the food to meet the protein need of the individual for whom the product is intended. Unfortunately, knowledge of protein/amino acid requirements of healthy human adults is inadequate and knowledge of protein/amino acid requirements of other stages of the life cycle under other physiological conditions is highly fragmentary. This means that nutritional value of a food cannot really be defined in terms of the chemical composition of the food itself or in terms of comparison of food composition and theoretical nutritional needs of the intended user. Interrelationships involving other components of the food, components of other items in the diet, total nutritional environment, food likes and dislikes, cultural values, and economic considerations affect protein consumption, digestion, absorption, and metabolism.

PROTEIN QUALITY ASSAY METHODS USING HUMANS

Expectations for human nutrition have not been fully defined. Most practical assay tools for judging adequacy of protein were derived for use in animal nutrition where greatest growth for least feed expenditure was the goal, at least at the time these research tools were developed. Is the goal for human nutrition to simply have bigger humans? Is it for expanded life span, or decrease in infant mortality, or what? Practical evaluation tools are difficult to develop without defined goals. In spite of its limitations, the nitrogen balance technique is still by far the most commonly used method for assaying adequacy of dietary protein in laboratory controlled studies involving human subjects.

Some simple questions on protein nutritive value of plant proteins for humans can be asked and answered in laboratory controlled studies using humans as biological models. Three basic approaches may be taken: (1) comparisons made by feeding the test foods at equal intakes of nitrogen or protein which essentially evaluates essential amino acid proportionality patterns and protein digestibility; (2) comparisons made by feeding at equal levels of test foods which evaluates both quantity and quality of protein as well as food digestibility; (3) comparisons made by allowing ad libitum feeding of the test foods as determined by subjects' desires which would most nearly duplicate field conditions but is an extremely difficult procedure to monitor accurately. The primary research emphasis of the Human Nutrition Research Laboratories, Department of Food and Nutrition, University of Nebraska, has been in investigating various factors affecting protein nutritive value of cereal, plant, and oilseed products to meet human protein needs. Some of these studies will be reviewed purely as examples of the possibilities of various approaches to evaluation of food products using a human biological assay technique. As illustrated by the recent bibliography compiled by Coons (1968) many researchers have been deeply involved in evaluation of cereal/plant/oilseed products using human biological models. Hence, this paper should be considered as a review of only a small portion of the total literature.

EVALUATION OF AMINO ACID PROPORTIONALITY PATTERNS

The first limiting amino acid of a food is that amino acid supplied in lowest amount in proportion to human need. When the product is fed in amounts insufficient to meet need, supplementation with the first limiting amino acid of that product should result in improvement in protein nutriture of the individual. More simply, increasing the amount of the first limiting amino acid of a food by any means reduces the total amount of the protein (and probably the total amount of the food) necessary to meet the protein needs of the individual. Theoretically, determination of the first limiting amino acid of a food could be accomplished by simple comparison of its amino acid proportionality pattern with an amino acid proportionality pattern reflecting human requirements. However, because of fore-mentioned limitations in knowledge on the amino acid requirements of even the normal, healthy human adult, this method can be said to give results that are only predictive. Since much information on first limiting amino acids of foods for humans has been determined by different investigators using different experimental approaches under widely different experimental conditions, comparison of results is difficult. A project was developed to determine the first limiting amino acids for human adults in a wide variety of plant products keeping experimental design and procedures as closely similar as possible. Standardization of procedures is a necessary basis of all testing approaches.

Each individual study making up the project was 33 days in length and consisted of an introductory 3-day nitrogen depletion period, a 5-day nitrogen adjustment period, and 5 experimental 5-day periods. Order of the 5 experimental periods was randomly arranged for each of the 8-10 men or men and women serving as subjects. During the experimental periods, the test food was fed in amounts to provide 4.0 gm N per subject per day (80% of the total dietary protein). Comparison of the amino acids provided by 4.0 gm test food nitrogen and the Rose provisional amino acid requirements for young men (Rose 1957) was used to predict the first 3 most limiting amino acids in the test food. During 3 of the 5 experimental periods, the test diets were singly fortified with each of the 3 selected amino acids (purified, crystalline, L-form) at that level necessary to make the total of food amino acid plus supplement equal the Rose recommended levels for that amino acid (Rose 1957). During 1 period, a combination of all 3 supplements were used (positive control) and during 1 period no supplements were used (negative control). Urea was used to maintain diets isonitrogenous at 5.0 gm N per subject per day. These levels of total nitrogen intake and total test protein intake have been selected in this laboratory as standards in studies designed to determine the first limiting amino acids in food proteins for human adults. If subjects are not in negative balance while receiving nonamino acid supplemented diets, the effects of amino acid additions are difficult to demonstrate. When diets are adequate in protein, supplementation with amino acids will not result in improved nitrogen balance. Obviously, for true adequate protein nutriture, higher level supplemented or unsupplemented protein diets are needed.

Caloric intake for each subject was kept constant during the experimental periods of each study at the level required for weight maintenance. A few low protein fruits and vegetables were part of the daily basal diets. Vitamin and mineral supplements were also included. Other procedures were basically those used in standard human metabolism studies of this type (Anon. 1958). Evaluation of protein nutriture of the subjects was primarily by the nitrogen balance technique. Amino acid composition (except tryptophan) of the test foods was determined via analysis with a Technicon or Beckman amino acid Auto-Analyzer. Handbook values were used for estimates of tryptophan content. Fasting blood samples were taken at the end of each experimental period during most studies. Individual serum amino acid constituents were determined. Various clinical blood values were also obtained to ascertain general maintenance of health and gross protein nutriture.

Using this approach, that amino acid which resulted in a decrease in body nitrogen loss was taken to be the first limiting amino acid in the test food for human adults. Thus far, seven different foods have been studied using this approach. Results are given in Table 18.1.

The following results were obtained when whole ground wheat flour was the test protein: wheat + no amino acid supplement = −0.68; wheat + L-lysine = +0.05; wheat + L-tryptophan = −0.64; wheat + L-methionine = −0.69; wheat + a combination of L-lysine, L-tryptophan, L-methionine = +0.10. The significant improvement in nitrogen retention resultant from lysine supplementation of the wheat diets indicates lysine as being the first limiting amino acid in this cereal grain for protein nutriture of the adult human (Kies and Fox 1970A).

Triticale is a new wheat-rye hybrid which has been suggested as being a potentially important food for humans. The lysine content of triticale protein has been found to be higher than that of wheat protein. Even so, the improvement in nitrogen retention with lysine supplementation indicates lysine as still being the first limiting amino acid in this cereal grain for maintenance of nitrogen equilibrium in human adults (Kies and Fox 1970A).

Recent information of development of a high-protein potato tuber (20% protein, dry weight basis) has created interest in the protein nutritive value of conventional potato products. When dehydrated potato flakes were the test food used in the experimental plan, mean nitrogen balance data indicated methionine as clearly being the first limiting amino acid (Kies and Fox 1970B).

Opaque-2 and *floury-2* mutant corns have created much excitement among both animal and human nutritionists because of the resultant lysine content increase in the cereal grain. In conventional corn, lysine is distinctly the first limiting amino acid for humans and for several domesticated animals such as swine as well. Unless a food has an absolutely perfect balance of amino acid content in proportion to human needs, obviously one or more amino acids must be first limiting. When a degerminated, ground product was used for evaluation, lysine was shown as being the first limiting amino acid. These results should *not* be interpreted as indicating that *opaque-2* shows no improvement over conventional corn grain in protein value. Other human feeding studies done in this and other laboratories clearly indicate the superiority of the *opaque-2* lines. However, a better interpretation would be that the higher lysine content of *opaque-2* still was lower in proportion to human need than was the content of any other essential amino acid (Kies and Fox 1971B).

Milo sorghum grain was included in our program of human biological testing of amino acid supplemented and unsupplemented cereals. Since lysine was indicated as being the first limiting amino acid in milo meal for human adults, improvement in the efficiency of milo protein for human feeding could be achieved by increasing the lysine content (Kies *et al.* 1968).

Comparison of the amino acid proportionality of rye grain with the Rose

TABLE 18.1

EFFECT OF AMINO ACID SUPPLEMENTATION OF CEREAL AND
PLANT PROTEIN DIETS FOR HUMANS

Study[1]	Test Protein Source[2]	Amino Acid Supplement	Mean N Balance[3] (Gm N/day)	Reference
1	Wheat (whole ground flour)	None	−0.68	(Kies and Fox 1970A)
		L-lysine	+0.05	
		L-tryptophan	−0.64	
		L-methionine	−0.69	
		Combined	+0.10	
2	Triticale (whole ground flour)	None	−0.59	(Kies and Fox 1970A)
		L-lysine	+0.11	
		L-tryptophan	−0.24	
		L-methionine	−0.41	
		Combined	+0.27	
3	Potatoes (de-hydrated flakes)	None	−1.18	(Kies and Fox 1970B)
		L-leucine	−0.83	
		L-methionine	−0.27	
		L-phenylalanine	−0.91	
		Combined	−0.30	
4	*Opaque-2* (ground, deger-minated meal)	None	−0.71	(Kies and Fox 1971B)
		L-methionine	−0.67	
		L-lysine	−0.06	
		L-tryptophan	−0.76	
		Combined	−0.13	
5	Milo (whole ground meal)	None	−1.15	(Kies *et al.* 1968)
		L-lysine	−0.34	
		L-methionine	−1.05	
		L-phenylalanine	−1.17	
		Combined	−0.48	
6	Rye (meal)	None	−1.28	(Kies and Fox 1971C)
		L-methionine	−1.21	
		L-tryptophan	−1.08	
		L-lysine	−0.58	
		Combined	−0.58	
7	Alfalfa (leaf meal)	None	−1.53	(Kies *et al.* 1968)
		L-lysine	−1.62	
		L-phenylalanine	−1.46	
		L-methionine	−0.69	
		Combined	−0.47	

[1] Separate studies using different subjects.
[2] Fed to provide 4.0 gm N/subject/day, 80% of total dietary nitrogen.
[3] Mean value based on values of 8–10 subjects per study. Each experimental period was 5 days in length.

amino acid requirements of young men indicates methionine as being first limiting. However, human biological testing results are quite different. On the basis of biological testing, lysine, not methionine, is clearly shown as being the first limiting amino acid in this cereal grain (Kies and Fox 1971C).

With the event of spun soya proteins, the possibility of conversion of other high-protein plant products to more palatable, digestible forms for human consumption is evident. Therefore, a high quality alfalfa leaf meal was investigated. Mean nitrogen balances demonstrated methionine as being the first limiting amino acid in this potentially important protein resource for humans (Kies *et al.* 1968).

Using slightly different experimental approaches, lysine was found to be the first limiting amino acid in both rice (Chen *et al.* 1967) and conventional corn grain meal (Kies *et al.* 1967A). Methionine was shown to be first limiting in "instant" oatmeal (a precooked, dehydrated product) and in a soybean textured product resembling ground beef (Kies and Fox 1970B, 1971A).

Serum amino acid patterns of subjects maintained on the previously described experimental regime have also been determined in some but not all cases. These have been rather disappointing. While changes in fasting level amino acid proportionality patterns can be demonstrated via the additions of purified amino acid dietary supplements, interpretation of these changes in meaningful terms reflecting state of protein nutriture are difficult. In other words, does a stated change indicate an improvement or the reverse in the individual's protein nutritive state? This is an unanswered question. Information on crude protein digestibility of the test products was also collected.

Another method of evaluation of protein value of food products, primarily on the basis of amino acid proportionality pattern, is to simply compare 2 or more products at 1 or more levels of nitrogen intake. The protein value of ground beef, an extruded soybean product resembling ground beef (TVP), and a 1% DL-methionine-fortified TVP product resembling ground beef were determined at 2 levels of nitrogen intake. Weighed, controlled, experimental diets were maintained adequate in vitamins, minerals, and calories for each of the 10 adult male subjects during the 62-day study. Mean nitrogen balances of subjects fed 8.0 gm N per day from beef, TVP, or methionine-enriched TVP were +0.74, +0.78, and +0.72 gm N per day, respectively. This would indicate that at this level of protein intake all three sources equaled or exceeded the protein requirements of the subjects. Mean nitrogen balances of subject fed 4.0 gm N per subject per day from beef, TVP, or methionine-enriched TVP were −0.30, −0.70, and −0.45 gm N per day, respectively. While none of the test protein sources fully met the protein needs of subjects at this level of nitrogen intake, beef was superior to TVP on the basis of nitrogen balance data. DL-methionine fortification at the 1% level of TVP was demonstrated as being partially effective in improvement of the protein value of TVP.

LEVEL OF PRODUCT PROTEIN

Total protein content as well as amino acid proportionality pattern is important in determining the value of a food as a protein resource for humans. Even a food source which is relatively poor in protein quality and quantity can meet human protein need if sufficient quantities of the food are fed.

In our laboratory, adult men were maintained on white, degerminated corn meal fed to provide 4.0, 6.0, and 8.0 gm N per subject per day during 3 randomly arranged experimental periods. Caloric intake was maintained constant for each subject at that level necessary for weight maintenance. Other aspects were basically as described in the previous project on first limiting amino acids. Mean nitrogen balances of the subjects while maintained on 4.0, 6.0 and 8.0 gm N per day from corn were —1.03, —0.48 and +0.50 gm per day, respectively. Thus, when subjects received 708 gm of dry corn meal providing 8.0 gm N per day, protein requirements were close to being met. This would be much more easily accomplished with a higher protein product (Kies *et al.* 1965A).

In a recent project, for which analyses are as of yet not completed, the protein nutritive value of 11 dry breakfast cereals for human adults was investigated. Comparisons were made by feeding equal quantities of the cereals (500 gm per subject per day, dry weight basis) during randomized experimental periods of 7 days each. Since the products ranged from 1.14 to 3.30% N (7.14–19.6% protein), nitrogen intake ranged from 5.72 to 16.51 gm per subject per day during the various experimental periods. The 500 gm cereal intake figure is obviously much, much higher than would normally be consumed. However, using this approach, preliminary results suggest that level of protein content and digestibility are far more important than amino acid proportionality pattern in determining protein nutritional value of a dry breakfast cereal product (Kies and Fox 1971C).

People generally eat in terms of portions of food rather than in terms of portion of nutrients. Hence, it seems reasonable to conclude that it is much better to compare on the basis of equal weights or portions of food rather than on the basis of equal intake of protein (or nitrogen). Unfortunately, this approach offers a disadvantage in that with increasing levels of dietary nitrogen, cumulative errors inherent in the nitrogen balance technique tend to favor a progressively greater tendency toward positive nitrogen balance, if great care is not observed.

TOTAL NUTRITION ENVIRONMENT

Among populations where dietary protein is marginal or inadequate, intake of other nutrients is also likely to be far from optimal. Laboratory studies designed to evaluate food proteins are usually conducted under conditions of optimal intake of other nutrients, a procedure which is justifiable in reducing experimental variables but which may result in misleading conclusions in terms of field action programs. Several short term, laboratory controlled, human metabolism studies have been conducted in this laboratory on the effects of overall nutritional value of cereal diets on their apparent protein value. Mean nitrogen balances of subjects consuming corn diets plus general vitamin/mineral supplementation, plus niacin, and plus no added supplements were —0.37, —0.61 and —1.07. These data plus other indices suggest that niacin (specifically) and overall nutriture are factors in determining adequacy of this protein. Nitrogen retention of subjects consuming corn diets plus niacin, plus niacin and lysine, plus tryptophan and lysine are higher than those achieved with no supplementation, with tryptophan alone, with lysine alone or with niacin and tryptophan. Under similar experimental conditions, lysine supplementation was just as effective as lysine/tryptophan supplementation when full vitamin/mineral supplementation was employed. Nitrogen balances of subjects consuming wheat diets plus no supplements, plus leucine, plus niacin, and plus leucine and niacin were —0.95, —1.20, —0.87, and —0.92. Effectiveness of niacin supplementation of corn diets may be related to the high leucine content rather than to the low tryptophan/low niacin availability of corn. The results stress the need to define total nutritional environment of individuals in indicating quantitative needs for a single nutrient (Kies and Fox 1971D).

OTHER FACTORS

The effect of total nitrogen intake, i.e., nitrogen from any metabolically useable nontoxic source on adequacy of cereal proteins has been investigated. It may include nitrogen from the essential amino acids, nonessential amino acids, proteins or nonamino acid sources such as urea or diammonium citrate. Supplementation of a variety of cereal products fed at suboptimal levels for adequate protein nutriture with a variety of nonspecific nitrogen sources seemingly results in a sparing of total protein requirements and a sparing of specific amino acid requirements (Kies *et al.* 1965A, B, 1966, 1967A, B, 1970C, 1971B).

Combining different dietary protein sources so as to hopefully improve the total amino acid proportionality pattern of the diet has been investigated in a study involving algae and fish meal concentrates (Lee *et al.* 1967). Mutual supplementary effects of wheat and rice combinations were studied in Hong Kong orphanage children (Fry and Fox 1968).

Product acceptability is primarily an area of research for the food scientist rather than for the nutritionist. From an application viewpoint, this separation of food science and human nutrition science is unfortunate. If a product is not eaten, it's not nutritious regardless of its favorable attributes. Palatability is only one quality involved in food acceptability. Methionine, the first limiting amino acid in several foods, is sometimes recommended as a supplement for improvement in protein quality. Regrettably, purified methionine has rather disagreeable taste/odor properties which may limit its usefulness as a food supplement. In an earlier described study, methionine was found to be the first limiting amino acid in "instant" oatmeal. Using standard sensory panel techniques, acceptability of fortifying this product with varying levels of L-methionine was judged. Decrease in product acceptability was found with increasing levels of methionine fortification; however, low level fortification resulted in a product having improved protein nutritional value with no discernable loss in product acceptability.

Product acceptability can be viewed from the food processor/producer's viewpoint as well as from that of the consumer. Again, although this is not an area of research emphasis in our department, the topic should not be ignored. Because of cost considerations, the DL-isomer of methionine is more acceptable as a possible nutritional supplement than is the L-isomer. In a recent project evaluating the comparative nutritional value of L-, D-, or DL-methionine-supplemented "instant" oatmeal, results indicated that L-isomer supplementation was more effective. Since level of methionine supplementation is definitely limited by palatability considerations, the greater nutritional effectiveness of L-methionine might justify its use in spite of its higher cost (Kies and Fox 1971C).

GROWTH VERSUS MAINTENANCE IN DETERMINING PROTEIN REQUIREMENTS

Most of the work in this laboratory has involved the use of adults as human biological models. Protein requirement for maintenance of adults is much less stringent than for growth and maintenance of young children. The adult, from legal, moral, and practical standpoints, offers advantages for initial evaluation of food products. If a particular alteration does not result in improvement in protein value for the adult, it is highly unlikely that it will do so for the child. Thus, the number of trials for child evaluation can be drastically reduced with prior evaluation using adults as biological models.

At present, early teenage boys (12–16 yr of age) in the rapid growth stage are being incorporated into our research format. Initial studies indicate that these young men may be excellent biological models for studying protein requirements for growth of humans and for the evaluation of protein value of food proteins. Many of the practical and moral problems inherent in the use of infants or very young children can be avoided by this approach to studying growth needs (Kies and Fox 1971C).

SUMMARY

Evaluation of the protein value of plant products using human biological assay techniques suggest the following factors to be of importance: (1) amino acid proportionality; (2) total protein content; (3) total dietary nitrogen; (4) digestibility; and (5) total nutritional environment. The use of the human as a biological model is expensive and presents unique problems in philosophy, in morality, in legality, in theory, and in practice. However, while small animal feeding studies and chemical/chromotography analysis are of great value in initial evaluation and elimination steps, the results are predictive but do not constitute true evidence of beneficial effects. Laboratory human biological evaluation as the final step before large-scale changes in food production/food consumption practices is highly recommended.

BIBLIOGRAPHY

ANON. 1958. Methods used for human metabolic studies in the North Central region. North Central Regional Publ. *80,* Univ. Minnesota Agr. Expt. Sta. Tech. Bull. *225,*.

CHEN, S. C. S., FOX, H. M., and KIES, C. 1967. Nitrogenous factors affecting the adequacy of rice to meet the protein requirements of human adults. J. Nutr. *92,* 429.

COONS, C. M. 1968. Selected references on cereal grains in protein nutrition: Human and experimental animal studies of major and minor cereals, 1910-1966. USDA Agr. Res. Serv. *ARS 61-5.*

FRY, P. C., and FOX, H. M. 1968. Effect of diet on serum protein fractions of Hong Kong Chinese children. J. Nutr. *95,* 129.

KIES, C., and FOX, H. M. 1970A. Determination of the first limiting amino acid of wheat and triticale grain for humans. Cereal Chem. *48,* 615.

KIES, C., and FOX, H. M. 1970B. Effect of amino acid fortification of "instant" potatoes and "instant" oatmeal on protein nutritive value for human adults. Proc. 3rd Intern. Congress Food Sci. Technol., Washington, D.C., Aug. 1970.

KIES, C., and FOX, H. M. 1970C. Effect of level of total nitrogen intake on second limiting amino acid in corn for humans. J. Nutr. *100,* 1275.

KIES, C., and FOX, H. M. 1971A. Comparative protein value for human adults of a soybean textured food, a methionine-fortified soybean textured food and ground beef. Proc. Soc. Exptl. Biol. *28,* 297 (Abstr.).

KIES, C., and FOX, H. M. 1971B. Studies on the nutritive value of *Opaque-2* corn for human adults. J. Nutr. (in press).

KIES, C., and FOX, H. M. 1971C. Unpublished data.

KIES, C., and FOX, H. M. 1971D. Interrelationships of leucine with lysine, tryptophan and niacin as they influence protein value of cereal grains for humans. Cereal Chem. (in press).

KIES, C., FOX, H. M., and WILLIAMS, E. R. 1966. Effect of quantitative-qualitative variations in "non-specific" nitrogen supplementation of several dietary proteins. Proc. Soc. Exptl. Biol. *25,* 300 (Abstr.).

KIES, C., FOX, H. M., and WILLIAMS, E. 1967A. Effect of non-specific nitrogen supplementation on minimum corn protein requirement and first limiting amino acid for adult men. J. Nutr. *92,* 377.

KIES, C., FOX, H. M., and WILLIAMS, E. R. 1967B. Time, stress, quality and quantity as factors in the non-specific nitrogen supplementation of corn protein for adult men. J. Nutr. *93,* 377.

KIES, C., JAN, J., and FOX, H. M. 1968. Determination of the first limiting amino acid in alfalfa leaf meal and in milo meal for meeting adult human need. Proc. Soc. Exptl. Biol. *25,* 299.

KIES, C., WILLIAMS, E., and FOX, H. M. 1965A. Determination of the first limiting nitrogenous factor in corn protein for nitrogen retention in human adults. J. Nutr. *86,* 350.

KIES, C., WILLIAMS, E., and FOX, H. M. 1965B. Effect of "non-specific" nitrogen intake in adequacy of cereal proteins for nitrogen retention in human adults. J. Nutr. *86,* 357.

LEE, S. K., FOX, H. M., KIES, C., and DAM, R. 1967. Supplementary value of algae protein in human diets. J. Nutr. *92,* 251.

ROSE, W. C. 1957. The amino acid requirements of adult men. Nutr. Abstr. Rev. *27,* 631.

CHAPTER 19

Melvin P. Klein
and
Leo N. Kramer

Determination of Protein
Quantity and Quality by X-Ray
Photoelectron Spectroscopy

X-ray photoelectron spectroscopy offers a method of high potential for determining the quantity and quality of grain protein. In this method a small quantity of material is irradiated with X-rays from a suitable target material and electrons from all constituent atoms may be photo-ejected. The kinetic energy (KE) of these photoemitted electrons is equal to the photon energy of the incident X-rays minus the binding energy (BE) of the level from which the electrons are emitted, $KE = h\nu - BE$. This kinetic energy, and hence the binding energy, are determined by a magnetic or electrostatic momentum analyzer, and are characteristic of the atom and level whence the electrons originated. All elements across the periodic table may be investigated. Calibration against compounds of known elementary composition permits a quantitative determination of the amount of each atomic species contained in the sample. The binding energies exhibit chemical shifts which permit a distinction between a given element in different chemical groupings. These experiments have demonstrated: (1) total protein may be evaluated by quantitative determination of the amide nitrogen peak; (2) amino nitrogen may be distinguished from amide nitrogen thus providing a measure of the basic amino acids lysine, arginine, and histidine; and (3) the sulfur content may be determined by observing the sulfur photoelectrons.

Instrumentation currently under development promises to reduce the time per sample from hours to minutes and thus render possible large-scale screening.

A convenient, rapid, economical, and nondestructive method for determining the quantity and quality of grain proteins is desired for mass screening operations. In this paper we present some results of a new method which can now satisfy some of these requirements and offers the potential of satisfying all of them.

METHOD[1]

When a sample is irradiated with light of sufficient energy, electrons may be ejected and their kinetic energy is given by the well known Einstein relation,

$$KE = h\nu - BE$$

---

[1] This work was performed under auspices of the US AEC. A preliminary account was reported in *Improving Plant Protein by Nuclear Techniques,* International Atomic Energy Agency, Vienna, 1970.

where KE is the kinetic energy of the electrons, h is Planck's constant, ν is the frequency of the light quantum, and BE is the binding energy of the electrons in the sample. If the photon energy is increased sufficiently, by using X-rays, electrons may be ejected from the inner or core levels of the constituent atoms of the sample. Since photon or X-ray energies are known to high accuracy, the binding energies may be determined with precision by measurement of the kinetic energies. The measurements may be performed with a variety of devices, but most commonly magnetic and electrostatic deflection analyzers are employed. The overwhelming majority of the energy levels of the elements across the periodic table have been determined in this manner. Although the complementary method of X-ray emission contains the same information, the former method, called X-ray photoelectron spectroscopy (XPS), offers so many direct advantages that we will confine the present discussion thereto.

The method is outlined schematically in Fig. 19.1. In this Figure are sketched the discrete energy levels of a particular atom in a compound as well as those levels which are a collective property of the compound as a whole and make up its valence band or molecular orbitals. An X-ray photon is shown lifting an electron from the 2S level of this atom into the continuum of energies or,

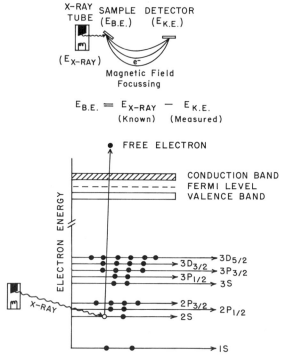

FIG. 19.1. ENERGETICS AND ANALYSIS OF X-RAY PHOTOELECTRONS

equivalently, removing it to infinity. At this point the electron enters the spectrometer wherein it is brought to a focus and impinges on a detector when the electron has the correct energy (more rigorously, the correct momentum). The energy range is scanned by varying the strength of the magnetic field and thus a spectrum is traced out by recording the number of electrons reaching the detector at each value of magnetic field. This method was developed and brought to its present state of refinement by the group at the Institute of Physics, University of Uppsala, Sweden (Hagstrom et al. 1968).

Since all atoms are constructed in the same manner, it is possible to photoeject their electrons, and by a suitable choice of exciting X-ray energy and particular atomic level one can distinguish among the different constituent atoms of the sample. It is apparent, then, that calibration against a sample of known elementary composition will permit a qualitative and quantitative analysis of the unknown sample. The absolute sensitivity of the method is very high although it is not especially suitable for detecting small amounts of one element in the presence of a large excess of other elements.

More recently, it has been shown that a given element in different chemical configurations exhibits chemical shifts of its binding energies (Siegbahn et al. 1967; Fadley et al. 1967). These chemical shifts thus extend the utility of the method so that not only the total quantity of a given element may be determined, but also the type or types of chemical bonding situation in which the atom is situated. The origins and theoretical foundations for these chemical shifts are rather well understood (Siegbahn et al. 1967; Fadley et al. 1968).

EXPERIMENTAL DESIGN

Since virtually all of the previous work in this field has been directed toward precision measurements of the energies of the photoelectrons, we were obliged to devise techniques to permit the quantitative determinations of the relative numbers of relevant atoms contained in the samples. We were further interested in establishing the feasibility of measuring not only total protein by nitrogen, but also of assessing the free amino nitrogen as a measure of lysine, arginine, and histidine, and the sulfur as a measure of cysteine, cystine, and methionine. Our principal research endeavors prior to our introduction to the present topic are related to iron and sulfur and the nonheme iron sulfur proteins known as the ferredoxins. Accordingly, we had some prior knowledge of sulfur spectra but no current nitrogen data were pertinent to the present problem (Kramer and Klein 1969).

The first task undertaken was the observation of the nitrogen and sulfur spectra of relevant amino acids, simple peptides, and known proteins. Figure 19.2 shows spectra of three such compounds. The cystine spectra show lines from both nitrogen and sulfur. The designations N1S and S2P indicate that the electrons arise from the 1S level of nitrogen and the 2P level of sulfur, respectively. The dipeptide, L-isoleucyl-L-alanine, exhibits a nitrogen peak

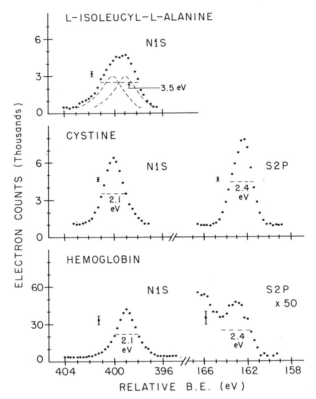

FIG. 19.2. X-RAY PHOTOELECTRON SPECTRA OF THREE COM-
POUNDS SHOWING NITROGEN AND SULFUR LINES

The dipeptide nitrogen line is decomposed into two components of
equal width. One component is isoenergetic with the cystine nitrogen
and is assigned to the amino nitrogen. The other component, isoenergetic with the hemoglobin, is assigned to the amide nitrogen.

considerably broader than that of the cystine. This peak may be decomposed
into 2 peaks of width equal to the cystine peak; 1 component is of the same
energy as the cystine peak and we conclude that it corresponds to the amino
nitrogen. The other peak occurs at lower binding energy and is attributed to the
amide nitrogen formed during the peptide bonding. The lower lines originate
from a sample of hemoglobin. The nitrogen peak occurs at the position of lower
binding energy and is therefore the amide nitrogen. The sulfur peak is shown
magnified by 50-fold and exhibits an increasing shoulder at higher binding
energies. We believe this shoulder arises from sulfur as $SO_4^=$ and that it most
probably entered the sample during ammonium sulfate precipitation of the
protein (Kramer and Klein, unpublished results).

Figure 19.3 shows the nitrogen and sulfur spectra from three protein samples,

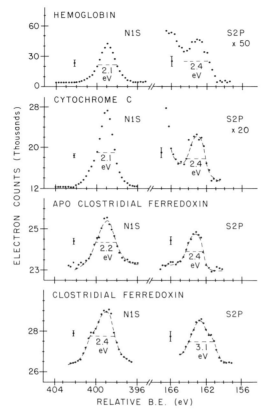

FIG. 19.3. NITROGEN AND SULFUR X-RAY PHOTOELECTRON
SPECTRA OF THREE PROTEINS

hemoglobin (repeated for continuity), equine cytochrome C, and the apo-ferredoxin and its native form from *Clostridium pasteurianum.* This latter protein has a molecular weight of 6000 and contains 8 iron atoms, 8 moles of cysteine, and 6—8 moles of acid labile sulfur (Malkin and Rabinowitz 1967). The cytochrome C spectrum is similar to that of hemoglobin but shows a higher sulfur content. It also contains $SO_4^=$. The apo-ferredoxin spectra are qualitatively similar to both the heme-proteins, except for the larger quantity of sulfur. The native ferredoxin spectra are significantly different from the heme-proteins; both lines are broader and exhibit rather obvious shoulders. The nitrogen shoulder at higher binding energy probably arises from nitrogen entering as NH_4^+ during ammonium sulfate precipitation of the protein (Kramer and Klein, unpublished results). The low energy sulfur shoulder definitely arises from sulfur bonded to iron and is characteristic of such bonding (Kramer and Klein 1969).

Figure 19.4 shows the nitrogen spectra of Barker barley seed and of a dark red kidney bean. The barley spectrum is very similar to the other proteins; the

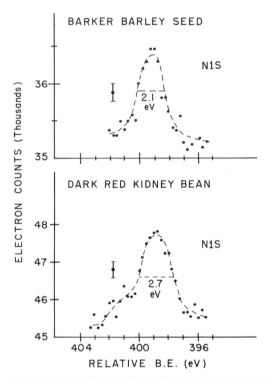

FIG. 19.4. NITROGEN X-RAY PHOTOELECTRON SPECTRA OF
TWO SEEDS

kidney bean spectrum is considerably broader and exhibits a high energy shoulder.

Figure 19.5 shows the nitrogen spectrum of a light red kidney bean and the curve has been decomposed into two components by computer. The major peak, at lower binding energy, represents the amide nitrogen, while the smaller peak at higher binding energy represents the amino nitrogen. Figure 19.6 shows the sulfur region of this same sample.

Figure 19.7 shows the nitrogen spectrum of a sample of Ramona wheat where, again, the experimental data have been fit by two peaks corresponding to the amide and amino nitrogen. Figure 19.8 shows the sulfur region of the wheat sample.

All of the foregoing spectra were excited by Mg Kα radiation at 1252 eV and analyzed in the Berkeley iron-free spectrometer.

SAMPLE PREPARATION

The nature of photoelectron spectroscopy imposes constraints on the form of the samples. Since the kinetic energy of the photoelectrons is very low, usually

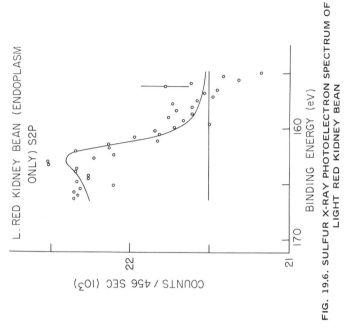

FIG. 19.6. SULFUR X-RAY PHOTOELECTRON SPECTRUM OF
LIGHT RED KIDNEY BEAN

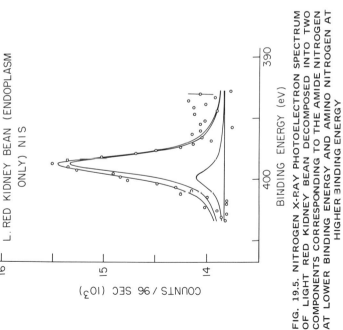

FIG. 19.5. NITROGEN X-RAY PHOTOELECTRON SPECTRUM
OF LIGHT RED KIDNEY BEAN DECOMPOSED INTO TWO
COMPONENTS CORRESPONDING TO THE AMIDE NITROGEN
AT LOWER BINDING ENERGY AND AMINO NITROGEN AT
HIGHER BINDING ENERGY

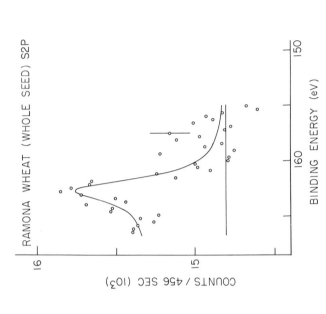

FIG. 19.8. SULFUR X-RAY PHOTOELECTRON SPECTRUM OF
RAMONA WHEAT SEED

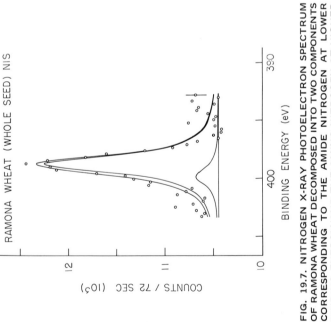

FIG. 19.7. NITROGEN X-RAY PHOTOELECTRON SPECTRUM
OF RAMONA WHEAT DECOMPOSED INTO TWO COMPONENTS
CORRESPONDING TO THE AMIDE NITROGEN AT LOWER
BINDING ENERGY AND THE AMINO NITROGEN AT HIGHER
BINDING ENERGY

less than 1.5 keV, they must be analyzed *in vacuo* to reduce scattering. Also, because of their low energy, they are able to penetrate a distance of only a few tens to a few hundreds of angstroms in solid material. Thus, only those electrons originating within this distance from the surface of the sample escape elastically and give rise to discrete photoelectron lines. Only the surface of a given sample is of use.

Accordingly, for these studies the samples were prepared as powders by grinding in a ball mill. For the quantitative determinations the biochemicals were mixed with NaC1 and then pulverized together. The amino acids, dipeptides, hemoglobin, and equine cytochrome C were obtained from commercial sources. The ferredoxin was prepared in the laboratory of Professor J. C. Rabinowitz in Berkeley. The seed samples were obtained from the seed station of the University of California, Davis, California.

QUANTITATIVE DETERMINATIONS

The basic relative sensitivity for nitrogen and sulfur was obtained from a sample of cystine mixed with NaC1. The areas under the N1S, S2P, and Na2S peaks were integrated by computer and were treated as follows:

Sulfur

$$(1) \quad \frac{\text{Area S2P}}{\text{Area Na2S}} \times \frac{\text{Moles NaC1}}{\text{Moles of compound}} = 14/1 \text{ atom equivalent}$$

$$(2) \quad \frac{\text{Weight of compound} \times \% \text{ S}}{32} = \text{Moles of compound}$$

$$(3) \quad \frac{\text{Area S2P}}{\text{Area Na2S}} \times \frac{\text{Moles NaC1}}{14} \times \frac{32}{\text{Weight of compound}} = \% \text{ S}$$

Nitrogen

$$(4) \quad \frac{\text{Area N1S}}{\text{Area Na2S}} \times \frac{\text{Moles NaC1}}{\text{Moles of compound}} = 9.4/1 \text{ atom equivalent}$$

$$(5) \quad \frac{\text{Weight of compound} \times \% \text{ N}}{14} = \text{Moles of compound}$$

$$(6) \quad \frac{\text{Area N1S}}{\text{Area Na2S}} \times \frac{\text{Moles NaC1}}{9.4} \times \frac{14}{\text{Weight of compound}} = \% \text{ N}$$

These sensitivity factors are specific for our instrument and for the particular sample geometry and method of data analysis.

Table 19.1 gives the quantitative estimates for several samples determined by

TABLE 19.1-A

QUANTITATIVE DETERMINATION OF NITROGEN AND SULFUR CONTENT

Compound	Element	Experimental Weight (%)	Calculated Weight (%)
Cysteine[1] (1)	Nitrogen	8.2	8.2
	Sulfur	18.0	19.3
Cysteine[1] (2)	Nitrogen	8.5	8.5
	Sulfur	20.0	19.3
Cystine (1)	Nitrogen	13.0	11.7
	Sulfur	26.3	26.7

[1] Cysteine-hydrochloride-monohydrate

TABLE 19.1-B

ELEMENTAL ANALYSIS

Seed	% N(XPS)	% N(anal)	% S(XPS)	% S(anal)
Barker barley	1.5±0.2	1.5±0.2	0.05±0.02	0.01±0.01
Ramona wheat	1.6±0.2	1.7±0.2	0.05±0.02	0.06±0.03
Rapida oats	2.2±0.2	1.9±0.2	0.1±0.03	0.02±0.02
Light red kidney bean	3.2±0.3	4.1±0.3	0.08±0.04	0.13±0.06

TABLE 19.1-C

SEED PROTEIN DISTRIBUTION (LIGHT RED KIDNEY BEAN)

Sample	% N(XPS)	% N(anal)	% Basic AA (XPS)	% Basic AA (anal)
Endoplasm	3.1±0.3	4.0±0.3	17±5	17.4
Embryo and endoplasm	3.1±0.3	4.2±0.3	18±5	

this method. The two cysteine samples were prepared independently and are shown as an indication of the reproducibility of the method. The cystine sample is different from that used as the calibration standard. For the seeds, the experimental weight percentage of nitrogen, determined from the photoelectron spectral amide nitrogen peak, was converted to percentage of protein by assuming a nitrogen:protein weight ratio of 0.15. Table 19.1 includes both nitrogen and sulfur analyses by both X-ray photoelectron spectroscopy and by conventional wet chemical analysis. Except for the light red kidney bean, the results are comparable. The apparent discrepancy for the light red kidney bean sample may be attributed to the fact that this material is extremely hard and may not have been milled adequately to produce a sufficiently fine powder.

Also shown in Table 19.1 is the distribution of protein between the embryo and the total seed for the light red kidney bean. To ascertain the validity of the hypothesis that the component of the nitrogen photoelectron line ascribed to the amino nitrogen did in fact correspond to the quantity of basic residues, an amino acid analysis was performed. The results, derived from such an analysis on the total acid hydrolysate, are given in Table 19.2, wherein the fraction of the basic residues is shown to be 17.4%. This value equals that derived from the photoelectron data, thus substantiating the assumption.

TABLE 19.2

AMINO ACID ANALYSIS OF LIGHT RED KIDNEY BEAN

Amino Acid	Quantity (μmole)	Amino Acid	Quantity (μmole)
Aspartic acid	0.163	Allo-isoleucine	0.007
Threonine	0.030	Isoleucine	0.060
Serine	0.016	Leucine	0.109
Glutamic acid	0.203	Tyrosine	0.011
Proline and cystine	0.063	Phenylalanine	0.057
Glycine	0.099	NH$_3$	0.369
Alanine	0.087	Basic Amino Acids	
Cystine	0.001	Lysine	0.093
Valine	0.086	Histidine	0.039
Methionine	0.005	Arginine	0.078

Total AA content = 1.107 μmole

$$\% \text{ Basic AA} = \frac{0.210}{1.207} \times 100 = 17.4\%$$

SUMMARY

This report demonstrates that the quantity and quality of grain protein may be determined by X-ray photoelectron spectroscopy. The quantity of protein may be determined by quantitation of the amide nitrogen photoelectron line. Two measures of the quality of the protein are available: (1) the basic amino acids including lysine, arginine, and histidine may be determined by quantitation of the amino nitrogen photoelectrons; and (2) the sulfur-containing amino acids, cystine, cysteine, and methionine, may be quantitated by measuring the sulfur photoelectron line.

The quantity of material required to perform the measurements need be only a few milligrams. Indeed, the active volume giving rise to the photoelectron lines is probably no greater than a few tens of micrograms of material of unit density. Thus, it is feasible to excise a small section of a seed for the analysis while retaining the remainder of the seed for future planting. We performed one such experiment and were able to bring the bulk of the seed to germination.

The experiments reported here were designed to test the feasibility of the method to measure protein quality and quantity. Thus, care was taken to measure the entire nitrogen and sulfur regions of the photoelectron spectrum. The times required for these nitrogen determinations were approximately 1 hr per sample. Such a large period of time clearly limits the utility of this method for mass screening operations. Were it satisfactory to perform only a total nitrogen—and thus total protein—analysis on a large number of samples, one could dispense with the high spectral resolution and measure only the integrated nitrogen intensity. Such a measurement could be accomplished in about 2 min per sample. This estimate is for the spectrometer observation time alone and does not include the time required for sample preparation. Since the seed material is analyzed in its native state, aside from the pulverizing operation, the time-consuming preparative steps requisite to and employed in purely chemical methods are eliminated, as are the separate and independent procedures required for each element of interest. Newer instrumentation promises to offer a significant further saving in time.

BIBLIOGRAPHY

FADLEY, C. S. et al. 1967. Chemical bonding information from photoelectron spectroscopy. Science 157, 1571-1573.

FADLEY, C. S., HAGSTROM, S. B. M., KLEIN, M. P., and SHIRLEY, D. A. 1968. Chemical effects on core-electron binding energies in iodine and europium. J. Chem. Phys. 48, 3779-3794.

HAGSTROM, S., NORDLING, C., and SIEGBAHN, K. 1968. Tables of electron binding energies. In Alpha-, Beta-, and Gamma-Ray Spectroscopy, Vol. 1, K. Siegbahn (Editor). North-Holland, Amsterdam.

KRAMER, L. N., and KLEIN, M. P. 1969. Chemical shifts in core electron binding energies of iron and sulfur in iron complexes determined by photoelectron spectroscopy. J. Chem. Phys. 51, 3618-3620.

MALKIN, R. M., and RABINOWITZ, J. C. 1967. Nonheme iron electron-transfer proteins. Ann. Rev. Biochem. 36, 113-147.

SIEGBAHN, K., et al. 1967. ESCA-atomic, Molecular and Solid State Structure Studied by Means of Photoelectron Spectroscopy. Almquist and Wiksells, Uppsala, Sweden.

CHAPTER 20

W. Emile Coleman

Automatic Amino Acid Analyses
For Determining the Amount and
Quality of Protein in Fungal
Protein and in Other Protein Sources

INTRODUCTION

The U.S. Environmental Protection Agency's Solid Waste Research Laboratory has, in its efforts to recycle starchy and cellulosic waste (Rogers *et al.* 1970, 1972), developed technology for producing fungal protein. Evaluating fungi for amino acid and, hence, for the amount of protein involves (1) sample preparation of the fungal growths, (2) hydrolysis of the fungal protein, (3) preparation of the amino acid calibration mixture for amino acid analysis, (4) automatic amino acid analyses of samples, and (5) calculations of individual amino acids to determine the protein content of the fungal mycelium. The quality of the protein, which is produced by fermentation, depends on the amino acid profile; this profile is determined by a quantitative analysis with the use of an Automatic Amino Acid Analyzer.[1] The amount of protein in the fungal mycelium is determined by totaling the amino acids liberated during the protein hydrolysis reaction. Results of analyses of 2 fungi (1 grown on starchy and 1 grown on cellulosic wastes) and of a commercial yeast are discussed; their amino acid compositions are compared with the Food and Agriculture Organization's (FAO) ideal pattern for amino acids in foodstuffs.

Calculation of percentage of protein must be treated with some care. Determining the amount of protein in products such as fungi is usually done by analyzing for total nitrogen and then multiplying by a factor of 6.25 to convert nitrogen to protein (Miller 1968). This gives a reasonable approximation for foods of known concentration; in the case of fungal protein, however, the determination is more complex. Not only do we not know if 6.25 is a proper factor for the nitrogen:protein ratio of the organism supplying the protein but we do not know how much nitrogen other sources contribute to the total nitrogen content. Nonprotein nitrogen materials such as nucleic acids, purines, pyrimidines, sugar amines, and others are known to occur in single cell organisms, and these contribute significantly to the total nitrogen content (Mateles and Tannenbaum 1968). Therefore, the protein content of the fungal protein is more accurately calculated by totaling only those amino acids liberated during the hydrolysis reaction.

---

[1] Mention of companies or products by name does not imply their endorsement by the U.S. Government.

EQUIPMENT

A Phoenix Biolyzer, Model B-9000, was used for quantitative amino acid analyses. The acidic, neutral, and basic amino acids were separated on a single 60 cm X 0.9 cm column of cation exchange resin (Spherix xx8-60-1) with the use of pH 2.91 citrate buffer and 0.25 M sodium citrate. The buffer system is referred to as the accelerated Piez-Morris System (Piez and Morris 1960; Phoenix Precision Instrument Co. 1968). A 4-channel Automatic Sample Injector incorporated into the analyzer provided 24-hr operation. A photoelectrically-controlled proportioning pumping system (Phoenix Varipump, Model 4000) produced the buffer gradients.

MATERIALS, REAGENTS, AND STANDARD SOLUTIONS

Ninhydrin and stannous chloride were dissolved in a solvent of half methyl Cellosolve and half propionate buffer to obtain ninhydrin reagent. Ninhydrin reagent, propionate buffer, and citrate buffers were volumetrically prepared as outlined in the Liquid Chromatography Handbook (Phoenix Precision Instrument Co. 1968). The pH 2.91 sodium citrate buffer contained 6% n-propanol to facilitate separation of threonine and serine. Each milliliter of the amino acids standard solution (Pierce Chemical Co., Rockford, Ill., Stock No. 200891) contained 1.25 μmol of cystine and 2.50 μmol each of the following: ammonia as $(NH_4)_2SO_4$, 1-lysine, 1-histidine, 1-arginine, 1-aspartic acid, 1-threonine, 1-serine, 1-glutamic acid, 1-proline, glycine, 1-alanine, 1-valine, 1-methionine, 1-isoleucine, 1-leucine, 1-tyrosine, and 1-phenylalanine. The protein samples were hydrolyzed with 6 N HCl in a vacuum hydrolysis tube (Phoenix Precision Instrument Co.). Prepurified nitrogen was used to deaerate the hydrolysis samples.

PROCEDURE FOR PROTEIN EVALUATION

The fungal protein was usually grown in a 500-ml flask containing 100 ml of mineral salts media (Rogers *et al.* 1972). Because the fungal growths were still active when presented for protein analysis, each flask was autoclaved for 15–20 min at 120°C.

Sample Preparation

After the fungal mass was autoclaved, the growth media was decanted and discarded; the mold was retained in the flask. The mold was washed by swirling approximately 200 ml of distilled water in the flask for about 5 min. The water was decanted, and this washing procedure was repeated twice. (This vital washing procedure ensures removal of the remaining mineral salts. If the metal ions in the growth media are not removed, they become bonded to the cation exchange resin in the chromatographic column, interfere with the analysis, and result in loss of resolution. Then, too, if the ammonium ion is not removed, it produces a huge peak that covers several amino acid peaks on the chromatogram.) After the last washings were decanted, the mold was transferred to an evaporating dish and freeze dried. The dried material was ground with mortar and pestle until uniform in particle size. (In the absence of freeze-drying equipment, the mold can be dried in a thermostatically-controlled oven at 75°C for 24 hr.)

Deaeration and Hydrolysis.—After 20 mg of the prepared protein material was weighed into a vacuum hydrolysis tube, 10 ml of 6 N HCl was added. The two sections of the tube were clamped securely together, and the tube was placed on an all-glass pressure system (Fig. 20.1). The sample was deaerated and evacuated as outlined in a procedure by Coleman (1971). While the sample was maintained *in vacuo,* the hydrolysis tube was removed from the pressure system and placed in a constant-temperature, thermostatically-controlled oven at 110°C ± 1° for 22 hr.

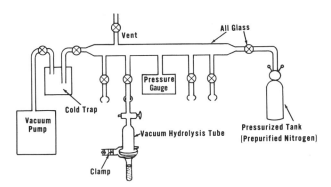

FIG. 20.1. PRESSURE SYSTEM FOR DEAERATING SAMPLES

Preparation of the Protein Hydrolyzate.—After the hydrolyzate was cooled to room temperature, the contents of the tube were filtered through a medium-porosity sintered-glass filter into an evaporating dish. The hydrolysis tube and the filter were carefully washed with distilled water; the volume of liquid in the evaporating dish did not exceed 20 ml. The HCl was removed by freeze drying the hydrolyzate. When dry, 10 ml of distilled water was added to the evaporating dish and the sample was again freeze dried. The freeze-dried material, representative of the original 20 mg of mycelium used for hydrolysis, was made up volumetrically to 5 ml with 0.1 N HCl. After this solution was refiltered through a medium-porosity sintered-glass filter into a small stoppered flask, it contained the equivalent of 4 mg of mycelium per milliliter. When the sample was not to be analyzed immediately, it was frozen.

Preparation of the Amino Acid Calibration Solution.—Enough 0.1 N HCl was added to 2 ml of the amino acid standard solution (Pierce Chemical Co.) to make 10 ml of amino acid calibration solution. A 0.5-ml injection of the calibration solution into the Amino Acid Analyzer results in a concentration of 0.125 μmol for cystine and 0.25 μmol for all other amino acids.

Amino Acid Analyses

An Automatic Sample Injector on the Amino Acid Analyzer can be loaded with 4 samples; with the use of a buffer programmer and a shutdown timer incorporated into the instrument, the 4 samples successively are injected every 6 hr and analyzed over a 24-hr period. An additional 1-1/2 hr is added to the analysis time to recondition the column. At the end of a predetermined time, the instrument is automatically shut down. A calibration standard of amino acids is always run in addition to the protein hydrolyzates. Customarily, the standard is placed in the first sample holder and the protein hydrolyzates in the other three. When the standard is analyzed first, instrument operation can be checked and malfunctions can be detected and corrected before the hydrolyzates are injected. The 2-ml sample reservoirs were half filled with ion exchange resin covered with buffer. The buffer above the resin was withdrawn, and 0.5-ml samples were injected onto the resins with a pipet or a syringe. (Our samples are stored in 25-ml Erlenmeyer flasks stoppered with silicone rubber. Extracting the liquid through the rubber stopper with a syringe was easy.) At the end of 6 hr, a complete amino acid chromatogram of the calibration solution was recorded.

The foregoing procedure has stressed analysis of fungal protein because our research is involved with biosynthesis of fungal protein from cellulosics; however, any other protein-bearing material is treated similarly.

Calculations

Peak Area Integration.—The amount of each amino acid component in a sample analyzed on the Automatic Amino Acid Analyzer is determined by measuring the area enclosed by the component's corresponding peak on the chromatogram. Because the height-times-width (HXW) method of integration is rapid and satisfactorily accurate, it was used to integrate all of the peaks. From the analysis of the standard amino acid solution, a constant (HXW per μmol) is calculated for each amino acid (Spackman et al. 1958). When analyzing for unknowns, HXW is divided by C to give the micromoles of each amino acid in the sample added to the column. For convenience in calculating the percentage of protein, the micromoles of each amino acid present in the unknown samples were converted to milligrams of amino acids.

Percentage of Protein.—In the case of fungal protein, a dry weight of 20 mg was hydrolyzed, freeze dried, and dissolved in 5 ml of 0.1 N HCl. A 0.5-ml aliquot of the hydrolyzate was, therefore, equivalent to 2 mg of sample (dry weight). The percentage of protein was then determined by:

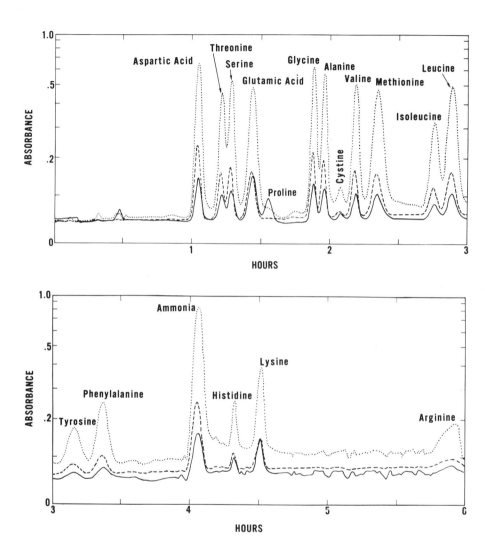

FIG. 20.2. CHROMATOGRAM OF HYDROLYZATE OF *ASPERGILLUS NIGER* PROTEIN GROWN ON POTATO WASTE

· · · · Corresponds to 570 mμ, ——— to 440 mμ, and — — — to 570 mμ (suppressed). The buffer flow was 80 ml/hr; the ninhydrin flow, 40 ml/hr; and the column temperature, 56° C.

$$\% \text{ protein } = \frac{\text{total weight amino acids (mg)}}{\text{dry weight sample (mg)}} \times 100$$

For example, if the summation of all amino acids from a protein hydrolyzate containing the equivalent of 2 mg dry weight of sample per 0.5 ml was 0.726 mg, then

$$\% \text{ protein } = \frac{0.726}{2} \times 100 = 36.3$$

RESULTS OF ANALYSES

The two fungi discussed in this paper were selected as a result of numerous screening experiments on the biodegradability of waste cellulosic and waste starchy substrates by fungi (Rogers *et al.* 1970, 1972). Cellulosic substrates included Solka Floc, kraft paper, and hydropulped refuse; starchy substrates included fruit and potato wastes. *Aspergillus fumigatus* best degraded the cellulose and *Aspergillus niger,* the starch. The yeast was selected solely because it is known to be approximately 80% protein and is commonly used as a food additive (Bressani *et al.* 1961).

TABLE 20.1

AMINO ACID COMPOSITION OF SELECTED FUNGI

Amino Acids	*Aspergillus fumigatus*	*Aspergillus niger*	Torula Yeast	Ideal FAO Amino Acid Pattern
		Gm Amino Acid/100 Gm Protein		
Alanine	6.4	6.3	7.1	—
Arginine	6.3	4.7	6.4	—
Aspartic acid	9.1	9.4	9.7	—
Cystine	0.4	0.8	Not detected	2.0
Glutamic acid	11.3	12.5	13.3	—
Glycine	4.8	5.0	4.1	—
Histidine	2.3	1.8	2.5	—
Isoleucine	7.2	5.3	6.3	4.2
Leucine	10.9	8.6	8.5	4.8
Lysine	6.8	5.9	8.5	4.2
Methionine	3.5	3.2	1.9	2.2
Phenylalanine	5.3	7.7	7.3	2.8
Proline	4.2	4.4	4.7	—
Serine	4.1	4.7	4.2	—
Threonine	4.4	4.8	4.8	2.8
Tryptophan	Not determined			
Tyrosine	3.0	4.4	4.5	2.8
Valine	8.3	9.2	6.1	4.2

The chromatogram of *Aspergillus niger* hydrolyzate shows all of the essential amino acids (Rose 1952; Morrison 1964) except tryptophan (Fig. 20.2).

Since tryptophan is destroyed by acid hydrolysis (Kimmel 1964), it may or may not have been present in the protein. Separate analyses will be conducted to determine its presence.

The amino acid content and protein composition of the fungi and yeast were determined (Table 20.1). Each amino acid is expressed as grams of amino acid per 100 gm protein and compared directly with the FAO ideal amino acid pattern (FAO 1957). With the use of the method of calculating percentage of protein discussed earlier, the percentage of protein of *Aspergillus fumigatus* was 13.3; of *Aspergillus niger*, 36.3; and of Torula yeast, 85.4.

BIBLIOGRAPHY

BRESSANI, R. *et al.* 1961. All-vegetable protein mixtures for human feeding. IV. Biological testing of INCAP vegetable mixture nine in chicks. J. Nutr. *74*, 209-216.

COLEMAN, W. E. 1971. Vacuum-acid hydrolysis of fungal protein and of other protein sources. Div. Res. Develop. Open-file Rept. *RS-03-68-17*. U.S. Environmental Protection Agency, Cincinnati.

FAO. 1957. Protein requirements. Bull. *16*, Food Agr. Organ. Rome, Italy.

KIMMEL, J. R. 1964. Some recent advances in techniques for protein degradation. *IN* Symposium on Foods: Proteins and Their Reactions, H. W. Schultz and A. F. Anglemier (Editors). Avi Publishing Co., Westport, Conn.

MATELES, R. I., and TANNENBAUM, S. R. 1968. Single-Cell Protein. M.I.T. Press, Cambridge, Mass.

MILLER, S. A. 1968. Nutritional factors in single-cell protein. *IN* Single-Cell Protein, R. I. Mateles and S. R. Tannenbaum (Editors). M.I.T. Press, Cambridge, Mass.

MORRISON, A. B. 1964. Some aspects of the nutritive value of proteins. *IN* Symposium on Foods: Proteins and Their Reactions, H. W. Schultz and A. F. Anglemier (Editors). Avi Publishing Co., Westport, Conn.

Phoenix Precision Instrument Company. 1968. Liquid Chromatography Handbook, Philadelphia.

PIEZ, K. A., and MORRIS, L. 1960. A modified procedure for the automatic analysis of amino acids. Anal. Biochem. *1*, 187-201.

ROGERS, C. J. *et al.* 1972. Production of fungal protein from cellulose and waste cellulosics. Environ. Sci. Technol. *6*, 715-719.

ROGERS, C. J., COLEMAN, W. E., SPINO, D. F., and PURCELL, T. C. 1970. Fungal biosynthesis of protein from potato waste. *IN* Abstracts of Papers. 160th National Meeting, ACS, Chicago, Sept. 14-18.

ROSE, W. C. 1952. Half-century of amino acid investigations. Chem. Eng. News *30*, 2385-2388.

SPACKMAN, D. H., STEIN, W. H., and MOORE, S. 1958. Automatic recording apparatus for use in the chromatography of amino acids. Anal. Chem. *30*, 1190-1206.

CHAPTER 21

Allen J. St. Angelo
and
Robert L. Ory

Investigations on Lipoxygenase and Associated Lipid-Oxidizing Systems in Dormant Peanuts

The development of rancidity in vegetable oils has been attributed to a number of causes; light, air, high temperature, trace metals, enzymes, micro-organisms, the presence of free fatty acids, and autoxidation by the unsaponifiable fraction in the oil. Triglycerides of vegetable oils contain unsaturated fatty acids which can undergo oxidation at the double bonds, catalyzed by metals or enzymes, particularly, lipoxygenase (E.C. 1.13.1.13).

Oxidative degradation of lipids during storage or processing is a major cause of the undesirable odors and flavors, of the destruction of essential fatty acids, aroma and flavor constituents, pigments, vitamins, and possibly of toxicity of food products. Roubal and Tappel (1966) reported damage to proteins, enzymes, and amino acids by peroxidizing lipids; methionine, histidine, cystine, and lysine were the most labile. Interestingly, the concentrations of these amino acids are generally low in oilseeds and cereal grains. More recently, lipid degradation products were shown to inactivate various proteolytic enzymes (Matsushita *et al.* 1970).

In general, the time required for rancidity to develop is associated with the relative proportions of various fatty acids in the oil, particularly those possessing high linoleic acid content. Peanuts contain from 45 to 55% oil, of which 20–37% is linoleic acid. The variety used in these investigations, Virginia 56-R, contains approximately 30% linoleic acid.

Lipoxygenase, the enzyme primarily responsible for biochemical peroxidation of unsaturated fatty acids, has been extensively studied in soybeans. However, there have been relatively few reports on this enzyme in peanuts; possibly because of an early report that peanuts are not a good source of lipoxygenase (Siddiqi and Tappel 1957), having less than 1% of the activity found in soybeans. Perhaps the low activity found in these early experiments was due to the presence of an inhibitor found in peanut testa (Narayanan *et al.* 1963). Since factors that affect aroma and flavor constituents and the solubility of proteins or protein quality can influence their use in food products, experiments were begun to determine the optimum conditions for lipoxygenase activity in raw peanuts, to determine the relative amounts of lipid oxidation due to causes other than lipoxygenase, and, ultimately, to investigate the effects of these oxidized lipids on protein quality of peanuts. This paper reports the partial purification of lipoxygenase and some properties of these sources of lipid oxidation in dormant unroasted peanuts.

MATERIALS AND METHODS

Materials

Peanut seeds, Virginia 56-R, hand selected for uniformity of size and quality by W. K. Bailey, USDA, Beltsville, were used throughout. Linoleic acid was purchased from Applied Science Laboratories[1] and soybean lipoxygenase from National Biochemical Corp.

Assay Methods

Lipoxygenase activity was determined spectrophotometrically by measuring the increase in absorption due to the formation of conjugated diene-hydroperoxides at 234 mμ. The assays were performed at room temperature with a Beckman DU spectrophotometer equipped with a Gilford Model 2000 Multiple Sample Absorbance Recorder. Solutions of the substrate, linoleic acid, were freshly prepared with Tween-20 as an emulsifier according to Ben-Aziz *et al.* (1970). This technique renders the Tween-solubilized linoleic acid optically clear over a broad range of pH and permits the measurement of initial rates of conjugated diene formation in the reaction mixture.

Linoleate hydroperoxide isomerase activity was assayed by the method of Gardner (1970). The substrate, linoleic acid hydroperoxide, was prepared similar to his procedure with the exception that 0.05 M Na_2HPO_4 solution was substituted for potassium borate. Final concentration of the substrate was 0.648 $\times 10^{-4}$ M. Isomerase activity was measured as the initial rate of decrease in conjugated diene absorption at 234 mμ and pH 6.5.

One unit of activity for lipoxygenase or for linoleic acid hydroperoxide isomerase was defined as the change in absorptivity per minute per milliliter of enzyme preparation.

Protein was determined according to Lowry *et al.* (1951).

Preparation of Enzymes

Ungerminated seeds (hulls and testa removed) were homogenized in 2 volumes of deionized water in a Serval Omnimix at 0°C for a total of 45 sec at alternating 15 sec intervals. The homogenate was squeezed through 8 layers of cheesecloth and centrifuged at 30,000 $\times$ g for 20 min at 4°C. The supernatant was removed and recentrifuged twice under the same conditions. The supernatant contained lipoxygenase and linoleic acid hydroperoxide isomerase, in addition to some other proteins, e.g., arachin and conarachin. Solid ammonium sulfate was added, at various concentrations from 30 to 85% saturation, and after a brief period, centrifuged. The precipitates were

---

[1] Mention of companies or products by name does not imply their endorsement by the U.S. Government.

resuspended in deionized water and dialyzed against deionized water overnight. All operations were conducted at 4°C.

RESULTS AND DISCUSSION

A typical curve illustrating the measurement of enzymic activities for the two enzymes is shown in Fig. 21.1. Curve A represents the initial rate of increase in conjugated diene formation for a lipoxygenase assay, using a partially purified enzyme preparation while curve B illustrates results obtained for isomerase activity. When peanut extracts were assayed with linoleate as substrate the results shown in curve C were obtained. This latter type curve suggested that the

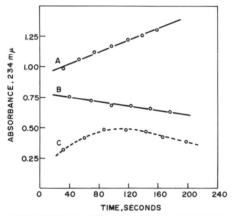

FIG. 21.1. LIPOXYGENASE AND LINOLEIC ACID HYDROPEROXIDE ISOMERASE ACTIVITIES IN PEANUTS

Activities were determined by measuring changes in absorptivity at 234 mμ. Curve A: partially purified lipoxygenase, 0.833×10^{-4} M linoleic acid as substrate. Curve B: hydroperoxide isomerase, 0.22×10^{-4} M linoleate hydroperoxide as substrate. Curve C: represents a mixture of the two enzymes with linoleic acid, 0.833×10^{-4} M as substrate.

TABLE 21.1

AMMONIUM SULFATE FRACTIONATION

Fraction	Total Protein (Mg)	Lipoxygenase		Isomerase	
		Total Activity	Specific Activity	Total Activity	Specific Activity
Supernatant	972	1293	1.64	195	0.25
30%	563	110	0.20	46	0.08
40%	168	414	2.47	63	0.38
50%	93	0	0	13	0.14

lipoxygenase and the isomerase were closely associated in the same fraction.

The two enzymes, lipoxygenase and linoleic acid hydroperoxide isomerase, could be separated from each other by ammonium sulfate fractionation of the supernatant as shown in Table 21.1.

Approximately 73% of the total lipoxygenase activity was recovered in the ammonium sulfate fractions. Of this total, 79% was recovered in the 40% precipitated fraction. After dialysis against deionized water overnight, there was a 30% loss in activity. These activity losses suggested that the enzyme is adversely affected by the conditions employed for separation of the two systems. However, although the lipoxygenase activity recovered was less than 50%, the specific activity increased from 1.64 in the supernatant to 2.47 in the 40% ammonium sulfate fraction. No lipoxygenase activity was found in the 50% ammonium sulfate fraction.

The isomerase recovered in each of the first 3 fractions was more widely distributed: the 30% ammonium sulfate fraction had 38% of the total enzyme activity, with a specific activity of 0.08; the 40% precipitate had 51% of the enzyme activity with a specific activity of 0.38; and the 50% precipitate had only 11% of the activity with 0.14 specific activity. The specific activity of the supernatant before addition of ammonium sulfate was 0.25. Since ammonium sulfate fractionation was the most effective way of separating lipoxygenase activity from most of the isomerase activity, this procedure was employed to prepare lipoxygenase for further studies.

Lipoxygenase isolated from the 40% ammonium sulfate precipitated fraction

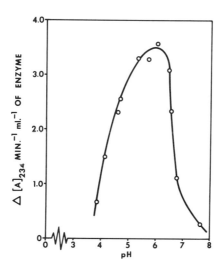

FIG. 21.2. EFFECT OF pH ON ACTIVITY OF AMMONIUM SULFATE
PRECIPITATED LIPOXYGENASE

Linoleic acid (0.83 × 10⁻⁴ M) substrate in Tween-20 (0.008%) solution.

was active over a pH range from 4 to 7, as shown in Fig. 21.2. The reaction medium consisted of enzyme (4.4 mg protein per milliliter of enzyme preparation), linoleic acid substrate (0.833×10^{-4} M), Tween-20 (0.0083%), and 0.1 M phosphate buffer, in a total volume of 3 ml, and was read against a blank consisting of each of these materials except enzyme. Additional controls with enzyme but no substrate also showed no activity. Maximum activity was observed at pH of about 6. Dillard *et al.* (1960) reported finding two pH optima for peanut lipoxygenase activity in whole peanut extracts with linoleic acid as substrate; a sharp peak at pH 8.1 and a slight hump at pH 6.0. No activity was found in our partially purified lipoxygenase preparation at pH 8.0. As Tappel (1963) pointed out in a study of pH optima for enzyme activities, the overall reaction is a function of the types of linoleate substrate solutions used, the method for solubilization, and the method used in preparing the lipoxygenase. For example, pH optima for soybean lipoxygenase have been reported ranging from 6.5 to 9.

Lipoxygenase prepared as described above loses activity even when stored below 0°C. The initial rate of linoleate oxidation during studies on effects of various temperatures showed that approximately 30% of the activity was destroyed by incubating the ammonium sulfate prepared enzyme for 1 hr at 30°C. At temperatures above 40°C, all lipoxygenase activity was lost, whereas the isomerase was still active at temperatures as high as 55°C.

As in soybeans (Holman 1948), lipoxygenase activity was high in the dormant peanut, but decreased with germination.

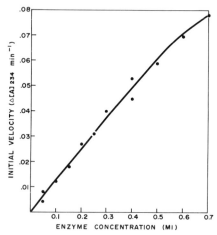

FIG. 21.3. EFFECT OF ENZYME CONCENTRATION ON OXIDATION
OF LINOLEIC ACID

Linoleate concentration, 0.83×10^{-4} M; Tween-20 concentration, 0.008%;
40% ammonium sulfate precipitate was used as source of enzyme; protein
concentration, 4.4 mg per ml.

Oxidation of linoleic acid by peanut lipoxygenase was linear with respect to time for several minutes without any lag period. However, readings were taken over the first 90 sec after adding enzyme to start the reaction in order to measure initial velocities over the linear portion of the curve only.

The effect of enzyme concentration is shown in Fig. 21.3; substrate concentration and temperature were kept constant. At lower enzyme concentrations, the initial rate of reaction was linear. For this particular preparation, when the concentration exceeded 0.22 mg of enzyme protein, the initial rate began to decrease.

A Lineweaver-Burk plot of the effect of substrate concentration on initial velocity is shown in Fig. 21.4. The apparent Km was 7.3×10^{-5} M. While no values of the Michaelis-Menton constant for peanut lipoxygenase have been reported in the literature, the value obtained here agrees well with that reported for soybean with Tween-20 added to the substrate solution.

The effect of various additives on the initial rate of linoleate oxidation showed the following: (a) incubating the enzyme with 10^{-3} M potassium cyanide solution did not effect lipoxygenase activity; (b) 10^{-3} M p-chloromercuribenzoate (PCMB) did not cause any inhibition; (c) 0.002 M 2,3-dimercapto-1-propanol (British Anti Lewisite, or BAL) yielded 59% inhibition; (d) incubation of the enzyme with dithiothreitol (DTT) at concentrations of 0.001 M and 0.01 M inhibited activity by 55 and 100% respectively. These additives were incubated with the enzyme preparation for 1 hr at 4°C before testing for activity. In a separate experiment, 0.25% Tween-20 showed almost total inhibition, even with the unfrac-

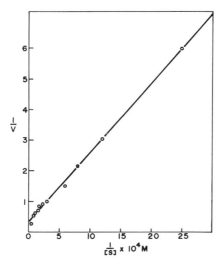

FIG. 21.4. LINEWEAVER-BURK PLOT FOR PARTIALLY PURIFIED PEANUT LIPOXYGENASE

Linoleic acid emulsified with 0.01% Tween-20, substrate; enzyme concentration, 0.4 mg enzyme protein per assay.

tionated supernatant as the source of enzyme. Lipoxygenase from peanut did not show any activity towards oleic acid as a substrate. It is interesting to note that PCMB, a sulfhydryl inhibitor, did not inhibit the enzyme, but BAL and DTT, sulfhydryl reducing agents, did show some inhibition. However, since there are no sulfhydryl groups located in the active site of the enzyme, BAL and DTT might be inducing a change in steric conformation by altering the tertiary structure of the protein. Soybean lipoxygenase activity is also unaffected by sulfhydryl reagents (Holman 1947).

The hydroperoxide isomerase is a newly reported enzyme of lipid metabolism in seeds, having been first identified in flaxseed by Zimmerman in 1966, more recently in corn germ by Gardner (1970), and in barley, wheat germ, mung beans, and soybeans according to Zimmerman and Vick (1970). Our studies with peanut isomerase indicated the optimum pH to be close to 6.1.

During assays on lipoxygenase, our results were assumed to be due to lipoxygenase activity only and not to any hemoprotein-catalyzed oxidation, as reported by Eriksson *et al.* (1970), for several reasons: (a) no induction period was necessary for linoleate oxidation; (b) cyanide did not inhibit the reaction; (c) the rate was not directly proportional to the square root of the enzyme concentration, a necessary criterion for such oxidation suggested by Tappel (1953) and Blain and Styles (1961); and (d) the lipoxygenase prepared from peanuts did not oxidize oleic acid. Furthermore, boiling the enzyme for 15 min destroyed all lipoxygenase activity. Eriksson *et al.* (1970) reported that hemo-proteins derived from catalase and peroxidase maintained their ability to oxidize linoleic acid even after being heated at 130°C to destroy catalase activity and at 140°C to destroy peroxidase activity.

Some investigators assay lipoxygenase activity by measuring the difference between the initial hydroperoxide content and the final concentration taken several minutes or even hours later. Since the lipoxygenase and isomerase are so closely associated in the seed, it is conceivable that a particular extract might contain a critical concentration of both enzymes, so that the isomerase would rapidly react with hydroperoxide formed by the lipoxygenase. The net results in such measurements would be low or would approach zero and might be recorded as little or no lipoxygenase activity. By following the initial formation of conjugated diene over the first 90 sec and plotting as many as 10 experimental points during that period, this particular problem was avoided.

BIBLIOGRAPHY

BEN-AZIZ, A., GROSSMAN, S., ASCARELLI, I., and BUDOWSKI, P. 1970. Linoleate oxidation induced by lipoxygenase and heme proteins: a direct spectrophotometric assay. Anal. Biochem. *34,* 88-100.

BLAIN, J. A., and STYLES, E. C. C. 1961. The unsaturated-fat oxidase activity of plant extracts. *In* Production and Application of Enzyme Preparations in Food Manufacture. Macmillian Company, New York.

DILLARD, M. G., HENICK, A. S., and KOCH, R. B. 1960. Unsaturated triglyceride and fatty acid lipoxidase activities of navy beans, small red beans, peanuts, green peas, and lima beans. Food Res. *25*, 544-553.

ERIKSSON, C. E., OLSSON, P. A., and SVENSSON, S. G. 1970. Oxidation of fatty acids by heat treated hemoproteins. Lipids *5*, 365-366.

GARDNER, H. W. 1970. Sequential enzymes of linoleic acid oxidation in corn germ: lipoxygenase and linoleate hydroperoxide isomerase. J. Lipid Res. *11*, 311-321.

HOLMAN, R. T. 1947. Crystalline lipoxidase. II. Lipoxidase activity. Arch. Biochem. *15*, 403-413.

HOLMAN, R. T. 1948. Lipoxidase activity and fat composition of germinating soybean. Arch. Biochem. *17*, 459-466.

LOWRY, O. H., ROSEBROUGH, N. J., FARR, L., and RANDALL, R. J. 1951. Protein measurements with the Folin-phenol reagent. J. Biol. Chem. *193*, 265-275.

MATSUSHITA, S., and KOBAYASHI, M. 1970. Effect of linoleic acid hydroperoxides on pepsin activity. Agr. Biol. Chem. (Tokyo) *34*, 825-829.

NARAYANAN, K. M., BAINS, G. S., and BHATIA, D. S. 1963. Lipoxidase inhibitor in groundnut (*Arachis hypogea* L.) testa. Chem. Ind. (London) 1588.

ROUBAL, W. T., and TAPPEL, A. L. 1966. Damage to proteins, enzymes, and amino acids by peroxidizing lipids. Arch. Biochem. Biophys. *113*, 5-8.

SIDDIQI, A. M., and TAPPEL, A. L. 1957. Comparison of some lipoxidases and their mechanism of action. J. Am. Oil Chemists' Soc. *34*, 529-533.

TAPPEL, A. L. 1953. Linoleate oxidation catalysts occurring in animal tissues. Food Res. *18*, 104-108.

TAPPEL, A. L. 1963. Lipoxidase. *In* The Enzymes, Vol. 8, P. D. Boyer, H. Lardy, and K. Myrback, (Editors). Academic Press, New York.

ZIMMERMAN, D. C. 1966. A new product of linoleic acid oxidation by a flaxseed enzyme. Biochem. Biophys. Res. Commun. *23*, 398-402.

ZIMMERMAN, D. C., and VICK, B. A. 1970. Specificity of flaxseed lipoxidase. Lipids *5*, 392-397.

CHAPTER 22

Determination of Lysine In Cereal Proteins

Jose Madrid Concon

There is need for a rapid and simple method for the determination of critical amino acids in cereal nutritional biochemistry and in screening and breeding programs in search for nutritionally superior grains. Current searches for grains high in lysine have underscored the need for such a method.

Present lysine methods—microbiological assay; enzymatic decarboxylation; high-voltage electrophoresis; paper, thin layer, ion exchange, and gas-liquid chromatography—are inherently slow partly because of prior hydrolysis of sample, removal of hydrolyzing agent, and concentration of hydrolyzates. Automation and computerization, however, have markedly speeded up the subsequent analytical steps after the preparatory treatment of samples. Unfortunately, the large capital investments required, and the service and maintenance costs of automated instruments are beyond the means of many laboratories. Therefore, a simple and rapid alternative method for the determination of lysine has been sought by many chemical analysts in cereal grain nutrition research and technology.

The requirements for simplicity and rapidity in the determination of lysine can be met by using unhydrolyzed proteins or samples. In addition, a requirement for sensitivity is also imposed because cereal grains are notoriously low in lysine. Furthermore, only limited amounts of samples may be available (i.e., single kernels or segments thereof).

PREVIOUS METHODS FOR LYSINE USING UNHYDROLYZED PROTEINS

Van Slyke Nitrous Acid Method

Lieben and Loo (1942) used the Van Slyke (1911, 1912) nitrous acid reaction to determine lysine in relatively purified proteins by measurements of the volume of nitrogen envolved in 30, 60 and 90 min. An empirical formula, was used:

$$\text{mg lysine} = (146/14) \, [c-3 \, (b-a)]$$

where a, b, and c are, respectively, milligrams of nitrogen evolved after 30, 60, and 90 min. But the Van Slyke reaction also produces nitrogen from a plethora of substances and chemical groups (Kainz *et al.* 1957) so that one can readily appreciate the difficulties with crude samples. Sure and Hart (1917) earlier found that specific deamination of α-amino groups occurs at $0°C$ and the ϵ-

amino groups deaminate completely at 32°C. However, with the more sensitive and specific reagent, trinitrobenzenesulfonic acid (TNBS) (Okuyama and Satake 1960; Habeeb 1966) the present author (unpublished observation) found that at 0°C the α-amino groups of lysine do not deaminate completely until after 9 min by which time part of the ε-amino groups has decomposed.

Hofmann Reaction and Iodometric Back Titration Method

Baraud (1957) back titrated the amount of iodine reacting with the dithio-carbamate derivatives of primary amino groups in order to determine lysine. With purified proteins, satisfactory results were obtained. Since the Hofman reaction using carbon disulfide is very unspecific for the ε-amino groups, this method may give doubtful results with crude protein extracts.

Dinitrofluorobenzene (DNFB) Method

Carpenter (Carpenter 1960; Carpenter *et al.* 1957; Carpenter and March 1961) developed a method for available lysine in crude proteins using DNFB. However, it is difficult to correlate chemical and biological availability since the latter could also involve protein digestibility which may not be related to the reactivity of the ε-amino groups of lysine. Some reactive lysine in the insoluble samples used in the methods may not react at all. The serious shortcomings of DNFB was pointed out by Carpenter *et al.* (1957). Because a major portion of the cereal proteins are insoluble in the DNFB reaction medium, variable and low results may be expected. DNFB reacts not only with primary amino groups but other groups as well.

Trinitrobenzenesulfonic Acid (TNBS) Method

Kakade and Liener (1969) used TNBS, a reagent more specific than Sanger's DNFB to determine "available lysine" in protein with good results. The method was used satisfactorily by Subramanian *et al.* (1970) on corn protein extracts. Some samples gave high results with the method (Mertz 1970). Some of the serious shortcomings of DNFB may also be expected of TNBS. c TNP-lysine is more unstable in boiling acid solution than the corresponding DNP derivative (Okuyama and Satake 1960); S-TNP derivatives are very stable in acid but are easily decomposed in alkaline medium (Kotaki *et al.* 1964). Since the rate of decomposition during the TNBS reaction could vary with the sample, serious errors could result if large amounts of S-TNP derivatives had not decomposed completely at the time of acidification. Small amounts of S-TNP will not cause serious errors because its extinction coefficient at 340 mμ is considerably lower than that of the NH-TNP derivatives (Kotaki *et al.* 1964).

PRELIMINARY STUDIES ON THE DINITROBENZENESULFONIC
ACID (DNBS) METHOD

Specificity of DNBS

The studies of Freedman and Radda (1968) on the reactivity of TNBS with amino acids and peptides showed that at room temperature, pH 7.4, the ϵ-amino groups were 30 times more reactive to TNBS than were the α-amino groups. It was also found that the α-amino group even becomes less reactive when the neighboring carboxyl group was amidated or joined in a peptide bond. These authors concluded that the decreased nucleophilicity of the amino group, as a result of amidation or peptide bond formation of the carboxyl radical, accounts for this decreased reactivity. The reactivity of the amino groups appears to increase with their pK values. (Steric and other factors may also influence this reactivity.) This is to be expected since TNBS has an acid strength close to that of perchloric acid (Pietrzyk and Belisle 1966).

The above considerations suggest that a less electrophilic reagent than TNBS might be more specific for the ϵ-amino groups. The decrease in electrophilicity must be such that the rate of reaction is not compromised to a point that it is useless from the practical standpoint. This requirement seems to be realized with dinitrobenzenesulfonic acid (DNBS) which is a weaker acid than TNBS (Pietrzyk and Belisle 1966). The absence of one extra ortho nitro group in DNBS is expected to diminish the corresponding electrophilicity of the sulfonate radical. Steric and other electrostatic factors may also influence the reactivity of both DNBS and primary amino groups. It follows, then, that under normal conditions a virtual absence of reactivity between DNBS and the α-amino groups in peptides may result. Since factors which influence the decreased reactivity of the amino groups with TNBS do not seem to affect the ϵ-amino groups (Freedman and Radda 1968) under certain conditions, a relatively specific reagent for protein ϵ-amino groups is seen in DNBS. (Based on the performance of fluoro-dinitrobenzene, it is evident that other factors, such as critical distance of approach between reactive centers, are operative in DNBS).

The specificity of DNBS to the ϵ-amino groups of proteins under specified conditions was in fact observed by previous investigators (Eisen et al. 1953; Li 1956; Ikenaka 1959; Hosoya 1960; Sugae 1960).

Factors Affecting the Specificity and Reactivity of DNBS

Although previous workers found DNBS to be specific for the ϵ-amino groups, as expected the reaction was slow. Therefore, attempts were made by the author to determine whether such specificity could be altered by speeding up the reaction. Experiments with pure lysine, other selected amino acids, and peptides were undertaken to determine the effect of pH, temperature, reaction time, concentration of reactants, and the presence of denaturing agents. The details of these experiments will be reported in another paper (Concon 1971).

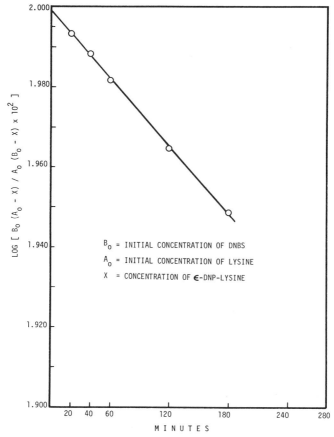

FIG. 22.1. KINETIC PLOT OF THE REACTION OF LYSINE ε-AMINO
GROUPS WITH DNBS AT pH 12 AND 40°C

But some of the most important results are discussed in the following sections.

Effect of pH.—Consistent with the results with TNBS (Goldfarb 1966A, B; Freedman and Radda 1968), the reactivity of ε-amino groups with DNBS increases almost linearly with pH at 40°C. No reaction is observed at pH 9. Maximum reaction appears to be between pH 12 and 13. However, at the latter pH significant destruction of chromophore seems to be indicated. No reaction was apparent at 40°C with the α-amino groups of lysine at all pH values tested. The reactivity of the α-amino group is readily determined from the absorbance at 360 mμ of the ether extractable chromophore dissolved in ethyl acetate. The absorbance of the remaining acid solution at the same wavelength represents the ε-amino-DNP derivative. There is evidence that the apparent absence of any reactivity of the α-amino groups at pH above 12 may be due to the instability of the α-NH-DNP derivative. That the ε-NH-DNP derivative is stable at pH 12 even after 3 hr at 40°C is shown by the excellent linearity of the second order kinetic plot shown in Fig. 22.1.

Effect of Temperature.—Figure 22.2 shows the remarkable difference in reaction rate between the α- and ϵ-amino groups with DNBS at pH 12 even at temperatures above 40°C. At 60°C, the absorbance of α,ϵ-bis-NH-DNP derivative is about 2% of the ϵ-NH-DNP form.

Effect of Reaction Time.—Figure 22.3 shows the differences in the amounts of the mono- and bis-DNP derivatives formed with time at different pH's and temperatures. The results at 45°C indicate that there is no advantage to be gained in terms of specificity by increasing the temperature to 45°C and at the same time, decreasing the pH to 11. Under these conditions there is a greater reactivity of the amino groups. The absence of the ether-soluble chromophore at pH 13, even after 3 hr, again indicates that the α-NH-DNP derivative is unstable

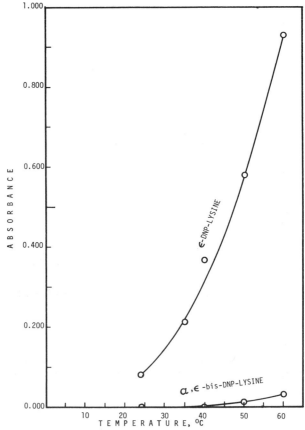

FIG. 22.2. REACTION OF LYSINE α- AND ϵ-AMINO GROUPS WITH DNBS FOR 1 HR AT pH 12 AS A FUNCTION OF TEMPERATURE

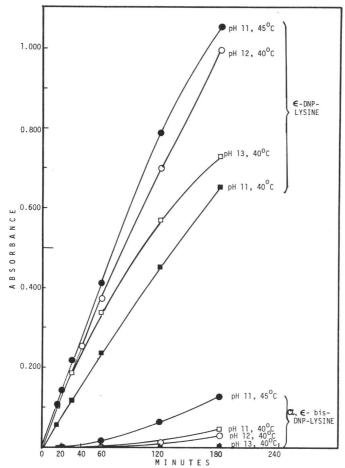

FIG. 22.3. REACTION OF LYSINE α- AND ε-AMINO GROUPS WITH DNBS
AS A FUNCTION OF TIME, pH, and TEMPERATURE

at high pH. The results in Fig. 22.3 also show that the ε-NH-DNP derivative is unstable at this pH. However, this instability at high pH can be tested more conclusively by determining the decrease in absorbance with time of a solution of pure chromophore at highly alkaline conditions.

Effect of DNBS Concentration.—The second order kinetic plot shown in Fig. 22.1 shows the dependence of the reaction to both the lysine and DNBS concentration. With 0.24% DNBS in the reaction medium, no ether-soluble chromophore can be detected at pH 12, 40°C for 1 hr. However, with a five-fold increase in DNBS concentration the absorbance of the ether-soluble chromophore dissolved in an equal volume of ethyl acetate was 23% of the absorbance of the aqueous phase.

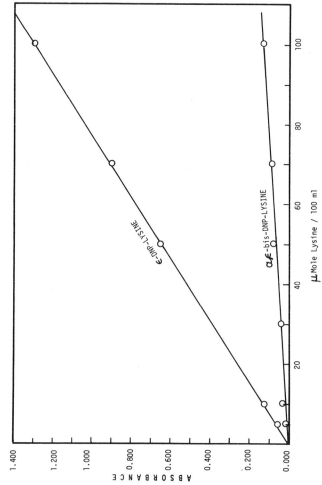

FIG. 22.4. CONFORMANCE OF THE FORMATION OF MONO-, ϵ-, AND BIS-Cl,ϵ-AMINO-DNP DERIVATIVES OF LYSINE TO BEER AND LAMBERT'S LAW AT 55°C, pH 10 and 2 HR REACTION TIME

Conformance to Beer and Lambert's Law.—Even though the reaction is incomplete, the formation of bis- α-, ϵ-amino- and mono- ϵ-amino-DNP derivatives conforms to Beer and Lambert's Law. For example, at 55°C, pH 10, and 2 hr, the reaction is far from complete but the formation of these chromophores follows Beer and Lambert's law as seen in Fig. 22.4.

Reactivities of Other Amino Acids and Peptides.—The reactivities of a number of selected amino acids to DNBS are indicated in Table 22.1. The reactivities of these amino acids at pH 12, 40°C, and 1 hr are variable. The reactivity of a particular α-amino group is expected to be modified by the corresponding side chain group. In contrast to histidine and arginine, tyrosine, tryptophan, and cysteine are the most reactive. As seen in Table 22.1, the interference of the latter amino acids can be eliminated by ether extraction. A major portion of the cysteine interference is eliminated by ether extraction However, the sulfhydryl group may be masked by mercuration with either $HgCl_2$ or organic mercurials, such as, phenylmercuric chloride. The slight reactivities of the other basic amino acids are reduced by mercuration. Mercury forms complexes with basic amino acids in alkaline medium (Kai 1967). Such complexation probably affects only the α-amino groups of lysine since no change in absorbance is observed when lysine is reacted with DNBS in the presence of mercuric chloride.

The terminal amino groups of peptides of these amino acids, however,

TABLE 22.1

REACTION OF AMINO ACIDS AND PEPTIDES WITH DNBS

Amino Acid or Peptide[1]	Optical Density		
	Before Ether Extraction	After Ether Extraction	With Mercuration
Lysine	0.215	0.215	0.215
Arginine	0.035	0.042	0.030
Histidine	0.071	0.080	0.015
Cysteine	1.400	0.070	0.005
Tyrosine	0.150	0.000	
Tryptophan	0.513	0.000	
Valine	0.085	0.000	
Tryptophanyl tryptophan (14.5 μmoles)	0.018		
Tryptophanyl histidine diacetate (14.3 μmoles)	0.017		
Histidyl leucine (15.1 μmoles)	0.028		

[1] Ten micromoles of amino acid per 25 ml solution.

become less reactive to DNBS as shown in Table 22.1. As noted before, similar decrease in reactivity of the terminal amino groups in peptides was observed with TNBS (Freedman and Radda 1968).

The studies on the reactivity of various amino acids and peptides to DNBS will be reported in detail in a separate paper (Concon 1971).

Determination of Lysine in Purified Proteins

Ikenaka (1959) found that at pH 10.7, 37°C, only 11 of 22 lysine residues in taka-amylase reacted with DNBS after 50 hr; similarly, 3 out of 25 lysine residues in bacterial amylase reacted after 74 hr (Sugae 1960). The low pH and calcium acetate (to minimize denaturation of enzyme) used by these authors could render many of the lysine residues unreactive by steric and electrostatic factors. Freedman and Radda (1968) showed that with TNBS, the unprotonated amino group is the reactive species. One would expect that same behavior with DNBS. The extent of dissociation of ϵ-amino groups in Ikenaka's taka-amylase (1959) as predicted by the Henderson-Hasselbalch equation can be correlated with their reactivity to DNBS but not in Sugae's bacterial amylase (1960).

Therefore, for DNBS to be valid as a specific reagent for lysine in unhydrolyzed proteins, all the ϵ-amino groups must be equally accessible and fully reactive. These requirements may be satisfied by conducting the dinitrophenylation reaction at high pH and in high concentration of urea. The details of the method are given in the Experimental section and is discussed more fully in a separate paper (Concon 1971). The validity of DNBS as a reagent for lysine determination in unhydrolyzed proteins is confirmed by the results with purified proteins shown in Table 22.2. These results also show the suitability of using pure lysine as standard. Among others, these results indicate that under the

TABLE 22.2

LYSINE RESIDUES IN PURIFIED PROTEINS DETERMINED
WITH SODIUM DINITROBENZENESULFONATE

Protein	Residues per Mole				Present[1]
	Found				
	1	2	3	Avg	
Insulin	1.04	0.98		1.01	1
Ribonuclease	10.30			10.30	10
Trypsin	14.50	14.30		14.40	14
Chymotrypsin	12.70	12.80	13.30	12.90	13
Beef liver catalase	142.10	153.2		147.6	152
Bovine serum albumin	60.70	61.3	58.50	60.2	60

[1] Haurowitz (1963).

conditions used all the lysine residues in these proteins are fully and equally reactive. The range of errors from the true values is from 0.8 to 3.6%.

Determination of Lysine in Crude Cereal Protein Extracts

Preliminary results show that there is little advantage to be gained by using powdered samples directly without preliminary extraction of proteins. This is due to the fact that many samples contain insoluble nonproteins which must be removed by centrifugation after color development and ether extraction. This extra step negates whatever time is gained by eliminating the protein extraction steps.

Extraction of Proteins.—The details of the extraction method are given in the Experimental section and are discussed more fully in a separate paper (Concon 1971).

Except for corn and sorghum the proteins of the other cereal grains are completely extractable with 0.2–0.5% NaOH solution. Corn requires a stepwise addition and mixing with 70% ethanol and dilute sodium hydroxide for quantitative extraction of proteins. In addition, the vitreous and floury endosperms must be ground to 105 μ and 74 μ particle size, respectively. The results of the quantitative extraction of corn proteins by the stepwise procedure are given in Table 22.3. The extraction efficiency is from 98 to 106%. The total extraction time is 15–20 min in contrast to several hours to several days with previous procedures (Mertz and Bressani 1957; Foster *et al.* 1950). The difficulties with single quantitative extractions of corn proteins are understandably due to the wide range of solubilities of these proteins. The success of the stepwise procedure is premised on the possibility of cosolubilization of the proteins. This is assumed to be brought about by effective interaction of zeins and glutelins by predispersion of the prolamins with 70% ethanol. With isolated mixtures of corn proteins, complete solubilization is not obtained even after 24 hr with dilute NaOH solution alone nor with a mixture of 70% ethanol and dilute NaOH solution in the same proportion as in the final mixture in the stepwise procedure. In contrast, the stepwise procedure brings about a relatively clear dispersion almost immediately with no precipitation even after 24 hr.

Some wheat samples are not quantitatively extracted with dilute sodium hydroxide solution. When subjected to the stepwise procedure, these samples gave quantitative yields. All the high prolamin grains such as wheat and barley (except sorghum) yield quantitative results with the stepwise procedure. Some other factors are involved in sorghum extraction and are currently being investigated. The low prolamin grains, such as rice and oats, do not give quantitative yields.

Rapid Nitrogen Determination.—The determination of nitrogen in the powdered samples as well as in the extracts is an important part of the entire methodology. The rapidity of obtaining the desired results is highly dependent

TABLE 22.3

EXTRACTION OF CORN ENDOSPERM PROTEINS BY STEPWISE ADDITION
OF 70% ETHANOL AND DILUTE SODIUM HYDROXIDE

Variety	Total Nitrogen (%)	Nitrogen Extracted (%)
TXS 102	1.15	98.4
Dekalb XL 315 (low protein fraction)	1.50	101.5
Dekalb XL 315 (high protein fraction)	2.32	105.0
Swenk 17 B (whole corn)	1.39	100.0
USDA normal (unknown)	1.43	100.8
Tolima 377	1.76	102.6
Cauca 343	1.55	101.1
Antigua grupo[1]	1.80	102.5
La Posta[1]	1.32	100.5
Brazil 70[1]	1.45	106.4
Brazil 2811[2]	2.07	97.7
Opaque-2[1]	1.27	98.1
Floury-2[1]	1.42	100.3

[1] These varieties have more than 30% floury endosperms. These were ground to $\leqslant 105\ \mu$ and extracted with 0.3% NaOH solution. All others were ground to $\leqslant 105\ \mu$ and extracted with 0.5% NaOH solution.
[2] This variety has only about 10% floury endosperms but was ground and extracted as the floury varieties.

on the nitrogen determination. No doubt the microKjeldahl procedure is the most popular method for nitrogen determination. However, as described in official handbooks, the method is by no means rapid. Previous attempts to speed up the method have been reviewed by Bradstreet (1965). Excessive frothing of sample and prolonged digestion time are the two major difficulties with the method.

The author controlled the frothing problem by addition of 1–2 drops of 50% palmitic or lauric acid in ethanol. The digestion time was shortened by addition of 30% H_2O_2. Some of the remarkable advantages in time with these modifications are as follows (the corresponding times by the standard method are in parenthesis): clearing or charring time: 0–1 min (0.5–1.5 hr); digestion time after clearing: 5–10 min (0.5–3 hr). It is necessary to control the heating to 1/2 the full capacity of the digester (Laboratory Construction Co.) during the initial vigorous reaction. The digester is then turned to full heating capacity when the reaction has subsided. The results obtained are equal if not better than the standard procedure. Previous methods using hydrogen peroxide are reviewed by Bradstreet (1965). Some of the details of this rapid procedure are described in the Experimental section of this chapter. The method and its results are discussed more fully in a separate paper to be published elsewhere (Concon 1971).

TABLE 22.4

DETERMINATION OF LYSINE IN CRUDE WHEAT
AND CORN PROTEIN EXTRACTS WITH DNBS

Sample	Gram Lysine per 16 Gm N			Amino Acid Analyzer[1]
	DNBS Method			
	1	2	Avg	
Corn[2]				
Tolima 377	1.10	1.22	1.16	1.19
Cauca 343	1.26	1.31	1.28	1.37
Antigua grupo[2]	1.52	1.46	1.49	1.46
Brazil 70	1.82	1.86	1.84	1.66
Opaque-2	3.18	2.99	3.09	2.99
Floury-2	2.68	2.76	2.72	2.64
Wheat				
1401[3]	2.43	2.55	2.49	2.69
1403[3]	2.48	—	2.48	2.63
1405[3]	2.70	2.68	2.69	2.87
Cjan 67 Y 68-69[4]	2.65	2.94	2.76	2.33[4]
Red spring, Illinois[4]	2.86	3.02	2.94	3.02[4]

[1] Average of two determinations.
[2] Analyses of endosperms only.
[3] Samples and amino acid analyzer values supplied by Dr. P. J. Mattern, University of Nebraska. Mercuration with $HgCl_2$.
[4] Single Values only. Mercuration with phenylmercuric chloride.

Lysine Determination of Crude Cereal Protein Extracts.—The determination of lysine with the new procedure was tried on corn and wheat samples. The results are shown in Table 22.4. Values are comparable with those obtained with amino acid analyzer. The steps in the procedure are described in the Experimental section of this chapter. A detailed discussion of the method and its results will be presented in a separate paper (Concon 1971). It is interesting to note however, that in spite of the crudeness of the extracts with varied possibilities for interference, DNBS under the conditions in which it is used, gave good results as a reagent for lysine. This fact gives further evidence of the high specificity of this reagent for the ϵ-amino groups under specified conditions.

EXPERIMENTAL[1]

Preparation of Sample

Dehulling.—Grains with loose hulls (1–5 gm of kernels in a tapered 50-ml tube) are dehulled by prodding repeatedly with a glass rod. The separated hulls

---

[1] Part of this work was done at North Star Research and Development Institute, Minneapolis.

are removed with a sieve or by winnowing. The intact kernels are handpicked and the dehulling operation repeated.

Corn kernels are dehulled by first soaking them in hot water at 75°–80°C for 3–4 min. The hulls are easily peeled off with tweezers. The hulls of intact kernels are kept moist during the dehulling operation.

Grinding.–Any suitable grinding equipment is used as long as the sample is ground rapidly with minimum exposure to excessive heat. The author used a water-cooled "micro mill" (Chemical Rubber Co.) to grind 5–20-gm quantities. In this case, the last few milligrams of residue are ground with a mortar and pestle, preferably with Dry Ice. The latter renders the samples brittle which, in this condition, are easier to grind.

With small amounts of sample, a dental amalgamator is suitable for very fine grinding. Floury corn samples require very fine grinding ($< 74\ \mu$, 200 mesh) for quantitative extraction. Ball milling in *dry* acetone or petroleum ether after preliminary grinding in a micro mill was used by the author for 5–20-gm samples. Moisture can adversely affect the extraction of the protein in ball-milled vitreous corn. Thus, as a general precaution, the moisture of the sample must be kept low (1% or less) before and after extraction. The ball-milled material is passed directly through a required sieve and the coarser particles reground by any convenient means; the sieved material is filtered rapidly, the residue freed of solvent in dry air or *in vacuo,* passed through the same sieve, and stored in airtight bottles in the cold.

Protein Extraction for corn, wheat, barley and other high-prolamin grains (except sorghum).–The sample is weighed (500 mg) into a heavy-duty poly-ethelene test tube; 1.5 ml of 70% ethanol (v/v) is pipeted. The tube is stoppered with a polyethelene cap and its content mixed (vortex mixer) vigorously for 3–5 min. Dilute sodium hydroxide (5.5 ml, 0.2–0.5%) is pipeted and vigorous mixing is resumed for another 3–5 min. In both mixing operations, the paste or slurry must be homogenous. Corn samples ground to $<74\ \mu$ require 0.3% NaOH solution; all samples ground to $<105\ \mu$ are added with 0.5% NaOH solution.

Rice, oats and other low-prolamin grains (except sorghum and corn) are extracted simply with 0.2–0.3% NaOH solution, 7 ml per 500 mg of sample. The sample and solvent are mixed vigorously (vortex mixer) for 3–5 min.

The slurry is centrifuged for 5 min at 8000 g (10,000 rpm, Sorvall Superspeed Automatic SS-3 centrifuge, Sorvall Inc. Newtown, Conn.) or 10–15 min at 3000 g. The clear supernatant is transferred to a clean test tube. Undefatted samples give turbid extracts. When these samples are used, the lipoproteins which float on the surface should be carefully shaken and redispersed back into the supernatant.

Rapid Nitrogen Determination.–About a 50 mg sample (containing 0.5–1 mg nitrogen) is weighed with a charge tube into a 30-ml micro Kjeldahl flask; 2.6 gm of K_2SO_4–HgO mixture (190:4, w/w), 2 ml concentrated H_2SO_4 (sp gr 1.84), 1 drop (about 10 mg solids) of 50% lauric acid in ethanol, 3 ml of 30% H_2O_2 and

2–3 boiling chips (10 mesh) are added. The mixture is heated on an electric digester (Laboratory Construction Company) at 1/2 full capacity, until vigorous reaction has subsided. If serious frothing occurs, another drop of lauric acid solution is added. When vigorous frothing has subsided, heat is applied at full capacity. Digestion is continued for 7–10 min after the initial fuming is observed or when all the water is evaporated. At termination of digestion, the digest must have ascended at least halfway through the neck of the tube if the optimum temperature of digestion has been attained. Some lauric acid which condensed on the neck during digestion will cause the digest to discolor when the former comes in contact with the hot melt. This does not affect the result. Cool, distill off, and determine the ammonia as in the usual procedure. (The author used a one piece model, micro nitrogen distilling apparatus, E. H. Sargent & Co.)

Determination of Lysine in Cereal Protein Extracts

Reagents.—*Urea–Buffer Solution.*—Dissolve 180 gm of urea, 35 gm of $Na_3PO_4 . 12H_2O$ and 500 mg of phenylmercuric chloride in 500 ml solution. Solution is hastened by warming. The pH is adjusted to 12.3 with 85% H_3PO_4. (Saturated $Ca(OH)_2$, pH 12.45 at 25°C is satisfactory for standardizing the pH meter.)

Sodium 2,4-Dinitrobenzenesulfonate (DNBS).—The commercial product (Eastman Organic Chemicals) is purified by treating 30 gm of DNBS dissolved in 100 ml water, with 10 gm of activated charcoal. The suspension is stirred for 30 min. Treatment with charcoal is repeated if the filtrate is not clear light yellow or has an orange tint. The purified filtrate may be lyophilized or used directly to prepare a 3% working solution. A 1:50 dilution of the working solution read against water at 25°C should have an absorbance at 360 mμ of about 0.510 at pH 5.3. The solution is stable in the dark at 5°–10°C.

Standard Lysine Solution.—A stock solution containing 50 μmoles lysine monohydrochloride per milliliter is prepared with reagent grade material dried to constant weight at 100°C. This is stored frozen. A working standard solution is prepared from the stock solution to contain 1 μmole lysine per milliliter. In order that the sample and standard solutions should have the same pH during color development, the working standard is prepared as follows: For samples extracted by the stepwise procedure with 70% ethanol and 0.5, 0.3, or 0.2% NaOH solution, the corresponding working standard is dissolved in 0.4, 0.2, or 0.1% NaOH solution, respectively. For those extracted with dilute NaOH solution only, the corresponding lysine working standard solution is prepared with the extractant of the same concentration.

Ether Extractant.—Peroxide-free diethyl ether is saturated with a solution containing concentrated HCl, urea-buffer, and water (4:1:1, v/v). The ether phase is filtered.

Stannous Chloride Solution.—A 1% $SnCl_2.2H_2O$ solution is prepared fresh each time. Concentrated HC1 is added drop by drop to dissolve $Sn(OH)_2$. The acidified solution is filtered.

Procedure.—Obtain 4, 20 X 125 mm pyrex test tubes, medium thickness with plastic lined caps. Put 1.0 ml of urea-buffer solution into all tubes. Pipet 0.50 ml each, of protein extract (containing 0.5–2.0 μmoles lysine[2]) into the third, and NaOH solution (the same concentration as in the lysine standard) into the fourth (assay blank). One of the protein extract tubes will serve as the "sample blank" to correct absorbance contribution, if any, from the sample. Mix and let all tubes stand a few minutes. At 1-min intervals, add to each tube (except the sample blank) starting with the assay blank, 0.50 ml 3% DNBS solution, mix, and immerse in a shaker water bath at 40°C. To the sample blank add 0.50 ml water instead of the DNBS solution before immersing in the water bath. After exactly 1 hr in the water bath, each tube is withdrawn and added rapidly with 4.0 ml concentrated HC1 (except the sample blank) successively at 1 min intervals to stop the reaction. The sample blank is added with 3.5 ml concentrated HC1 followed by 0.50 ml of 3% DNBS solution. Cool, mark the level of solution, and extract three times with (equal volume for the first extraction, carefully measured) specially prepared ether. Remove the separated ether phase by gentle suction (water aspirator). If necessary, centrifuge at the highest speed possible for a few minutes to break any emulsion or thick gel which forms at the interphase. Add 0.1 ml (about 3 drops) $SnC1_2$ solution to each tube after the last extraction step. Mix and remove the dissolved ether in the test solutions by heating in a water bath at 50°–55°C with occasional *cautious* shaking to hasten its removal. Removal is adequate when the original volume before extraction is obtained and little frothing is observed when the tube is shaken. Cool in a water bath, mix any droplets that adhere on the side of the tube, and read the absorbance of the solution at 360 mμ against the assay blank. Results are calculated as follows.

$$\frac{\text{Gram lysine}}{16 \text{ gm N}} = \frac{A-B}{C} \cdot \frac{16 \text{ L}}{N}$$

A = absorbance of sample
B = absorbance of sample blank (zero for negative values)
C = absorbance of standard lysine solution
L = gm lysine in standard solution
N = gm nitrogen in sample

---

[2] The extracts of samples containing more than 4 μmoles lysine per milliliter should be diluted to contain the desired amounts with the same dilute NaOH solution used in the lysine standard.

Comments

The protein extraction procedure is an important aspect of the lysine method. Since the nitrogen recovery may be over 100% because of withdrawal of part of the solvent during gelation of starch, it is best to calculate the result based on the nitrogen content of the extract rather than the original sample.

A proportionate amount of sample should be weighed so that diluting the extract to bring the lysine content of the aliquot used within range is obviated. However, quantitative extraction of corn samples by the stepwise procedure is adversely affected if the liquid:solid ratio is too high. In general, with the same amounts of liquids, the weight of sample may be halved without significant negative effects. No such dependence on the liquid:solid ratio is observed with samples extracted with dilute NaOH solution only.

Preliminary results indicate that aging of sample may decrease the extractability of the proteins. In the case of corn, quantitative extraction of the protein of aged samples is obtained if the NaOH solution used is 0.5% in sodium dodecyl sulfate (SDS). The interference of SDS at pH below 10 in the reactivity of the ϵ-amino groups has been mentioned.

An important source of error is the ether extraction step. If the volume of ether added is not controlled, the precision of the method is adversely affected. Since water is soluble in ether, the final volume of test solution is reduced in proportion to the total amount of ether used in the extraction. Thus, the concentration of the ether-insoluble chromophore would vary with the total volume of ether used. This problem may be minimized by presaturating the ether with a solution identical to the assay blank without DNBS and by carefully controlling the first volume of ether added. An equal volume of ether is found adequate for the first extraction. A smaller volume is used in the next two extractions.

During removal of ether by heating, loss of sample may result if heating is applied too rapidly, such as at temperatures above 55°C. The samples tend to bump at higher temperatures. In addition, precipitates may collect on the surface of the solution and obstruct the escape of ether. The resulting buildup of pressure may cause serious bumping. Such pressure buildup is prevented by shaking the solutions more frequently when the conditions demand, at the beginning of the heating process.

Because the absorbance of the solution during color development increases with time, greater precision will be realized if the time of reaction is observed carefully. For example, one should expect a change of about 0.006 absorbance units per minute per milliliter with a lysine concentration of 0.4 μmoles per ml at pH 12 and 40°C. This value may vary depending on several factors. Results in the author's laboratory indicate some inconsequential errors with a variation of 1–2 min in reaction time. Reagents may be conveniently and rapidly dispensed with burets rather than pipets. Accurate self-filling burets with elongated tips are preferable.

If an insoluble precipitate is formed during addition of acid when the reaction is terminated, analysis of that particular sample should be repeated with increased amounts of acid or decreased amounts of sample. Parallel changes in conditions should be done with the standard and blanks. Precipitates tend to form when the initial amounts of acid are added. Such precipitates eventually dissolve if the amount of acid is adequate and the concentration of urea and salts in the reaction medium is not excessive. Some samples develop nonprotein turbidity on acidification. This is usually eliminated during ether extraction and by centrifugation. At any rate, the solution must be free of turbidity before the absorbance is read.

Although Pataki (1968) and Russell (1962, 1963) state that the ε-DNP-lysine is not light sensitive, Selim (1965) conducted the dinitrophenylation with DNFB and sample manipulations after color development in subdued light. Therefore, as a general precaution in the present method, the manipulations during and after color development are conducted in subdued light. Samples kept in the dark for several days at room temperature show little change in absorbance. It would be beneficial to determine the effects of ordinary laboratory light on the color stability of the solutions.

Certain cereal samples give high blanks. This is most often encountered in undefatted materials or those grains with highly pigmented outer layers. These samples, especially rice and oats, frequently develop purple sample blanks on standing at 40°C or above after addition of HCl. This is observed especially during removal of ether at 50°–55°C. These purple pigments have significant absorbance at 360 mμ and may cause serious errors. The corresponding DNBS-treated sample has a brownish yellow solution which would be expected when the colors, yellow and purple are combined. Some oxidative mechanism appears to be responsible since the color is intensified by addition of a few drops of 30% H_2O_2. Therefore, the addition of a reducing agent such as $SnCl_2$ to the acidified solution prior to heating prevents the development of the purple solution. It should be emphasized, however, that the addition of excess amounts of $SnCl_2$ should be avoided since this reagent may promote the formation of a yellow or brown pigment which absorbs at 360 mμ.

The sample blank frequently gives lower absorbance than the assay blank. In this case the negative reading should be ignored. Obviously, no absorbance is contributed by the sample without dinitrophenylation. The higher absorbance of the assay blank may come from the reaction of DNBS with impurities in the reagents or simply contaminations. Certain batches of urea give higher blank readings than others. This should not materially affect the results.

It may be necessary at times to interrupt the determination at some point for various reasons. The analysis should not be interrupted for prolonged periods at any time after addition of the urea-buffer solution and before addition of HCl. If the samples, such as corn, are allowed to stand for several hours after the urea-buffer solution is added, insoluble precipitates may form. The sulfhydryl concentrations of alkaline protein solutions also increase on standing (Concon 1966). However, no serious errors can be anticipated from these sources since the concentration of the masking agent, phenylmercuric chloride, is in large excess.

The rapidity of any analytical method is very dependent on the weighing of samples which occupy a significant block of time. For survey or screening purposes a rapid balance accurate to the third decimal place should be satisfactory. It is suggested that the analyst assess his requirements and needs and use time-saving devices, techniques, and innovations.

ACKNOWLEDGEMENTS

This work is supported by Contract No. 12-14-100-9507 (71) from the USDA, Agricultural Research Service, Northern Utilization Research and Development Division, Peoria, Illinois, 61604. Mention of firm names or trade products does not imply endorsement by USDA.

The author wishes to thank Dr. R. J. Dimler, Northern Marketing and Nutrition Research Division, USDA, Peoria, Illinois and Dr. E. L. Kendrick, Cereal Crops Research Branch, Plant Industry Station, USDA, Beltsville, Maryland, for making this work possible; and Mr. C. H. VanEtten, Northern Marketing and Nutrition Research Division, USDA, Peoria, Illinois for helpful advice and review of manuscript, and supplying some of the samples.

The following have been very generous in supplying the samples and the author thanks them: Dr. E. Villegas and Dr. M. G. Gutierrez, International Maize and Wheat Improvement Center, Londres, Mexico 6, D.F.; Dr. Willis Skrdla, North Central Regional Plant Introduction Station, ARS, Ames, Iowa; Dr. C. R. Adair, Plant Science Research Division, ARS, Beltsville, Maryland; Dr. T. H. Johnston, University of Arkansas Rice Branch Experiment Station, Stuttgart, Arkansas; Dr. P. J. Mattern, Department of Agronomy, University of Nebraska, Lincoln, Nebraska; Dr. R. C. Pickett, Department of Agronomy, Purdue University, Lafayette, Indiana; Dr. V. C. Finkner, Dept. of Agronomy, University of Kentucky, Lexington, Kentucky; and Dr. B. Juliano, International Rice Research Institute, Los Banos, Laguna, Philippines. The technical assistance of Miss Maxine Sinkler, Miss Diane Soltess and especially Miss Gertrude Skerski is gratefully appreciated.

BIBLIOGRAPHY

BARAUD, J. 1957. Iodometric analysis of free amino nitrogen in proteins by means of their dithiocarbamate derivatives. Bull Soc. Chim. France, 948-951 (French).

BRADSTREET, R. B. 1965. The Kjeldahl Method for Organic Nitrogen. Academic Press, New York.

CARPENTER, K. J. 1960. The estimation of available lysine in animal protein foods. Biochem. J. 77, 604-610.

CARPENTER, K. J., ELLINGER, G. M., MUNRO, M. I., and ROLFE, E. J. 1957. Fish products as protein supplements to cereals. Brit. J. Nutr. 11, 162-173.

CARPENTER, K. J., and MARCH, B. E. 1961. The availability of lysine in groundnut biscuits used in the treatment of kwashiorkor. Brit. J. Nutr. *15*, 403-410.

CONCON, J. M. 1966. The proteins of *opaque-2* maize. *In* Proc. High-Lysine Corn Conference, E. T. Mertz and O. E. Nelson (Editors). Corn Ind. Res. Found., Washington, D. C.

CONCON, J. M. 1971. Unpublished data.

EISEN, H., BELMAN, S., and CARSTEN, M. E. 1953. The reaction of 2,4-dinitrobezenesulfonic acid with free amino groups of proteins. J. Am. Chem. Soc. *75*, 4583-4585.

FOSTER, J. F., YANG, J. T. and YUI, N. H. 1950. Extraction and electrophoretic analysis of proteins of corn. Cereal Chem. *27*, 477-487.

FREEDMAN, R. B., and RADDA, G. K. 1968. The reaction of 2,4,6-trinitrobenzenesulfonic acid with amino acids, peptides and proteins. Biochem. J. *108*, 383-391.

GOLDFARB, A. R. 1966A. A kinetic study of the reaction of amino acids and peptides with trinitrobenzenesulfonic acid. Biochemistry *5*, 2470-2574.

GOLDFARB, A. R. 1966B. Heterogeneity of amino groups in proteins. I. Human serum albumin. Biochemistry *5*, 2574-2578.

HABEEB, A. F. S. A. 1966. Determination of free amino groups in proteins by trinitrobenzenesulfonic acid. Anal. Biochem. *14*, 328-336.

HAUROWITZ, F. 1963. The Chemistry and Function of Proteins. Academic Press, New York.

HOSOYA, T. 1960. Turnip peroxidase. V. The effect of several chemical reagents upon the activity of turnip peroxidase D. J. Biochem. *48*, 375-381.

IKENAKA, T. 1959. Chemical modification on taka-amylase A. I. Dinitrophenylation of taka-amylase A. J. Biochem. *46*, 177-183.

KAI, F. 1967. Reaction between mercury (II) and organic compounds. III. Separation and quantitative determination of basic amino acids with mercury (II) salt precipitants. Bull. Chem. Soc. Japan *40*, 2297-2302.

KAINZ, G., HUBER, H., and KASLER, F. 1957. Microdetermination of primary amino groups with nitrosylbromide. Mikrochim. Acta *5*, 744-750 (German).

KAKADE, M. L., and LIENER, I. E. 1969. Determination of available lysine in proteins. Anal. Biochem. *27*, 273-280.

KOTAKI, A., HARADA, M., and YAGI, K. 1964. Reaction between sulfhydryl compounds and 2,4,6-trinitrobenzene-1-sulfonic acid. J. Biochem. *55*, 553-561.

LI, C. H. 1956. Preparation and properties of dinitrophenyl-NH (ϵ)-insulin. Nature *178*, 1402.

LIEBEN, F., and LOO, Y. C. 1942. On the liberation of free amino nitrogen from proteins in the Van Slyke apparatus. J. Biol. Chem. *145*, 223-228.

MERTZ, E. T. 1970. Private communication. Dept. Biochemistry, Purdue Univ., Lafayette, Ind.

MERTZ, E. T. and BRESSANI, R. 1957. Studies on corn proteins. I. A new method of extraction. Cereal Chem. *34*, 63-69.

OKUYAMA, T., and SATAKE, K. 1960. On the preparation and properties of 2,4,6-trinitrophenyl amino acids and peptides. J. Biochem. *47*, 454-466.

PATAKI, G. 1968. Techniques of Thin Layer Chromatography in Amino Acid and Peptide Chemistry. Ann Arbor Science Publishers, Ann Arbor, Michigan.

PIETRZYK, D. J., and BELISLE, J. 1966. Behavior of 2,4-dinitrobenzenesulfonic acid as an acid catalyst in acetylation reactions. Anal. Chem. *38*, 1508-1512.

RUSSELL, D. W. 1962. Photolysis of α-N-2,4-dinitrophenyl amino acids in dilute aqueous sodium hydrogen carbonate. Biochem. J. *83*, 8P-9P.

RUSSELL, D. W. 1963. Photochemical behavior of 2,4-dinitrophenyl derivatives of some amino acids and peptides. Biochem. J. *87*, 1-4.

SELIM, A. S. M. 1965. Microdetermination of lysine in protein hydrolyzates. J. Agr. Food Chem. *13*, 435.

SUBRAMANIAN, S. S., JAMBUNATHAN, R., CONCON, J. M., and MERTZ, E. T. 1970. Simple methods for the determination of lysine and tryptophan in high lysine and normal maize. Federation Proc. *29*, 761.

SUGAE, K. 1960. Studies on bacterial amylase. V. Chemical modification of bacterial amylase and coupling of taka-amylase A with *p*-sulfo-benzene-diazonium chloride. J. Biochem. *48*, 790-802.

SURE, B., and HART, E. B. 1917. The effect of temperature on the reaction of lysine with nitrous acid. J. Biol. Chem. *31*, 527-532.

VAN SLYKE, D. D. 1911. A method for the quantitative determination of aliphatic amino groups. J. Biol. Chem. *9*, 185-203.

VAN SLYKE, D. D. 1912. Quantitative determination of aliphatic amino groups. J. Biol. Chem. *12*, 275-284.

CHAPTER 23

Aaron M. Altschul

Concluding Commentary on
Seed Proteins

One could look at this symposium simply as a series of recitations of interesting scientific achievements and of new vistas of scientific knowledge. In that respect the papers in this symposium can stand on their own feet, and the symposium itself is justified. But I think that there was more in the mind of Dr. Inglett, who organized the symposium, and in the minds of many of the speakers than simply a recitation of scientific achievements. Throughout there was an undercurrent of trying to relate what was being done in this field to the problems of feeding people who do not have enough protein. So there is a very definite social factor in these discussions: What can science and technology do to solve the world protein problem?

I would like to talk a bit about the protein problem. Most considerations of the world's food problem by economists generally neglect the protein problem entirely and consider the problem primarily or exclusively one of total food supply. Insofar as grain supply is an important component of the total food supply, protein considerations are included since grains are the major source of protein in the world. Nutritionists consider the problem of total food and of protein as separate. Only if there is evidence of specific protein malnutrition is there considered to be a protein problem. If there is evidence of famine then it is a total food problem.

I would suggest that you might look at the world food supply from the joint view of economics and nutrition. It turns out that you cannot divorce protein problems from total food problems because the way in which people decide to get their protein determines how much food they have or how much food they need. Also, the way people get protein is determined more by aesthetic considerations than by nutritional needs.

It is well known that the amount of animal protein in the people's diet is increased with increase in income. People with very low incomes, such as $100.00 per capita per year, eat less than 10 gm of animal protein per person per day and more often less than 5 gm. People with higher incomes eat much more. In order to compare both quality and quantity of diet on the same scale, we can convert the total food eaten per day into grain equivalent calories and this includes the plant sources eaten directly added to the plant sources eaten by the animals which, in turn, are eaten by man. It turns out that the average American requires about 11,000 plant equivalent calories per day. He actually eats about 2500 or so calories, but if he takes into account the amount of animal protein eaten per day it turns out to be equivalent to 11,000 grain equivalent calories. In Sweden they eat about an average of about 8900 calories. The average for the so-called developed countries is about 8000; the average for the under-developed countries is about 3000; and India has an expenditure of little over 2500 calories which includes about 300 grain equivalent calories to supply the animal protein. This kind of an approach is a simple and perhaps dramatic way of showing the interrelationship between proteins and calories. If we in the United States decided to change the average diet and reduce the amount of animal protein this would be reflected in the total amount of calories that we need. I would venture that if we were to reduce our equivalent caloric intake by 2000 calories (from 11,000 to 9000) there would be a lot of unhappy people. I'm not sure that it was much more than that, or perhaps less, which was happening in Poland when the riots were precipitated because of increase in the price of meat.

I would conclude that there is no separate protein and calorie food problem. Any way in which the equivalent of animal protein can be replaced by plant protein reduces the cost of the food supply and makes available more funds for total food. That is why it is entirely appropriate for us to consider the protein aspects of the major calorie food in the world: the cereals.

Another point I would like to comment on is the relationship between genetic and cultural evolution. Many of the presentations have been directed primarily at genetic evolution, that is, the modification of plants in such a way that their nutritional impact will be increased at no additional cost in terms of yield or at a minimum additional cost. I would assume that there are some amongst those who spoke who think that this can all be done by genetic evolution. I would imagine that there are others who think that none of this can be done by genetic evolution, that major changes in food processing, such as by fortification, will be required. Both sides are not entirely correct, I would think. There never was a time in history when we abided by either one or the other type of evolution very long. It is highly improbable that that time will come now. Anybody who thinks that the plant breeder in wheat has been the sole hero in the development of wheat as a major source of food in the world has only to go around the world and look at the different ways in which wheat has been used and consider the enormous contribution of bread as a concept in utilizing wheat. The invention of bread was an aspect of cultural evolution. There are any number of examples of that kind of an interrelationship between processing and genetics.

The ultimate solution to any problem of exacting the maximum from any of the foods being grown and eaten will be the one that costs the least and is most practical. This will require a judicious combination of genetic and cultural evolution.

Index